Optimal Long-Term Operation of Electric Power Systems

MATHEMATICAL CONCEPTS AND METHODS IN SCIENCE AND ENGINEERING

Series Editor: **Angelo Miele**
Mechanical Engineering and Mathematical Sciences
Rice University

Recent volumes in this series:

A Continuation Order Plan is available for this series. A continuation order will bring delivery of each new volume immediately upon publication. Volumes are billed only upon actual shipment. For further information please contact the publisher.

Optimal Long-Term Operation of Electric Power Systems

G. S. Christensen

University of Alberta
Edmonton, Alberta, Canada

and

S. A. Soliman

Ain Shams University
Cairo, Egypt

PLENUM PRESS • NEW YORK AND LONDON

Library of Congress Cataloging in Publication Data

Christensen, G. S. (Gustav S.)
  Optimal long-term operation of electric power systems.

  (Mathematical concepts and methods in science and engineering; 38)
  Includes bibliographical references and index.
  1. Hydroelectric power plants — Management — Mathematical models. 2. Mathematical optimization.I. Soliman, S. A. II. Title. III. Series.
TK1081.C49   1988   621.31′2134′068   88-15268
ISBN 0-306-42875-X

© 1988 Plenum Press, New York
A Division of Plenum Publishing Corporation
233 Spring Street, New York, N.Y. 10013

Printed in the United States of America

To our wives
Penelope and Laila

Preface

This book deals with a very important problem in power system planning for countries in which hydrogeneration accounts for the greatest part of the system power production. During the past thirty years many techniques have been developed to cope with the long-term operation of hydro reservoirs. These techniques have been discussed in a number of publications, but they have not until now been documented in book form.

This book is intended as the foundation for a special graduate course dealing with aspects of electrical engineering, operational research, water resource research, and applied mathematics. It may also be used for self-study by practicing personnel involved in the planning and operation of hydroelectric power systems for utilities, consulting groups, and government regulatory agencies.

The book consists of eight chapters. Chapter 1 reviews the historical developments in the field, discusses briefly all techniques used to solve the problem, and summarizes the modeling of hydroplants for long-term operation studies. At the end of the chapter we present in detail an outline of the book.

Some optimization techniques are presented in Chapter 2; these include the calculus of variations, dynamic programming, and the maximum principle. Recent results obtained by the authors are based on the powerful minimum norm approach, which is also reviewed in this chapter. The chapter starts with a mathematical review, especially of matrix algebra and some statistics definitions. Emphasis is placed on clear, methodical development rather than on rigorous formal proofs. A number of solved examples illustrating the techniques are included.

Chapter 3 is devoted to the operation of multireservoir power systems connected in series on a river; in this chapter the problem is solved both by using dynamic programming with the decomposition approach, and by using the minimum norm formulation of functional analysis to obtain maximum benefits from the system. A computational example is provided to compare the performance of both approaches.

Chapter 4 is devoted to maximizing the total benefits from a multichain hydroelectric power system. This chapter compares different techniques used to solve this problem, including dynamic programming, the maximum principle, and recent work by the authors using the minimum norm formulation. Computational examples are presented to explain the main features of each approach. Different models are used for each technique, depending on the operation of the system.

There is a period in water management during which the inflow to the reservoirs is the lowest on record, and the reservoirs must then be drawn down from full to empty. This period is referred to as the "critical period," and the stream flows that occur during the critical period are called the critical stream flows. Chapter 5 is concerned with the optimal operation of a multireservoir power system during such a critical period. The first part of the chapter is devoted to maximizing the total benefits from the system during this period; in this part the load shape on the system is not taken into account. The second part is devoted to maximizing the total generation from the system, with the load shape on the system taken into account. This load is a variable load depending on the total generation from the system at the end of each year of the critical period. The minimum norm formulation technique is used in this chapter, and a practical example for a real system in operation is provided.

Chapter 6 is concerned with the operation of a hydro system for maximum firm hydro energy capability during the critical period with a fixed load on the system.

Most electric utility power systems utilize a combination of hydro and thermal power plants to supply the load on the system. The aim of Chapter 7 is to study the optimal long-term operation of this combination, where the objective function is to minimize the thermal fuel cost and at the same time satisfy the hydro constraints on the system. In this chapter we discuss different techniques used to solve the problem, including the stochastic dynamic programming, aggregation–decomposition, nonlinear programming, discrete maximum principle approaches, and, finally, the minimum norm approach developed by the authors. Practical examples are provided for each technique to explain its main features.

In the final chapter we summarize the results obtained and offer suggestions for future work in the field.

Each chapter in this book is concluded by comments and references for further reading.

Most of the material in this text originated from work done in the past, and a number of *Water Resources Research* and IEEE papers have been used as primary sources. These are cited at appropriate points in the text.

The application of the minimum norm formulation of functional analysis as an optimization technique to the problem of long-term operation of hydro reservoirs was conceived by the authors after several years of research in this field, and several discussions with utility companies in Canada and the United States.

G. S. Christensen
Edmonton, Alberta, Canada
S. A. Soliman
Cairo, Egypt

Acknowledgments

During the planning and writing of this book, we have incurred indebtedness to many people. We wish to acknowledge continuing encouragement received from Dr. A. M. Elarabatie, Chairman of the Department of Electrical Power and Machines at Ain Shams University, and Dr. C. R. James, Chairman of the Department of Electrical Engineering at the University of Alberta. We are grateful to our many friends and colleagues, and in particular to Dr. M. A. Morsy of Ain Shams University and Dr. D. H. Kelly of the University of Alberta. The second author wishes to thank his father-in-law, Dr. Mousa Shaheen Lasheen, former vice president of Al-Azhar University, Cairo, his brothers-in-law, Eng. N. M. Shaheen and Dr. S. M. Shaheen, and his brother, Eng. R. A. Soliman, for taking care of his family during his absence to write this book. We also wish to express our thanks to Ms. Barbara J. Peck, Director, Canterbury Executive Services, for a neatly done job in typing many drafts of the manuscript. This work would not have been possible without the patience and understanding of our wives, Dr. Penelope Christensen and Laila Soliman, and our children, Lynne and Neil Christensen and Rasha, Shady, and Samia Soliman.

G.S.C.
S.A.S.

Contents

Optimal Long-Term
Operation of Electric
Power Systems

1

Introduction

1.1. A Historical Survey

The hydro optimization problem involves planning the use of a limited resource over a period of time. The resource is the water available for hydro generation. Most hydroelectric plants are multipurpose. In such cases, it is necessary to meet certain obligations other than power generation. These may include a maximum forebay elevation, not to be exceeded because of the danger of flooding, and a minimum plant discharge and spillage to meet irrigational navigational commitments. Thus, the optimum operation of the hydro system depends upon the conditions that exist over the entire optimization interval (Ref. 1.1).

Other distinctions among power systems are the number of hydro stations, their location, and special operating characteristics. The problem is quite different if the hydro stations are located on the same stream or on different ones. An upstream station will highly influence the operation of the next downstream station. The latter, however, also influences the upstream plant by its effect on the tail water elevation and effective head. Close coupling of stations by such a phenomenon is a complicating factor (Refs. 1.1–1.8).

The problem of determining the optimal long-term operation of a multireservoir power system has been the subject of numerous publications over the past forty years, and yet no completely satisfactory solution has been obtained, since in every publication the problem has been simplified in order to be solved.

Aggregation of the multireservoir hydroplant into a single complex equivalent reservoir and solution by stochastic dynamic programming (SDP) was one of the earlier approaches used. Obviously, such a representation of the reservoir cannot take into account all local constraints on the contents of the reservoir, water flows, and hydroplant generation. This method can

1

perform satisfactorily for systems where reservoirs and inflow characteristics are sufficiently "similar" to justify aggregation into a single reservoir and hydroplant model.

Turgeon has proposed two methods for the solution of the problem (Refs. 1.2 and 1.5). The first is really an extension to the aggregation method, and it breaks the problem down into two levels. At the second level the problem is to determine the monthly generation of the valley. This problem is solved by dynamic programming. The problem at the first level is to allocate that generation to the installation; this is done by finding functions that relate the water level of each reservoir to the total amount of potential energy stored in the valley. The second method is the decomposition method by combining many reservoirs into one reservoir for the purpose of optimization of a multireservoir power system connected in series on a river, and using dynamic programming for solving $n - 1$ problems of two state variables each. The solution obtained by this method is a function of the water content of that reservoir and the total energy content of the downstream reservoirs. The main drawback is that the approach avoids answering basic questions as to how the individual reservoirs in the system are to be operated in an optimal fashion. Also the inflows to some reservoirs may be periodic in phase with the annual demand cycle, while other reservoirs have an inflow cycle that lags by a certain time.

Stochastic dynamic programming with successive approximation (DPSA) has been proposed to solve the problem of a parallel multireservoir hydroelectric power system. The successive approximation involves a "one-at-a-time" stochastic optimization of each reservoir. The major drawback of this approach is that it ignores the dependence of the operation of one reservoir on the actual energy content of other reservoirs.

In the aggregation–decomposition (AD) approach, the optimization of a system of n reservoirs is broken down into n subproblems in which one reservoir is optimized knowing the total energy content of the rest of the reservoirs. For each subproblem one of the reservoir-hydroplant models is retained and the remaining $n - 1$ are aggregated into an equivalent reservoir hydroplant model. A comparison of the last two approaches, DPSA and AD, has been made on a simulation basis for a six-complex-reservoir system. The results indicate that the AD approach gives better results and the computational effort increases only linearly with the number of reservoirs. More precisely, for each new reservoir added to the system, only one additional dynamic programming problem of two-state variables has to be solved. The computing time for each of the last two approaches was 150 min in CPU units (Refs. 1.2 and 1.5).

Linear programming and dynamic programming have been applied to the optimization of the production of hydroelectric power. The solution was obtained in two steps with linear and dynamic programming methods.

The models that have been used are deterministic. The first step in the solution was the long-term optimization problem; this problem was solved by a linear programming (LP) method. The variation in the efficiency of the turbines, the variation of the water heads, and the time delays are neglected. The second step is the optimal short-term run of the turbine-generator units; this is determined by dynamic programming (DP). The total computation time on a typical minicomputer was 1–3 min per power station.

Successive linear programming, an optimal control algorithm, and a combination of linear programming and dynamic programming (LP–DP) are employed to optimize the operation of multireservoir hydro systems given a deterministic inflow forecast. The alogrithm maximizes the total benefits from the system (maximization of energy produced, plus the estimated value of water remaining in storage at the end of the planning period). The LP–DP algorithm is the least satisfactory: it takes longer to find a solution and produces significantly less hydro power than the other two procedures. Successive linear programming (SLP) appears to find the global maximum and is easily implemented. The optimal control algorithm can find the optimum in about one fifth the time required by SLP for small systems but is more difficult to implement. The computing costs were reasonable with SLP and the optimal control algorithm, and increase only as the square of the number of reservoirs in contrast to the exponential growth of dynamic programming (Ref. 1.7).

Marino and Loaiciga applied a quadratic optimization model to a large-scale reservoir system to obtain operation schedules. The model they used has the minimum possible dimensionality, treats spillage and penstock releases as decision variables, and takes advantage of system-dependent features to reduce the size of the decision space. They compared the quadratic model with a simplified linear model.

1.2. Hydro Plant Modeling for Long-Term Operation

Hydro power plants are classified into pumped storage plants and conventional hydro plants.

1.2.1. Pumped Storage Plants

A pumped storage plant is associated with upper and lower reservoirs. During light load periods water is pumped from the lower to the upper reservoirs using the available energy from other sources as surplus energy. During peak load the water stored in the upper reservoir is released to generate power to save fuel costs of the thermal plants. The pumped storage plant is operated until the added pumping cost exceeds the saving in thermal costs due to the peak shaving operation.

The conventional hydroplants are classified into run-of-river plants and storage plants.

1.2.2. Run-of-River Plants

The run-of-river plants have little storage capacity, and use water as it becomes available; water not utilized is spilled. The MWh generated from a run-of-river plant is equal to a constant times the discharge through the turbines

$$G_k = hu_k \text{ MWh} \qquad (1.1)$$

where h is a constant measured in MWh/m^3 and referred to as the water conversion factor; u_k is the discharge through the turbine during a period k in m^3; and k is an index used for the period—this period may be a week or a month.

1.2.3. Storage Plants

Storage plants are associated with reservoirs with significant storage capacity. In periods with low power requirements water can be stored and then released when the demand is high.

Modeling of storage plants, for a long-term study, depends on the water head variation. For hydro plants in which the water head variation is small, the MWh generated from the plants can be considered as a constant times the discharge, as given in equation (1.1), and this constant is equal to the average number of MWh generated during a period k by an outflow of $1\,\text{m}^3$. But for the power systems in which the water heads vary by a considerable amount, this assumption is not true, and the water conversion factor, MWh/m^3, varies with the head, which itself is a function of the storage. For a long-term study, the MWh generated can be written as

$$G_k = \alpha u_k + \tfrac{1}{2}\beta u_k(x_k + x_{k-1}) + \tfrac{1}{4}\gamma u_k(x_k + x_{k-1})^2 \text{ MWh} \qquad (1.2)$$

where α, β, and γ are constants—these can be obtained by least-squares curve fitting to typical plant data available; and x_k is the storage at the end of period k in m^3.

Equation (1.2) is a function of the discharge through the turbines and the average storage between two successive months, $k - 1$ and k, to avoid underestimation for rising water levels and overestimation for falling water levels in the MWh generated.

1.2.4. Reservoir Models

Modeling of reservoirs is of a great importance for long-term operation of hydroelectric power systems. Reservoir models contain the storage and the release in a well-known equation, called the continuity equation, or the water conservation equation. For long-term study, the reservoir dynamics may be adequately described by the discrete difference equation

$$x_k = x_{k-1} + I_k - u_k - s_k + z_k \qquad (1.3)$$

where I_k is the natural inflow adjusted for evaporation and seepage losses during a period k in m^3, s_k is the spillage during a period k in m^3. Water is spilt when the discharge u_k is greater than the maximum discharge and the reservoir is filled to capacity, and z_k is the outflow from upstream reservoirs.

The variation of storage of a reservoir of regular shape with the elevation can be computed using the formulas for the volumes of solids. In practice, natural factors will change the reservoir configuration with time. An example is sediment accumulation. It is important to update the reservoir model periodically. A mathematical model for the storage-elevation curve may be obtained by using Taylor's expansion

$$h_k = \sum_{p=0}^{N} \alpha_p (x_k)^p \text{ m} \qquad (1.4)$$

where h_k is the net head at the end of period k in meters and N is the highest order of the approximation. In equation (1.4), we assume that the tailwater elevation is constant, independent of the discharge. α_p's can be obtained by using curve fitting to data available from typical reservoirs.

1.2.5. Operational Constraints

Many dams and associated reservoirs are multipurpose developments. Irrigation, flood control, water supply, stream flow augmentation, navigation, and recreation use are among the possible purposes of water resource development. For these purposes the reservoir is regulated so that full requirements for each element are available under the design drought conditions.

The operational reservoir constraints are

$$x^m \leq x_k \leq x^M \qquad (1.5)$$

$$u_k^m \leq u_k \leq u_k^M \qquad (1.6)$$

The first set of inequality constraints simply states that the reservoir storage (or elevation) may not exceed a maximum level, nor be lower than a minimum level. For the maximum level, this is determined by the elevation of the spillway crest or the top of the spillway gates. The minimum level may be fixed by the elevation of the lowest outlet in the dam or by conditions of operating efficiency for the turbines. The second set of the inequality constraint is determined by the discharge capacity of the power plant as well as its efficiency.

1.3. Outline of the Book

The body of this book consists of Chapters 2, 3, 4, 5, 6, and 7. The final chapter, Chapter 8, is a concluding one in which we summarize the results. We shall now review, in detail, the contents of each chapter.

Chapter 2 reviews some optimization techniques used throughout this book. This chapter starts with some mathematical notions, especially matrix algebra. Here concepts of vectors, matrices, and quadratic forms are briefly reviewed. Some statistical definitions used throughout the book are also considered. The second section reviews the static optimization results. This includes unconstrained optimization via calculus of variations; equality and inequality constaints are considered, and Lagrange and Kuhn–Tucker multipliers are introduced.

The third section focuses on the dynamic programming approach and the principle of optimality. The following two sections deal with Pontryagin's maximum principle and the functional analytic optimization technique. Here we review the mathematical concepts of functional analysis, especially for random variables, and follow that with a statement of a minimum norm problem and its solutions. Emphasis will be given to clear, methodical developments rather than rigorous, formal proofs. A number of solved examples to illustrate the techniques are included in this chapter.

Chapter 3 is devoted to solving the long-term operating problem of a multireservoir power system connected in series on a river. The first section of this chapter deals with the optimal long-term operation of these reservoirs for maximum total benefits. The system described has a constant water conversion factor. The problem is solved by using both the dynamic programming with decomposition approach, and the minimum norm formulation of functional analysis. A computational example is provided to compare the performance of each approach.

The next section deals with a more complicated problem for the same system; the water conversion factor used in this section has a nonlinear relation with storage, and the generated function used for each power house is a nonlinear function of the discharge and the average storage. Here, the minimum norm formulation of the functional analysis optimization technique is used to solve the problem.

Chapter 4 is devoted to maximizing the total benefits from the hydroelectric generation plus benefits from the water left in storage at the end of the planning period, from a multichain hydroelectric power system. This chapter compares different techniques used to solve this problem, including dynamic programming, the maximum principle, and the minimum norm formulation of functional analysis. Computational examples are presented to explain the main features of each approach. Different models have been used for each approach.

In water management there is a period during which the inflow to the reservoirs is the lowest on record, and at that point the reservoirs should be drawn down from full to empty. This period is referred to as the "critical period," and the stream flows that occurred during the critical period are called the critical stream flows.

Chapter 5 is concerned with the optimal operation of a multireservoir power system during this critical period. The first section is devoted to maximizing the total benefits from the system during this period; the load shape on the system in this part is not taken into account. The second section is devoted to maximizing the total generation from the system. Here the load shape on the system is taken into account; this load is a variable load and depends on the total generation at the end of each year of the critical period. The minimum norm formulation technique is used in this chapter, and a practical example of a real system in operation is provided.

Chapter 6 is concerned with the operation of a hydro system for a maximum firm hydro energy capability during the critical period with a fixed load on the system. The firm energy is defined as the difference between the total generation during a certain period and the load on the system during that period. In this chapter the minimum norm approach is used.

Most electric utility power systems contain a combination of hydro and thermal power plants to supply the required load on the system. The aim of Chapter 7 is to study the optimal long-term operation of this combination, where the objective function is to minimize the thermal fuel cost and at the same time satisfy the hydro constraints on the system. In this chapter we discuss different techniques used to solve the problem, including the stochastic dynamic programming, aggregation–decomposition, nonlinear programming, and discrete maximum principle approaches, and finally the minimum norm approach developed by the authors. Practical examples are provided for each technique to explain its main features.

The final chapter, Chapter 8, summarizes the result and briefly outlines certain directions for future research needs. Only the titles and the broad outline of the problems considered in this book are mentioned here. A more detailed description of each of the problems and its relationship to the previous work in this area will be found in each chapter.

References

1.1. EL-HAWARY, M. E., and CHRISTENSEN, G. S., *Optimal Economic Operation of Electric Power Systems*, Academic Press, New York, 1979.
1.2. TURGEON, A., "A Decomposition Method for the Long-Term Scheduling of Reservoirs in Series," *Water Resources Research* 17(6), 1565–1570 (1981).
1.3. ARVANITIDIES, N. V., and ROSING, J., "Composite Representation of a Multireservoir Hydroelectric Power System," *IEEE Transactions on PAS* **PAS-89**(2), 319–326 (1970).
1.4. ARVANITIDIES, N. V., and ROSING, J., "Optimal Operation of Multireservoir Systems Using a Composite Representation," *IEEE Transactions on PAS* **PAS-89**(2), 327–335 (1970).
1.5. TURGEON, A., "Optimal Operation of Multireservoir Power Systems with Stochastic Inflows," *Water Resources Research* 16(2), 275–283 (1980).
1.6. GRYGIER, J. C., and STEDINGER, J. R., "Algorithms for Optimization Hydropower System Operation," *Water Resources Research* 21(1), 1–10 (1985).
1.7. OLCER, S., HARSA, C., and ROCH, A., "Application of Linear and Dynamic Programming to the Optimization of the Production of Hydroelectric Power," *Optical Control Application and Methods* **6**, 43–56 (1985).
1.8. MARINO, M. A., and LOAICIGA, H. A., "Dynamic Model for Multireservoir Operation," *Water Resources Research* 21(5), 619–630 (1985).

2

Mathematical Optimization
Techniques

2.1. Introduction

We present in this chapter a brief review of some optimization techniques that are considered to be background material for the text. We start by introducing the basic concepts of matrix algebra, and some basic statistical definitions in Section 2.2. Following this, in Section 2.3, we introduce the discrete variational calculus. Here we determine a discrete version of Euler's equation and the transversality conditions. In Section 2.4, we discuss the discrete maximum principle and its applications to a stochastic discrete process (Refs. 2.2–2.11).

Dynamic programming and the principle of optimality are introduced in Section 2.5. A brief review of some basic concepts of the functional analytic techniques of formulating the problems in the minimum norm form is dealt with in Section 2.6. Here, we discuss and construct Hilbert spaces of random variables; at the end of this section we introduce a powerful version of minimum norm problems (Refs. 2.1 and 2.8). Emphasis will be given in this chapter to clear, methodical development rather than rigorous formal proofs.

2.2. A Review of Matrices (Ref. 2.2)

For large-scale systems, unless we adjust the notation, it is difficult to see what is occurring and to understand the structure of the optimal control and of the optimal policy. To overcome these difficulties, we use the vector matrix notation to handle multidimensional matters. Our objective here is to briefly review some concepts useful in the text.

9

2.2.1. Vectors

Let $x_1, x_2, \ldots, x_n$ be real, complex numbers. A column of these numbers

$$x = \begin{bmatrix} x_1 \\ x_2 \\ \vdots \\ x_n \end{bmatrix} \quad \text{or} \quad x = \text{col}(x_1, x_2, \ldots, x_n)$$

is called a vector; the order of this vector is n. Throughout this text, we will deal with a functional vector, which is a function of another function, say time. We have the n-dimensional discrete vector

$$x(k) = \text{col}(x_{1,k}, x_{2,k}, \ldots, x_{n,k})$$

or in short

$$x(k) = \text{col}(x_{i,k}; i = 1, \ldots, n)$$

where $x_{i,k}$ are called the components of the vector.

Two vectors x and y are equal if their respective components are equal. In this case we can write $x_i = y_i$; $i = 1, \ldots, n$.

Addition of two vectors $x + y$ is carried out by adding the corresponding elements in each vector, $x_i + y_i$, $i = 1, \ldots, n$, and is given by

$$x + y = \text{col}(x_i + y_i; i = 1, \ldots, n)$$

In general, given the vectors x, y, and z, the following laws apply:

1. $x + y = y + x$ (commutative law)
2. $(x + y) + z = x + (y + z) = x + y + z$ (associative law)

Multiplication of the vector x by the scalar α is a vector and is defined by

$$x\alpha = \alpha x = \text{col}(\alpha x_i, i = 1, \ldots, n)$$

2.2.2. Matrices

A matrix is a rectangular array of elements; these elements may be real or complex numbers. The matrix A can be written as

$$A = \begin{bmatrix} a_{11} & a_{12} & \cdots & a_{1n} \\ a_{21} & a_{22} & \cdots & a_{2n} \\ \vdots & \vdots & a_{ij} & \cdots \\ a_{m1} & a_{m2} & \vdots & a_{mn} \end{bmatrix} \leftarrow \text{row } i$$

$$\uparrow$$
$$\text{column } j$$

In shorthand the elements of the matrix A are

$$A = [a_{ij}], \qquad i = 1, \ldots, m; j = 1, \ldots, n$$

the order of the matrix A in this case is $m \times n$.

If $m = n$, the matrix A in this case is a square matrix of order $n \times n$. Special matrices in our analysis can be specified as follows:

1. *The Diagonal Matrix.* This is a square matrix in which all the diagonal elements have a certain value, and all the off-diagonal elements are "zero," that is

$$A = \text{diag}(a_{ii}), \qquad i = 1, 2, \ldots, n$$

2. *Identity Matrix I.* An identity matrix I is a square matrix in which all the diagonal elements are "one" and all the off-diagonal elements are "zero," that is,

$$a_{ii} = 1, \qquad i = 1, 2, \ldots, n$$

$$a_{ij} = 0 \qquad \text{for } i \neq j$$

3. *Upper and Lower Triangular Matrices.* An upper triangular matrix is a square matrix in which all the lower-diagonal elements are "zero," that is

$$a_{ij} = 0 \qquad \text{for } j < i$$

and a lower triangular matrix is a square matrix in which all the upper-diagonal elements are "zero," that is

$$a_{ij} = 0 \qquad \text{for } j > i$$

Two matrices A and B are equal if and only if they are of the same order and all their corresponding elements are equal:

$$[A] - [B] = [0]$$

The transpose of matrix A, A^T, can be obtained by interchanging rows and columns. The transpose of the matrix A is

$$A^T = \begin{bmatrix} a_{11} & a_{21} & a_{31} & a_{n1} \\ a_{12} & & & \\ \vdots & & & \\ a_{1n} & & & a_{nm} \end{bmatrix}$$

For any two matrices A and B

$$(AB)^T = B^T A^T$$

We will define a matrix A as an orthogonal matrix if

$$A^T A = AA^T = I$$

2.2.2.1. Addition

Two matrices A and B can be added together only if they have the same dimensions and their elements in the corresponding positions are added, i.e., if $C = A + B$, then

$$c_{ij} = a_{ij} + b_{ij} \qquad \text{for all possible } i \text{ and } j$$

It is clear that

$$A + B = B + A \qquad \text{(commutative law)}$$

$$A + (B + C) = (A + B) + C \qquad \text{(associative law)}$$

2.2.2.2. Multiplication by a Scalar

If α is a constant, then the product αA is obtained by multiplying every element of A by α. In our shorthand notation this can be written

$$\alpha[a_{ij}] = [\alpha a_{ij}]$$

2.2.2.3. Multiplication of Two Matrices

Given two matrices A and B, our objective is to define AB. The product AB can be defined if, and only if, the number of columns of A is equal to the number of rows of B. A and B are then said to be conformable for multiplication. The dimensions of the resulting product can be found by the simple rule

$$
\begin{array}{ccccc}
A & \cdot & B & = & C \\
(m \times n) & & (n \times p) & & m \times p
\end{array}
$$

It is clear that a matrix is a symbol for a linear transformation. Using this idea, if $z = Ay$ and $y = Bx$ denote, respectively, the linear transformation by

$$z_i = \sum_{j=1}^{N} a_{ij} y_i, \qquad y_i = \sum_{j=1}^{N} b_{ij} x_j, \qquad i = 1, 2, \ldots, N$$

It is clear by direct substitution that z_i are linear functions of the x_j. Since $z = A(Bx)$, it is natural to define AB as the matrix of this resultant linear transformation. Hence, we define the ijth element of AB to be

$$\sum_{k=1}^{N} a_{ik} b_{kj}$$

Notice that, in general $AB \neq BA$ even if BA is defined, i.e., matrix multiplication is not commutative. If $AB = BA$, in this case A and B are said to commute with each other.

Matrix multiplication has the following general properties

$$(AB)C = A(BC)$$

$$C(A + B) = CA + CB$$

$$(A + B)C = AC + BC$$

$$\alpha(AB) = (\alpha A)B = A(\alpha B)$$

$$(ABC)^T = C^T B^T A^T$$

2.2.2.4. The Inverse of a Matrix

The inverse B of an $n \times n$ matrix A is defined to be a matrix satisfying

$$AB = BA = I$$

and it is denoted by A^{-1}. The classical method of computing A^{-1} involves determinants and factors of A. Three important results can be proved for nonsingular matrices:

1. If A and B are nonsingular n-square matrices, then

$$(AB)^{-1} = B^{-1}A^{-1}$$

2. If A is nonsingular, then $AB = AC$ implies that $B = C$.
3. If A is a nonsingular matrix with complex elements, then

$$(A^*)^{-1} = (A^{-1})^*$$

2.2.2.5. Partitioned Matrices

Partitioning is useful when applied to large matrices since manipulations can be carried out on the smaller blocks. More importantly, when multiplying matrices in partitioned form the basic rule can be applied to the blocks as though they were single elements.

For example, the following 3×4 matrix is partitioned into four blocks:

$$A = \begin{bmatrix} 2 & 3 & 1 & 5 \\ 0 & -2 & 2 & 0 \\ 1 & 0 & -4 & 10 \end{bmatrix} = \begin{bmatrix} B & C \\ D & E \end{bmatrix}$$

where B, C, D, and E are the arrays indicated by dashed lines. The matrix entries of such a partitioned matrix are called submatrices. The main matrix is sometimes referred to as a supermatrix.

If A is square, and its only nonzero elements can be partitioned as principal submatrices, then it is called block diagonal. A convenient notation that generalizes is to write A as

$$A = \text{diag}(A_1, A_2, \ldots, A_k)$$

where the submatrices $A_1, A_2, \ldots, A_k$ are square matrices not necessarily of equal dimension, which appear on the major diagonal. The inverse A^{-1} of $A = \text{diag}(A_1, \ldots, A_k)$ is

$$A^{-1} = \text{diag}(A_1^{-1}, A_2^{-1}, \ldots, A_k^{-1})$$

Another advantage of partitioned matrices is that if the partitioned matrix A is multiplied by the matrix X, which is given by

$$X = \begin{bmatrix} X_1 \\ X_2 \end{bmatrix}$$

then

$$AX = \begin{bmatrix} B & C \\ D & E \end{bmatrix} \begin{bmatrix} X_1 \\ X_2 \end{bmatrix}$$

$$= \begin{bmatrix} BX_1 + CX_2 \\ DX_1 + EX_2 \end{bmatrix}$$

The only restriction is that the blocks must be conformable for multiplication, so that all the products BX_1, CX_2, etc. exist. This requires that in a product AX the number of columns in each block of A must equal the number of rows in the corresponding block of X.

2.2.2.6. Partitioned Matrix Inversion

For matrices of high dimension, it is difficult to obtain the inverse of these matrices by using the classical method. In this case the partitioned form is useful to obtain the inverse of these matrices. Suppose F is a matrix in partitioned form as

$$F = \begin{bmatrix} A_{n \times m} & B_{n \times m} \\ C_{m \times n} & D_{m \times m} \end{bmatrix}$$

and

$$F^{-1} = \begin{bmatrix} W_{n \times n} & X_{n \times m} \\ Y_{m \times n} & Z_{m \times m} \end{bmatrix}$$

By definition

$$FF^{-1} = I$$

so

$$\begin{bmatrix} A & B \\ C & D \end{bmatrix}\begin{bmatrix} W & X \\ Y & Z \end{bmatrix} = \begin{bmatrix} I_n & 0 \\ 0 & I_m \end{bmatrix}$$

and applying the rules of partitioned multiplication produces

$$AW + BY = I_n$$

$$AX + BZ = 0$$

$$CW + DY = 0$$

$$CX + DZ = I_m$$

By solving the above equations, one can obtain

$$W = A^{-1} - A^{-1}BY$$

$$Y = -(D - CA^{-1}B)^{-1}CA^{-1}$$

$$Z = (D - CA^{-1}B)^{-1}$$

$$X = -A^{-1}B(D - CA^{-1}B)^{-1}$$

provided that the matrix A is nonsingular.

2.2.3. Quadratic Forms and Definiteness (Ref. 2.3)

One of the most useful forms that plays a central role in the solution of many problems treated in this work is quadratic form. We will define a quadratic form in X as

$$Q(X) = X^T A X = \sum_{i=1}^{n} \sum_{j=1}^{n} a_{ij} x_i x_j$$

where the matrix A can always be assumed symmetric.

A quadratic form is said to be

1. Positive definite if $Q(X) > 0$ for every $X \neq 0$;
2. Positive semidefinite if $Q(X) \geq 0$ for every X, and there exists $X = 0$ such that $Q(X) = 0$;
3. Negative definite if $-Q(X)$ is positive definite;
4. Negative semidefinite if $-Q(X)$ is positive semidefinite;
5. Indefinite if $Q(X)$ satisfies none of the above conditions.

A necessary and sufficient test for a matrix A to be a positive definite matrix is that the following inequalities be satisfied:

$$|a_{11}| > 0, \qquad \begin{vmatrix} a_{11} & a_{12} \\ a_{21} & a_{22} \end{vmatrix} > 0, \qquad \begin{vmatrix} a_{11} & a_{12} & a_{13} \\ a_{21} & a_{22} & a_{23} \\ a_{31} & a_{32} & a_{33} \end{vmatrix} > 0, \qquad \text{etc.}$$

We can test for definiteness independently of the vector X. If λ_i are the eigenvalues of A, then

$$Q(X) > 0 \qquad \text{if all } \lambda_i > 0$$

$$Q(X) \geq 0 \qquad \text{if all } \lambda_i \geq 0$$

$$Q(X) < 0 \qquad \text{if all } \lambda_i < 0$$

$$Q(X) \leq 0 \qquad \text{if all } \lambda_i \leq 0$$

2.3. Discrete Variational Calculus

The calculus of variations is a powerful method for the solution of problems in optimal control. In this section we introduce the subject of variational calculus through a derivation of the Euler equation and associated transversality conditions.

2.3.1. Unconstrained Discrete Optimization

Given a certain discrete cost function in the form

$$J = \sum_{k=k_0}^{k_f-1} F(x_k, x_{k+1}, k) = \sum_{k=k_0}^{k_f-1} F_k \tag{2.1}$$

the problem is find the function x_k that maximizes (or minimizes) the cost functional in equation (2.1).

We shall introduce an arbitrary discrete function η_k, and we assume that

$$x_k = \hat{x}_k + \varepsilon\eta_k, \qquad x_{k+1} = \hat{x}_{k+1} + \varepsilon\eta_{k+1} \tag{2.2}$$

where $\hat{x}_k$ is the optimal solution to the problem in equation (2.1). The parameter ε is a real positive parameter. With the above equations, the cost functional in equation (2.1) becomes

$$J(\varepsilon) = \sum_{k=k_0}^{k_f-1} F(\hat{x}_k + \varepsilon\eta_k, \hat{x}_{k+1} + \varepsilon\eta_{k+1}, k) \tag{2.3}$$

It follows that the function $J(\varepsilon)$ must take on its minimum at $\varepsilon = 0$, where its derivative must vanish, so that

$$\left.\frac{dJ(\varepsilon)}{d\varepsilon}\right|_{\varepsilon=0} = 0 \tag{2.4}$$

η_k is introduced with the stipulation that if x_{k_0} is specified as part of the problem, then η_{k_0} must vanish, and similarly if x_{k_f} is specified, then η_{k_f} must vanish.

The cost functional in equation (2.3) is a function of several variables. From the principles of calculus of variations, we have the following chain rule:

$$\frac{dJ(\varepsilon)}{d\varepsilon} = \left(\frac{\partial J}{\partial x_k}\right)^T\left(\frac{\partial x_k}{\partial \varepsilon}\right) + \left(\frac{\partial J}{\partial x_{k+1}}\right)^T\left(\frac{\partial^x k + 1}{\partial \varepsilon}\right) + \left(\frac{\partial J}{\partial k}\right)^T\left(\frac{\partial k}{\partial \varepsilon}\right) \quad (2.5)$$

The transpose, T, in equation (2.5) may be dropped if x_k is a function of one variable only.

Performing the differentiations we obtain

$$\left.\frac{dJ(\varepsilon)}{d\varepsilon}\right|_{\varepsilon=0} = \sum_{k=k_0}^{k_f-1}\left[\eta_k^T\left(\frac{\partial F_k}{\partial x_k}\right) + \eta_{k+1}^T\left(\frac{\partial F_k}{\partial x_{k+1}}\right)\right] \quad (2.6)$$

The second term in equation (2.6) can be written as

$$\sum_{k=k_0}^{k_f-1}\eta_{k+1}^T\left(\frac{\partial F_k}{\partial x_{k+1}}\right) = \left.\eta_k^T\frac{\partial F(x_{k-1}, x_k, k-1)}{\partial x_k}\right|_{k=k_0}^{k_f}$$
$$+ \sum_{k=k_0}^{k_f-1}\eta_k^T\frac{\partial F(x_{k-1}, x_k, k-1)}{\partial x_k} \quad (2.7)$$

Substituting from equation (2.7) into equation (2.6), and by using equation (2.4), we obtain

$$0 = \left.\eta_k^T\frac{\partial F(x_{k-1}, x_k, k-1)}{\partial x_k}\right|_{k_0}^{k_f} + \sum_{k_0}^{k_f-1}\eta_k^T\left[\frac{\partial F(x_k, x_{k+1}, k)}{\partial x_k}\right.$$
$$\left. + \frac{\partial F(x_{k-1}, x_k, k-1)}{\partial x_k}\right] \quad (2.8)$$

Equation (2.8) consists of a boundary part and a discrete integral part, which are independent of each other. For equation (2.8) to be equal to zero for arbitrary variations, each part must be equal to zero. Thus, we obtain the following vector difference equations:

$$\left.\eta_k^T\frac{\partial F(x_{k-1}, x_k, k-1)}{\partial x_k}\right|_{k=k_0}^{k_f} = 0 \quad (2.9)$$

and

$$\frac{\partial F(x_k, x_{k+1}, k)}{\partial x_k} + \frac{\partial F(x_{k-1}, x_k, k-1)}{\partial x_k} = 0 \quad (2.10)$$

Equation (2.9) is the well-known transversality condition, and equation (2.10) is known as the discrete Euler equation. For a free end point, we have

$$\left.\frac{\partial F(x_{k-1}, x_k, k-1)}{\partial x_k}\right|_{k=k_f} = 0, \qquad \eta_{k_f} \neq 0 \quad (2.11)$$

If x_k is specified at $k = k_0$, then η_k at $k = k_0$ must be equal to zero. Also if x_k is specified at $k = k_f$, then η_k at $k = k_f$ must be zero.

2.3.2. Constrained Discrete Optimization (Ref. 2.6)

Most of the practical problems have constraints relationships on the state variables. The problem with these constraints can be formulated as follows.

Find the function $\hat{x}_k$ that minimizes the following index:

$$J = \sum_{k=k_0}^{k_f-1} F(x_k, x_{k+1}, k) \qquad (2.12)$$

Subject to satisfying the constraints

$$\sum_{k=k_0}^{k_f-1} g(x_k, x_{k+1}, k) = 0 \qquad (2.13)$$

We can now form an unconstrained cost function, by adjoining the constraints in equation (2.12) to the cost functional in equation (2.12) via Lagrange multipliers

$$\tilde{J} = \sum_{k=k_0}^{k_f-1} \hat{F}(x_k, x_{k+1}, \lambda_{k+1}, k) \qquad (2.14)$$

where

$$\hat{F}(\cdot) = F(\cdot) + \lambda_{k+1}^T g(x_k, x_{k+1}, k) \qquad (2.15)$$

as a result we obtain the following modified Euler equation:

$$\frac{\partial \hat{F}(x_k, x_{k+1}, \lambda_{k+1}, k)}{\partial x_k} + \frac{\partial \hat{F}(x_{k-1}, x_k, \lambda_k, k-1)}{\partial x_k} = 0 \qquad (2.16)$$

Furthermore, the transversality conditions become

$$\left. \eta_k^T \frac{\partial \hat{F}(x_{k-1}, x_k, \lambda_k, k-1)}{\partial x_k} \right|_{k=k_0}^{k_f} = 0 \qquad (2.17)$$

Example 2.3.1. In this example, we will consider a simple problem. It is required to find the optimal equations for a hydro reservoir, such that the total MWh generated is maximum. The cost function to be maximized is

$$J = \sum_{k=k_0}^{k_f-1} (au_k + bu_k x_k)$$

Subject to satisfying the reservoir's continuity equation

$$x_{k+1} = x_k + y_k - u_k, \qquad x(k_0) = x(0)$$

where a and b are constants and y_k is the natural inflow. For this example we have

$$\hat{F}(x_k, u_k, x_{k+1}, \lambda_{k+1}, k) = au_k + bu_kx_k + \lambda_{k+1}(-x_{k+1} + x_k + y_k - u_k)$$

$$\frac{\partial \hat{F}(x_k, u_k, x_{k+1}, \lambda_{k+1}, k)}{\partial x_k} = bu_k + \lambda_{k+1},$$

$$\frac{\partial \hat{F}(x_{k-1}, u_{k-1}, x_k, \lambda_k, k-1)}{\partial x_k} = -\lambda_k$$

$$\frac{\partial \hat{F}(x_k, u_k, x_{k+1}, \lambda_{k+1}, k)}{\partial u_k} = a + bx_k - \lambda_{k+1},$$

$$\frac{\partial \hat{F}(x_{k-1}, u_{k-1}, x_k, \lambda_k, k-1)}{\partial u_k} = 0$$

Thus the discrete Euler equations are

$$bu_k + \lambda_{k+1} - \lambda_k = 0$$

$$a + bx_k - \lambda_{k+1} = 0$$

Since the final state is free

$$\lambda_{k_f} = 0$$

Now, the optimal equations with the original continuity equation are given by

$$x_{k+1} = x_k + y_k - u_k$$

$$\lambda_{k+1} - \lambda_k - bu_k = 0, \qquad \lambda_{k_f} = 0$$

$$a + bx_k - \lambda_{k+1} = 0$$

2.4. Discrete Maximum Principle

This section addresses the necessary conditions using the maximum principle approach for a discrete problem having equality constraints of the form

$$x_{k+1} = f(x_k, u_k, k), \qquad k = k_0, \ldots, k_f - 1 \tag{2.18}$$

where k_0 and k_f are fixed. The problem considered is to find an admissible sequence u_k, $k = k_0, \ldots, k_f - 1$ for all k, in order to minimize the cost function

$$J = G(x_k, k)\Big|_{k=k_0}^{k=k_f} + \sum_{k=k_0}^{k_f-1} \phi(x_k, u_k, k) \tag{2.19}$$

The necessary conditions for u_k to be an optimal are as follows.

(1) There exists a function or vector λ_k such that x_k and λ_k are the solutions of the following equations:

$$x_{k+1} = \frac{\partial H^k}{\partial \lambda_{k+1}} \tag{2.20}$$

$$\lambda_k = \frac{\partial H^k}{\partial x_k} \tag{2.21}$$

Subject to the boundary conditions given by

$$x_{k_0} = x(0) \tag{2.22}$$

$$\lambda_{k_f} = \frac{\partial G(x_k, k)}{\partial x_k} \quad \text{at } k = k_f \tag{2.23}$$

The function H^k is a scalar function, which is called the Hamiltonian and is given by

$$H^k(x_k, u_k, \lambda_k, k) = \phi(x_k, u_k, k) + \lambda_{k+1}^T f(x_k, u_k, k) \tag{2.24}$$

(2) If the admissible control vector u_k is unrestricted the functional $H^k(x_k, u_k, \lambda_k, k)$ has a local minimum at

$$\frac{\partial H^k(x_k, u_k, \lambda_k, k)}{\partial u_k} = 0 \tag{2.25}$$

In practice most of the problems have inequality constraints on the control vector u_k and states x_k; in this case we are not free to apply equation (2.25). The maximum principle addresses this difficulty. Instead of equation (2.25), the necessary condition is that the Hamiltonian function $H^k(x_k, u_k, \lambda_k, k)$ has an absolute minimum as a function of u_k over the admissible region Ω for all k in the interval (k_0, k_f). This can be expressed by the following inequality:

$$H^k(x_k, \hat{u}_k, \lambda_k, k) \leq H(x_k, u_k, \lambda_k, k) \tag{2.26}$$

where $\hat{u}_k$ is the optimal control vector in the admissible region Ω.

Example 2.4.1. Find the discrete canonic equations whose solutions minimize

$$J = \tfrac{1}{2} \sum_{k=k_0}^{k_f - 1} (x_k^2 + u_k^2)$$

Subject to the equality constraints

$$x_{k+1} = x_k + \alpha x_k^3 + \beta u_k, \qquad x_{k_0} = x_0$$

Define the Hamiltonian H^k as

$$H^k(x_k, u_k, \lambda_k, k) = \tfrac{1}{2}(x_k^2 + u_k^2) + \lambda_{k+1}(x_k + \alpha x_k^3 + \beta u_k) \tag{2.27a}$$

$$\lambda_k = \frac{\partial H^k}{\partial x_k} \tag{2.27b}$$

or

$$\lambda_k = x_k + \lambda_{k+1}(1 + 3\alpha x_k^2) \tag{2.27c}$$

Since there is no restriction on u_k

$$\frac{\partial H^k}{\partial u_k} = 0$$

$$0 = u_k + \beta\lambda_{k+1} \tag{2.27d}$$

or

$$u_k = -\beta\lambda_{k+1} \tag{2.27e}$$

Use of equation (2.23) yields

$$\lambda_{k_f} = 0 \tag{2.27f}$$

Now, the system equations are given by

$$x_{k+1} = x_k + \alpha x_k^3 - \beta^2\lambda_{k+1}, \qquad x_{k_0} = x_0 \tag{2.27g}$$

$$\lambda_k = \lambda_{k+1}(1 + 3\alpha x_k^2) + x_k, \qquad \lambda_{k_f} = 0 \tag{2.27h}$$

2.4.1. Stochastic Discrete Maximum Principle

This section addresses the necessary conditions for a stochastic discrete problem. The problem is to find an admissible sequence u_k, $k = k_0, \ldots, k_f$ in order to minimize the cost function

$$J = E\left\{ G_f[x(k_f)] + G_0[x(k_0)] + \sum_{k=k_0}^{k_f-1} \phi[x(k), u(k), \xi(k), k] \right\} \tag{2.28}$$

subject to satisfying the difference equality constraint

$$x(k + 1) = f[x(k), u(k), \xi(k), k] \tag{2.29}$$

where $x(k)$ is a generalized state vector that includes all unknown parameters, and $\xi(k)$ is a vector of a stochastic process with presumed known probability density function $P[\xi(k)]$. We may take a particular sample of the stochastic process $\xi^i(k)$ and formulate a deterministic optimization problem. We define the Hamiltonian

$$H^i = \phi[x(k), u(k)\xi^i(k), k] + \lambda^T(k + 1)f[x(k), uy(k), \xi^i(k), k] \tag{2.30}$$

The canonic equations and their associated two-point boundary conditions are

$$x(k + 1) = \frac{\partial H^i}{\partial\lambda(k + 1)} \tag{2.31}$$

$$\lambda(k) = \frac{\partial H^i}{\partial x(k)} \tag{2.32}$$

$$\frac{\partial H^i}{\partial u_k} = 0 \tag{2.33}$$

and the boundary conditions are

$$\lambda(k_0) = \frac{\partial G_0[x(k_0)]}{\partial x(k_0)} \tag{2.34}$$

$$\lambda(k_f) = \frac{\partial G_f[x(k_f)]}{\partial x(k_f)} \tag{2.35}$$

Now, this problem occurs with probability P_i, for $i = 1, 2, \ldots, M$. Thus solving the original stochastic problem is equivalent to solving a weighted sum of problems with appropriate weighting coefficients P_i. The resulting necessary conditions are

$$\frac{\partial H}{\partial \lambda(k+1)} = x(k+1) \tag{2.36}$$

$$\sum_{i=1}^{M} P_i \frac{\partial H^i}{\partial u(k)} = 0 \tag{2.37}$$

$$\sum_{i=1}^{M} P_i \left\{ \frac{\partial H^i}{\partial x(k)} - \lambda(k) \right\} = 0 \tag{2.38}$$

$$\sum_{i=1}^{M} P_i \left\{ \lambda(k_0) + \frac{\partial G_0[x(k_0)]}{\partial x(k_0)} \right\} = 0 \tag{2.39}$$

$$\sum_{i=1}^{M} P_i \left\{ \lambda(k_f) - \frac{\partial G_f[x(k_f)]}{\partial x(k_f)} \right\} = 0 \tag{2.40}$$

We pass formally from the discrete distribution of P_i to a continuous distribution, and then have the relations for the discrete time stochastic maximum principle

$$H = \phi[x(k), u(k), \xi(k), k] + \lambda^T(k+1)f[x(k), u(k), \xi(k), k] \tag{2.41}$$

$$\frac{\partial H}{\partial \lambda(k+1)} = x(k+1) \tag{2.42}$$

$$E\left[\frac{\partial H}{\partial u(k)}\right] = 0 \tag{2.43}$$

$$E\left[\frac{\partial H}{\partial x(k)} - \lambda(k)\right] = 0 \tag{2.44}$$

$$E\left[\lambda(k_0) + \frac{\partial G_0[x(k_0)]}{\partial x(k_0)}\right] = 0 \qquad (2.45)$$

$$E\left[\lambda(k_f) - \frac{\partial G_f[x(k_f)]}{\partial x(k_f)}\right] = 0 \qquad (2.46)$$

where E stands for the expectation. The expectation in (2.41) is taken over the random variables $\xi(k)$.

Example 2.4.2. Find the control vector $u(k)$ that minimizes

$$J = E\left[\tfrac{1}{2}x^2(k_f) + \tfrac{1}{2}\sum_{k_0}^{k_f-1} u^2(k)\right]$$

Subject to the system dynamics

$$x(k+1) = x(k) + u(k) + w(k), \qquad x(k_0) = x_0$$

where $w(k)$ in an unknown random input with known probability density function.

We define the Hamiltonian

$$H = \tfrac{1}{2}u^2(k) + \lambda(k+1)[x(k) + u(k) + w(k)]$$

The stochastic maximum principle of equations (2.42)-(2.46) tends to the canonic equations

$$x(k+1) = x(k) + u(k) + w(k), \qquad x(k_0) = x_0$$

$$E[u(k) + \lambda(k+1)] = 0$$

$$E[\lambda(k+1) - \lambda(k)] = 0, \qquad E[\lambda(k_f) - x(k_f)] = 0$$

The solution to these equations is

$$\bar{\lambda}(k) = x(k_f)$$

$$\bar{x}(k) = x(0) + \sum_{k=k_0}^{k-1} [\bar{u}(k) + w(k)]$$

2.5. Dynamic Programming (Ref. 2.1)

In the previous two sections, we found that for nonlinear systems with coupled variables the state and costate equations are hard to solve; on the other hand, the presence of constraints on the variables makes the problem more complicated.

The aim of this section is to present a brief introduction to dynamic programming, which is an alternative to the variational calculus and the maximum principle approaches discussed in the previous two sections for discrete problems.

The foundation of this technique is based on Bellman's principle of optimality:

> An optimal policy has the property that whatever the initial
> state and initial decision are, the remaining decisions must   (2.47)
> constitute an optimal policy with regard to the state resulting
> from the first decision.

Let the system dynamics be described by

$$x_{k+1} = f^k(x_k, y_k) \tag{2.48}$$

and associated with this system the performance index

$$J_i(x_i) = G(k_f, x_f) + \sum_{k=1}^{k_f-1} \phi(x_k, u_k) \tag{2.49}$$

To explain the principle of optimality, which is the basis for dynamic programming, suppose that we calculated $J_{k+1}^*(x_{k+1})$ from stage $k+1$ to k_f for all possible states x_{k+1}, and also that we calculated the corresponding optimal control sequences for the same period, $k+1$ to k_f. The optimal control results when the optimal control sequence $u_{k+1}^*, u_{k+2}^*, \ldots, u_{k-1}^*$ is applied to the system with a state of x_{k+1}.

If we apply any arbitrary control u_k at time k and then use the known optimal control sequence from $k+1$ onwards, the resulting cost function will be

$$\phi_k(x_k, u_k) + J_{k+1}^*(x_{k+1}) \tag{2.50}$$

where x_k is the state at time k, and x_{k+1} is given by equation (2.48). According to Bellman's principle of optimality, the optimal cost from stage k on is equal to

$$J_k^*(x_k) = \min_{u_k}[\phi_k(x_k, u_k) + J_{k+1}^*(x_{k+1})] \tag{2.51}$$

and the optimal control u_k^* at period k is the one that achieves this minimum. The above equation is the principle of optimality for a discrete system and allows us to optimize over only one control vector at a time by working backward from k_f.

The main disadvantages of dynamic programming are the large storage requirements and large computing time in the case of large-scale systems, which has been termed the "curse of dimensionality" by Bellman.

2.6. Functional Analysis Optimization Technique (Refs. 2.5 and 2.8)

In contrast with other optimization techniques, which generally approach problems of a particular dynamic nature, functional analysis with its geometric character provides a unified framework for optimization problems of discrete, continuous, distributed, or composite nature.

Our object in this section is to state one important minimum norm result that plays an important part in the solution of problems treated in this work. Before we do this, a brief discussion of relevant concepts from functional analysis is given.

2.6.1. Norms and Inner Products

A norm, commonly denoted by $\| \cdot \|$, is real valued and positive definite. The norm satisfies the following axioms:

1. $\|x\| \geq 0$ for all $x \in X,$ $\|x\| = 0$ if and only if $x = 0$;

2. $\|x + y\| \geq \|x\| + \|y\|$ for each $x, y \in X$;

3. $\|\alpha x\| = |\alpha| \cdot \|x\|$ for all scalars α and each $x \in X$.

A normed linear (vector) space X is a linear space in which every vector x has a norm (length). The norm functional is used to define a distance and a convergence measure

$$d(x, y) = \|x - y\|$$

For example let $[1, K]$ be a closed bounded interval. The space of the discrete function $x(k)$ on $[1, K]$ can have one of the following norms:

$$\|x\|_1 = \sum_{k=1}^{K} |x(k)|$$

$$\|x\|_2 = \left[\sum_{k=1}^{K} \|x(k)\|^2 \right]^{1/2}$$

A very important concept in Euclidean geometry is the concept of orthogonality of two vectors. Two vectors are orthogonal if their inner product is zero. Extension of this concept to more generalized spaces leads to powerful results. The concept of orthogonality is not present in all normed spaces and leads to the definition of an inner product defined on $X \times X$ denoted by $\langle \cdot \rangle$. The inner product satisfies

1. $\langle x, y \rangle = \langle y, x \rangle$
2. $\langle \alpha x + \beta y, z \rangle = \alpha \langle x, z \rangle + \beta \langle y, z \rangle$
3. $\langle x, x \rangle \geq 0,$ $\langle x, x \rangle = 0 \leftrightarrow x = 0$
4. $\langle x, x \rangle = \|x\|^2$

2.6.2. Hilbert Space

A linear space X is called a Hilbert space if X is an inner product space that is complete with respect to the norm induced by the inner product.

Equivalently, a Hilbert space is a Banach space whose norm is induced by an inner product. Let us now consider some specific examples of Hilbert spaces. The space E^n is a Hilbert space with inner product as defined by

$$\langle x, y \rangle = x^T y$$

or

$$\langle x, y \rangle = \sum_{i=1}^{n} x_i y_i$$

The space $L_2(k_0, k_f)$ is a Hilbert space with inner product

$$\langle x, y \rangle = \sum_{k=k_0}^{k_f} x^T(k\tau) y(k\tau)$$

An extremely useful Hilbert space is used in this study. The elements of the space are vectors whose components are functions of time over the interval $[k_0, k_f]$. We can define the Hilbert space $L_{2B}^n(k_0, k_f)$:

$$\langle V(k\tau), U(k\tau) \rangle = \sum_{k=k_0}^{k_f} V^T(k\tau) B(k\tau) U(k\tau)$$

for every $V(k\tau)$ and $U(k\tau)$ in the space.

2.6.3. Hilbert Space of Random Variables

Let $P(\xi)$ be the probability that the random variable x assumes a value less than or equal to the number ξ. The expected value of a discrete random variable g, denoted by μ_x, is defined by

$$E[x] = \mu_x = \sum_x g(\xi) P(\xi)$$

where $\sum_x$ means the sum over all x values.

Given a finite collection of real random variables $[x_1, x_2, \ldots, x_n]$, we define their joint probability distribution P as

$$P(\xi_1, \xi_2, \ldots, \xi_n) = \text{Prob}(x_1 \le \xi_1, x_2 \le \xi_2, \ldots, x_n \le \xi_n)$$

i.e., the probability of the simultaneous occurrence of $x_i \le \xi_i$ for all i. The expected value of any discrete function g of x_i's is defined by

$$E[g(x_1, x_2, \ldots, x_n)] = \sum_x g(\xi_1, \xi_2, \ldots, \xi_n) P(\xi_1, \ldots, \xi_n)$$

Two random variables x_i, x_j are said to be uncorrelated if

$$E(x_i x_j) = E(x_i) E(x_j)$$

With this elementary background material, we now construct a Hilbert space of random variables. Let $\{y_1, y_2, \ldots, y_m\}$ be a finite collection of random variables with $E[y_i^2] < \infty$ for each i. We define a Hilbert space H consisting of all random variables that are linear combinations of the y_i's. The inner product of two elements x, y in H is defined as

$$\langle x, y \rangle = E(xy)$$

If $x = \sum \alpha_i y_i$, $y = \sum \beta_i y_i$, then

$$E(xy) = E\left[\left(\sum_i \alpha_i y_i\right)\left(\sum_j \beta_j y_j\right)\right]$$

The space H is a finite-dimensional Hilbert space with dimension equal to at most m. If in the Hilbert space H each random variable has an expected value equal to zero, then two vectors x, z are orthogonal if they are uncorrelated:

$$\langle x, z \rangle = E(x)E(z) = 0$$

The concept of a random variable can be generalized in an important direction. An n-dimensional vector-valued random variable x is an ordered collection of n scalar-valued random variables. Notationally, x is written as a column vector

$$x = \mathrm{col}(x_1, x_2, x_3, \ldots, x_n)$$

(the components being random variables). x in the above equation is referred to as a random vector.

A Hilbert space of random vectors can be generated from a given set of random vectors in a manner analogous to that for random variables. Suppose $\{y_1, y_2, \ldots, y_m\}$ is a collection of n-dimensional random vectors. Each element y_i has n components y_{ij}; $j = 1, 2, \ldots, n$, each of which is a random variable with finite variance. We define the Hilbert space $\mathcal{H}$ of n-dimensional random vectors as consisting of all vectors whose components are linear combinations of the components of the y_i's. Thus an arbitrary element y in this space can be expressed as

$$y = K_1 y_1 + K_2 y_2 + \cdots + K_n y_n$$

where the K_i's are real $n \times n$ matrices.

If x and z are elements of $\mathcal{H}$, we define their inner product as

$$\langle x, z \rangle = E\left(\sum_{i=1}^{n} x_i z_i\right)$$

which is the expected value of the n-dimensional inner product. A convenient notation is

$$\langle x, z \rangle = E(x^T z)$$

The norm of an element x in the space of n-dimensional random vectors can be written as

$$\|x\| = \{\mathrm{Tr}[E(xx^T)]\}^{1/2}$$

where

$$E(xx^T) = \begin{bmatrix} E(x_1x_1) & E(x_1x_2) & \cdots & E(x_1x_n) \\ E(x_2x_1) & E(x_2x_2) & \cdots & E(x_2x_n) \\ E(x_nx_1) & E(x_nx_2) & \cdots & E(x_nx_n) \end{bmatrix}$$

is the expected value of the random matrix dyad xx^T. Similarly, we have

$$\langle x, z \rangle = \mathrm{Tr}[E(xz^T)]$$

2.6.4. A Minimum Norm Theorem

With the preliminary definitions and concepts outlined above, we are now in a position to introduce one powerful result in optimization theory. Our result is a generalization of the idea that the shortest distance from a point to a line is given by the orthogonal to the line from the point.

Theorem. Let B and D be Hilbert spaces. Let T be a bounded linear transformation defined on B with values in D. Let $\hat{u}$ be a given vector in B. For each ξ in the range of T, there exists a unique element $u_\xi \in B$ that satisfies

$$\xi = Tu$$

while minimizing the objection functional

$$J(u) = \|u - \hat{u}\|$$

The unique optimal $u_\xi \in B$ is given by

$$u = T^\dagger[\xi - T\hat{u}] + \hat{u}$$

where the pseudoinverse operator $T^\dagger$ is given by

$$T^\dagger \xi = T^*[TT^*]^{-1}\xi$$

provided that the inverse of TT^* exists.

We can obtain the above result by using a Lagrange multiplier argument. Here we consider an augmented objective

$$\tilde{J} = \|u - \hat{u}\|^2 + \langle \lambda, (\xi - Tu) \rangle$$

where λ is a multiplier (in fact, $\lambda \in D$) to be determined so that the constraint $\xi = Tu$ is satisfied. By utilizing properties of the inner products we can write

$$\tilde{J} = \|u - \hat{u} - T^*(\lambda/2)\|^2 - \|T^*(\lambda/2)\|^2 + \langle \lambda, \xi \rangle$$

Only the first norm depends explicitly on the choice of u. To minimize $\tilde{J}$ we therefore require that

$$u_\xi = \hat{u} + T^*(\lambda/2)$$

The vector $(\lambda/2)$ is obtained using the constraint as the solution to

$$\xi = T\hat{u} + TT^*(\lambda/2)$$

or

$$(\lambda/2) = [TT^*]^{-1}[\xi - T\hat{u}]$$

It is therefore clear that with an invertible TT^* we write

$$u_\xi = T^*[TT^*]^{-1}[\xi - T\hat{u}] + \hat{u}$$

which is the required result.

The theorem as stated is an extension of the fundamental minimum norm problem where the objective functional is

$$J(u) = \|u\|$$

the optimal solution for this case is

$$u_\xi = T^*[TT^*]^{-1}\xi$$

In applying this result to our physical problem we need to recall two important concepts from ordinary constrained optimization. These are the Lagrange multiplier rule and the Kuhn–Tucker multipliers. An augmented objective functional is formed by adding to the original functional terms corresponding to the constraints using the necessary multipliers. The object in these cases is to ensure that the augmented functional can indeed be cast as a norm in the chosen space.

References

2.1. BELLMAN, R., *Introduction to the Mathematical Theory of Control Process*, Vol. 1, Academic Press, New York, 1967.

2.2. BARNETT, S., *Matrix Methods for Engineers and Scientists*, McGraw-Hill, New York, 1979.

2.3. DENN, M. M., *Optimization by Variational Methods*, McGraw-Hill, New York, 1969.

2.4. DORNY, C. N., *A Vector Space Approach to Models and Optimization*, John Wiley & Sons (Intersciences), New York, 1975.

2.5. EL-HAWARY, M. E., and CHRISTENSEN, G. S., *Optimal Economic Operation of Electric Power Systems*, Academic Press, New York, 1979.

2.6. KIRK, D. E., *Optimal Control Theory: An Introduction*, Prentice-Hall, Englewood Cliffs, New Jersey, 1970.

2.7. LUENBERGER, D. G., *Optimization by Vector Space Methods*, John Wiley & Sons, New York, 1969.

2.8. PORTER, W. A., *Modern Foundations of Systems Engineering*, MacMillan, New York, 1966.

2.9. SAGE, A., *Optimum System Controls*, Prentice-Hall, Englewood Cliffs, New Jersey, 1968.

2.10. LEWIS, F. L., *Optimal Control*, John Wiley & Sons, New York, 1986.

2.11. BRYSON, A. E., and HO, YU-CHI, *Applied Optimal Control*, John Wiley & Sons, New York, 1975.

3

Long-Term Operation of Reservoirs in Series

3.1. Introduction

In this chapter we discuss the optimal long-term operation of multireservoir power systems connected in series on a river for maximum total benefits from the system. This chapter begins with the problem formulation, where the problem is posed as a mathematical problem. The second section, Section 3.3.1, is concerned with the applications of dynamic programming with the decomposition approach to solve the problem. For a large-scale power system, the use of full stochastic dynamic programming to solve the problem is computationally infeasible for a system greater than three or four reservoirs. In Section 3.3.2, we develop a method to solve the problem using the minimum norm formulation in the framework of functional analysis optimization technique. We compare, for the same system, the results obtained using dynamic programming with the decomposition approach with those obtained using the minimum norm formulation approach (Refs. 3.7–3.10).

In Sections 3.3.1 and 3.3.2, we use a linear storage-elevation curve and a constant water conversion factor, MWh/m^3, for modeling the hydroelectric generation and the amount of water left in storage at the end of the optimization interval. However, for a hydro power system in which the water heads vary by a considerable amount, these two assumptions are not true; for this reason, Section 3.3.3 is concerned with the solution of the same problem, but the model used for both the storage-elevation curve and the water conversion factor is a nonlinear model. The minimum norm formulation is used to solve the problem in that section (Refs. 3.15 and 3.16).

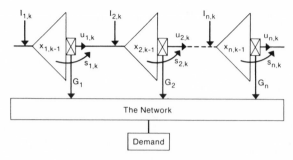

Figure 3.1. The power system.

3.2. Problem Formulation

3.2.1. The System under Study

The system under study consists of n hydroelectric power plants in series on a river, Figure 3.1. We will number the installations from upstream to downstream.

3.2.2. The Objective Function

The problem of the power system of Figure 3.1 is to determine the discharges $u_{i,k}$, $i = 1, \ldots, n$, $k = 1, \ldots, K$ as a function of time under the following conditions:

(1) The expected value of the water left in storage at the end of the last period studied is a maximum.

(2) The expected value of the MWh generated during the optimization interval is a maximum.

(3) The water conservation equation for each reservoir is adequately described by the following difference equations:

$$x_{i,k} = x_{i,k-1} + I_{i,k} - u_{i,k} - s_{i,k} + s_{(i-1),k}, \qquad i = 1, \ldots, n, \qquad k = 1, \ldots, K \tag{3.1}$$

(4) To satisfy the multipurpose stream use requirements the following operational constraints should be satisfied:

$$x_i^m \le x_{i,k} \le x_i^M, \qquad i = 1, \ldots, n, \qquad k = 1, \ldots, K \tag{3.2}$$

$$u_{i,k}^m \le u_{i,k} \le u_{i,k}^M, \qquad i = 1, \ldots, n, \qquad k = 1, \ldots, K \tag{3.3}$$

where x_i^M is the capacity of the reservoir, x_i^m is the minimum storage, $u_{i,k}^m$ is the minimum discharge through the turbine, and $u_{i,k}^M$ is the maximum discharge through the turbine. If $u_{i,k} > u_{i,k}^M$ and $x_{i,k}$ is equal to x_i^M then $u_{i,k} - u_{i,k}^M$ is discharged through the spillways.

In mathematical terms, the long-term problem for the power system of Figure 3.1 is to determine $u_{i,k}$ that maximizes

$$J = E\left[\sum_{i=1}^{n} V_i(x_{i,K}) + \sum_{i=1}^{n}\sum_{k=1}^{K} c_k G_i(x_{i,k-1}, u_{i,k})\right] \tag{3.4}$$

Subject to satisfying the equality constraints given by equation (3.1) and the inequality constraints given by equations (3.2) and (3.3). Where $V_i(x_{i,K})$ is the value of water left in storage in reservoir i at the end of the last period studied, $G_i(x_{i,k-1}, u_{i,k})$ is the generation of plant i during period k in MWh and c_k is the value (in dollars) of a MWh produced anywhere on the river in month k.

3.3. The Problem Solution

3.3.1. Turgeon Approaches (Refs. 3.1–3.6)

Turgeon proposed two methods for solving the problem formulated in equation (3.4). The first method is stochastic dynamic programming; this method is used only for power systems with a small number of reservoirs ($n \leq 4$). The second method is the decomposition approach, which combines many reservoirs into one reservoir and the solution is obtained using dynamic programming.

3.3.1.1. Global Feedback Solution

In this section we discuss how to use stochastic dynamic programming to obtain the global feedback solution. The global feedback solution can be obtained by solving the following functional equation, recursively, starting in period K and going backwards in time:

$$Z_{0,k}(x_{1,k-1}, \ldots, x_{n,k-1}) = \max_{u_k \in U_k} E[Z_{0,K+1}(x_{1,K}, \ldots, x_{n,K})$$

$$+ \sum_{i=1}^{n}\sum_{k=1}^{K} c_k H_i(x_{i,k-1}, u_{i,k})] \tag{3.5}$$

with

$$Z_{0,K+1}(x_{1,K}, x_{2,K}, \ldots, x_{n,K}) = V_1(x_{1,K}) + \cdots + V_n(x_{n,K}) \tag{3.6}$$

Here $u_k = \{u_{1,k}, \ldots, u_{n,k}\}$ and U_k is the set of u_k that satisfy constraints (3.1)–(3.3). The expectation E is taken over the random variables $I_{1,k}, \ldots, I_{n,k}$.

3.3.1.2. Decomposition Approach

The key to this approach is that the optimum discharge $\hat{u}_{i,k}$ is approximately a function of $x_{i,k-1}$ and $c_{i+1,k-1}$ only, where $c_{i+1,k-1}$ is the total energy content of the downstream reservoirs $i + 1, i + 2, \ldots, n$. In other words $c_{i+1,k-1}$ is the total energy content of the composite reservoir of the downstream reservoirs $i + 1, \ldots, n$. Using this approximation, the release policies $\hat{u}_{1,k}(x_{1,k-1}, c_{2,k-1})$, $\hat{u}_{2,k}(x_{2,k-1}, c_{3,k-1})$, $\ldots$, $\hat{u}_{n,k}(x_{n,k-1})$ are determined separately and in the implied order. This requires solving $n - 1$ dynamic programming problems, each with two state variables.

The procedure to find these optimal release policies is first to build a composite reservoir for the downstream reservoirs $i + 1, \ldots, n$ as described in the next section and to represent the whole power system in Figure 3.1 by two reservoirs, the reservoir i and the composite reservoir. Then we determine

$$\{\hat{u}_{i,k}(x_{i,k-1}, c_{i+1,k-1}), \hat{r}_{i+1,k}(x_{i,k-1}, c_{i+1,k}), i = 1, \ldots, n, k = 1, \ldots, K\}$$

the optimal operating policy of this system of two reservoirs by dynamic programming. Next, the conditional probability distribution of the inflow to reservoir $i + 1$, which equals $I_{i+1,k} + \hat{u}_{i,k}(x_{i,k-1}, c_{i+1,k})$, is computed.

3.3.1.2.1. Composite Model (Figure 3.2). The aim of this section is to explain the procedure for building a composite model of reservoirs $i + 1$ to n. This procedure is that of Arvanitidis and Rosing (Ref. 3.1).

First, a fixed conversion factor MWh/m^3 is assigned to each hydroplant. This factor is equal to the average MWh produced in a period k by an outflow of 1 m^3.

Second, the water stored in each reservoir is converted into energy by multiplying it by the sum of the conversion factors of the downstream plants.

Using the above two steps, the total potential energy ($c_{i+1,k}$) stored in reservoirs $i + 1$ to n at the end of the period k is given by

$$c_{i+1,k} = \sum_{j=i+1}^{n} \sum_{m=j}^{n} h_m x_{j,k} \qquad (3.7)$$

where h_m is the water conversion factor measured by MWh/m^3 at site m.

Similarly, the inflow of potential energy to reservoirs $i + 1$ to n in period k is

$$q_{i+1,k} = \sum_{j=i+1}^{n} \sum_{m=j}^{n} h_m I_{j,k} \qquad (3.8)$$

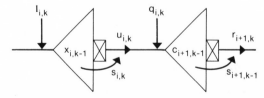

Figure 3.2. Reservoir i and reservoirs $i + 1$ to n combined.

The outflow of potential energy from reservoirs $i + 1$ to n in period k is

$$r_{i+1,k} = \sum_{j=i+1}^{n} h_j u_{j,k} \tag{3.9}$$

Equations (3.7)–(3.9) completely define a composite model for reservoirs $i + 1$ to n (Figure 3.3).

In practice, the MWh generated from the composite reservoir is not equal to the actual MWh generated from each reservoir. The first reason is that the actual water conversion factor, MWh/m^3, at a plant depends heavily on the net head, which may vary considerably. The second reason is that there may be spilllages at some plants. Therefore a generation function $G_{i+1}(c_{i+1,k-1}, r_{i+1,k})$, must be constructed that relates the actual generation to the MWh outflow and the energy content of the composite reservoir.

Let $W_{i+1}(c_{i+1,K})$ denote the expected value of the potential energy stored in the composite reservoir at the end of month K. Then

$$\{u_{i,k}(x_{i,k-1}, c_{i+1,k-1}), r_{i+1,k}(x_{i,k-1}, c_{i+1,k-1}), k = 1, \ldots, K\}$$

the optimal operating policy of reservoir i and reservoirs $i + 1$ to n combined, can then be determined by solving the following problem:

$$\max E\left\{ V_i(x_{i,K}) + W_{i+1}(c_{i+1,K}) + \sum_{k=1}^{K} [c_k G_i(x_{i,k-1}, u_{i,k}) \right.$$

$$\left. + c_k G_{i+1}(c_{i+1,k-1}, r_{i+1,k})] \right\}, \qquad i = 1, \ldots, n \tag{3.10}$$

Subject to the constraints below:

$$x_{i,k} = x_{i,k-1} + I_{i,k} + \hat{u}_{i-1,k} - u_{i,k}, \qquad k = 1, \ldots, K \tag{3.11}$$

$$c_{i+1,k} = c_{i+1,k-1} + q_{i+1,k} + \sum_{j=i+1}^{n} h_j u_{j,k} - r_{i+1,k}, \qquad k = 1, \ldots, K \tag{3.12}$$

$$x_i^m \le x_{i,k} \le x_i^M, \qquad k = 1, \ldots, K \tag{3.13}$$

$$u_{i,k}^m \le u_{i,k} \le u_{i,k}^M, \qquad k = 1, \ldots, K \tag{3.14}$$

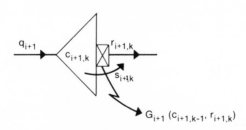

Figure 3.3. Composite model of reservoirs $i + 1$ to n.

$$c_i^m \le c_{i+1,k} \le c_i^M, \qquad k = 1, \ldots, K \tag{3.15}$$

$$r_{i+1,k}^m \le r_{i+1,k} \le r_{i+1,k}^M, \qquad k = 1, \ldots, K \tag{3.16}$$

In equations (3.10) and (3.11), Turgeon assumed that no spillages take place at all. The above equations have been solved by Turgeon using his algorithm mentioned in his famous paper, Ref. 3.5. In this paper, he gave an example; we repeat this example here to make it easy for the reader to compare these results and the results obtained using the algorithm of the next section.

3.3.1.2.2. Example. The system described in this section consists of four reservoirs connected in series on a river. The characteristics of these reservoirs are given in Table 3.1. The optimization is done on a monthly time basis for a period of a year. If d_k is the number of days in a month k, then the maximum and minimum discharges in Mm³ (1 Mm³ = 10⁶ m³) are given by

$$u_{i,k}^M = 0.0864 d_k \qquad \text{(maximum effective discharge in m}^3\text{/sec)}$$

$$u_{i,k}^m = 0.0864 d_k \qquad \text{(minimum effective discharge in m}^3\text{/sec)}$$

Turgeon chose the functionals $G_i(x_{i,k-1}, u_{i,k})$, $V_i(x_{i,K})$, and $W_{i+1}(c_{i+1,k})$ as

$$G_i(x_{i,k-1}, u_{i,k}) = f_i(x_{i,k-1}) \min(u_{i,k}, u_{i,k}^M) \tag{3.17}$$

$$V_i(x_{i,K}) = \sum_{j=i}^{n} h_j x_{i,K} \tag{3.18}$$

and

$$W_{i+1}(c_{i+1,K}) = c_{i+1,K} \tag{3.19}$$

where $u_{i,k}$ is in Mm³, h_j the average monthly production capability of plant j; this is given in the last column of Table 3.1. Selected values of the function f_i are given in Table 3.2. Furthermore, he assumed that

$$G_{i+1}(c_{i+1,k-1}, r_{i+1,k}) = \min(\hat{r}_{i+1,k}) \qquad \forall c_{i+1,k-1} \tag{3.20}$$

where

$$\hat{r}_{i+1,k} = \sum_{j=i+1}^{n} h_j \hat{u}_{j,k} \tag{3.21}$$

The expected natural inflow to the sites in the year of high flows, which we call year 1, and the cost of the energy are given in Table 3.3. In Table 3.4, we report the results for this year. Turgeon has simulated the operation

Table 3.1. Characteristics of the Installations

Site	Capacity of the reservoir (Mm³)	Maximum effective discharge (m³)	Average monthly productibility (MWh/Mm³)
1	9628	400	18.31
2	570	547	234.36
3	50	594	216.14
4	3420	1180	453.44

Table 3.2. Value of $f(m\bar{x}_i)$ for Different Values of m and for $i = 1, \ldots, 4$

m	Plant 1	2	3	4
0	11.28	231.53	215.82	431.75
0.2	14.41	232.68	215.95	441.99
0.4	17.02	233.81	216.08	449.94
0.6	19.54	234.89	216.20	456.54
0.8	21.97	235.94	216.33	462.32
1.0	24.32	236.96	216.46	467.52

Table 3.3. Monthly Inflows to the Four Sites in Year 1

Month k	$I_{1,k}$ (Mm³)	$I_{2,k}$ (Mm³)	$I_{3,k}$ (Mm³)	$I_{4,k}$ (Mm³)	c_k ($/MWh)
1	828	380	161	1798	0.78
2	829	331	82	1201	0.93
3	578	224	41	810	1.03
4	394	146	32	486	1.30
5	265	95	18	302	1.42
6	233	82	14	258	1.34
7	193	68	6	194	1.21
8	219	127	279	1485	0.98
9	1101	614	181	3239	0.83
10	1887	781	205	2560	0.73
11	1150	491	146	1583	0.70
12	824	363	132	1454	0.76

Table 3.4. Monthly Releases from the Four Reservoirs in Year 1 and the Profits Realized When the Local Feedback Solution is Implemented

Month k	$u_{1,k}$ (Mm^3)	$u_{2,k}$ (Mm^3)	$u_{3,k}$ (Mm^3)	$u_{4,k}$ (Mm^3)	Profits ($)
1	0	368	528	2260	980,300
2	0	331	413	1617	858,930
3	1071	1295	1338	2145	1,672,310
4	1071	1217	1250	2112	2,040,840
5	968	1323	1334	2855	2,771,260
6	1071	1463	1524	3161	2,856,200
7	1037	1104	1111	1749	1,552,640
8	659	566	844	2333	1,306,010
9	0	835	1016	3059	1,440,070
10	306	1087	1289	3161	1,424,850
11	0	150	251	310	161,000
12	0	134	266	1717	677,750

Value of the water remaining in the reservoirs at the end of the year	10,403,170
Total profits	28,145,330

of the system for widely different water conditions. In Table 3.5 we give the expected natural inflows to the sites in the year of low flow, which we call year 2. In Table 3.6 we report the corresponding results for that year. We began both years with

$$x_{1,0} = 6688.5, \qquad x_{2,0} = 557.9, \qquad x_{3,0} = 48.9, \quad \text{and} \quad x_{4,0} = 3347.4 \text{ Mm}^3$$

Table 3.5. Monthly Inflows to the Four Sites in Year 2

Month k	$I_{1,k}$ (Mm^3)	$I_{2,k}$ (Mm^3)	$I_{3,k}$ (Mm^3)	$I_{4,k}$ (Mm^3)
1	568	227	29	708
2	442	173	150	772
3	460	171	22	608
4	505	187	22	601
5	324	116	22	272
6	305	113	39	307
7	208	81	13	219
8	161	110	259	1327
9	498	301	64	1539
10	642	274	51	886
11	424	169	22	605
12	527	229	75	895

Table 3.6. Monthly Releases from the Four Reservoirs in Year 2 and the Profits Realized When the Local Feedback Solution is Implemented

Month k	$u_{1,k}$ (Mm^3)	$u_{2,k}$ (Mm^3)	$u_{3,k}$ (Mm^3)	$u_{4,k}$ (Mm^3)	Profits ($)
1	0	215	243	885	402,850
2	0	173	323	1097	580,040
3	1071	1242	1264	1689	1,509,300
4	1071	1259	1281	2258	2,149,750
5	968	1323	1348	2855	2,771,320
6	1071	1465	1552	3161	2,863,470
7	1037	1167	1181	1895	1,667,460
8	136	25	284	1614	751,270
9	0	522	586	2125	967,810
10	0	274	275	1162	455,890
11	0	0	22	0	3,330
12	0	0	75	0	12,370

Value of the water remaining in the reservoirs at the end of the year	7,020,170
Total profits	21,155,030

The processing time required to determine the monthly operating policy for a period of a year of the system just described was 0.199 min on an IBM 370/233 computer.

3.3.1.2.3. Comments. In Section 3.3.1, we reviewed the dynamic programming approach with the decomposition technique to solve the long-term optimal operating problem of a multireservoir power system connected in series on a river. The decomposition approach is justified only for similar reservoirs; otherwise, the solution obtained by this method is not a global feedback solution, but it is rather a suboptimal operating policy, because it is not possible yet to obtain global feedback solutions for large-scale systems. The control law supposes that the river flows for period k are known in probability terms only when the policy is chosen.

If the river flows are known with certainty, a control law of the form $\hat{u}_{i,k}(x_{i,k-1}, c_{i+1,k-1}, I_{i,k}, q_{i,k})$ must be sought. This can be done with the proposed method, but at the expense of more computer time and more memory space.

3.3.2. A Minimum Norm Approach

In this section we offer a formulation to the problem of a multireservoir power system connected in series on a river, which is based on the minimum norm approach. We repeat here the objective function for the reader for ease of understanding.

The problem of the power system shown in Figure 3.1 is to determine the discharge $u_{i,k}$, $i = 1, \ldots, n$, $k = 1, \ldots, K$ that maximizes

$$J = E\left[\sum_{i=1}^{n} V_i(x_{i,K}) + \sum_{i=1}^{n} \sum_{k=1}^{K} c_k G_i(u_{i,k}, x_{i,k-1})\right] \qquad (3.22)$$

Subject to satisfying the following constraints:

$$x_{i,k} = x_{i,k-1} + I_{i,k} + u_{(i-1),k} - u_{i,k} - s_{i,k} + s_{(i-1),k} \qquad (3.23)$$

($s_{i,k}$ denotes spill from reservoir i during month k)

$$x_i^m \leq x_{i,k} \leq x_i^M \qquad (3.24)$$

$$u_{i,k}^m \leq u_{i,k} \leq u_{i,k}^M \qquad (3.25)$$

3.3.2.1. Modeling of the System

The conventional approach for obtaining the equivalent reservoir and hydroplant is based on the potential energy concept. Each reservoir on a river is mathematically represented by an equivalent potential energy balance equation. The potential energy balance equation is obtained by multiplying both sides of the reservoir balance-of-water equation by the water conversion factors of at-site and downstream hydroplants. We may choose the following for the function $V_i(x_{i,K})$:

$$V_i(x_{i,K}) = \sum_{j=1}^{n} h_j x_{i,K} \qquad (3.26)$$

where h_j is the average water conversion factor (MWh/Mm^3) at site j. In the above equation we assumed that the cost of this energy is one dollar/MWh (the average cost during the year).

The generation of a hydroelectric plant is a nonlinear function of the water discharge $u_{i,k}$ and the net head, which itself is a function of the storage. In this section, we will assume a linear relation between the storage and the head (the storage-elevation curve is linear and the tailwater elevation is constant independently of the discharge). We may choose

$$G_i(u_{i,k}, x_{i,k-1}) = a_i u_{i,k} + b_i u_{i,k} x_{i,k-1} \text{ MWh} \qquad (3.27)$$

where a_i and b_i are constants for the reservoir i. These were obtained by least-squares curve fitting to typical plant data available.

Now, the cost functional in equation (3.22) becomes

$$J = E\left[\sum_{i=1}^{n} \sum_{j=i}^{n} h_j x_{i,K} + \sum_{i=1}^{n} \sum_{k=1}^{K} (A_{i,k} u_{i,k} + u_{i,k} B_{i,k} x_{i,k-1})\right] \qquad (3.28)$$

Subject to satisfying the constraints given by equations (3.23)–(3.25) where

$$A_{i,k} = a_i c_k, \qquad i = 1, \ldots, n$$

$$B_{i,k} = b_i c_k, \qquad i = 1, \ldots, n$$

3.3.2.2. Formulation

The reservoir dynamic equation is added to the cost functional using the unknown Lagrange multiplier $\lambda_{i,k}$, and the inequality constraints (3.24) and (3.25) are added using the Kuhn–Tucker multipliers, so that a modified cost functional is obtained:

$$J_0(u_{i,k}, x_{i,k-1}) = E\left[\sum_{i=1}^{n} \sum_{j=i}^{n} h_j x_{i,K} + \sum_{i=1}^{n} \sum_{k=1}^{K} \{A_{i,k} u_{i,k} + u_{i,k} B_{i,k} x_{i,k-1} \right.$$

$$+ \lambda_{i,k}(-x_{i,k} + x_{i,k-1} + u_{(i-1),k} - u_{i,k}) + (e_{i,k}^{M} - e_{i,k}^{m}) x_{i,k}$$

$$\left. + (f_{i,k}^{M} - f_{i,k}^{m}) u_{i,k} \} \right] \tag{3.29}$$

Here terms explicitly independent of $u_{i,k}$ and $x_{i,k}$ are dropped. In the above equation $e_{i,k}^{M}, e_{i,k}^{m}, f_{i,k}^{M}$, and $f_{i,k}^{m}$ are Kuhn–Tucker multipliers. These are equal to zero if the constraints are not violated and greater than zero if the constraints are violated.

Let R denote the set of n reservoirs, and define the $n \times 1$ column vectors

$$H = \text{col}[H_i, i \in R] \tag{3.30}$$

where

$$H_i = \sum_{j=i}^{n} h_j \tag{3.31}$$

$$x(k) = \text{col}[x_{i,k}, i \in R] \tag{3.32}$$

$$u(k) = \text{col}[u_{i,k}, i \in R] \tag{3.33}$$

$$\lambda(k) = \text{col}[\lambda_{i,k}, i \in R] \tag{3.34}$$

$$\mu(k) = \text{col}[\mu_{i,k}, i \in R] \tag{3.35}$$

$$\psi(k) = \text{col}[\psi_{i,k}, i \in R] \tag{3.36}$$

where

$$\mu_{i,k} = e_{i,k}^{M} - e_{i,k}^{m}, \qquad i \in R \tag{3.37}$$

and

$$\psi_{i,k} = f_{i,k}^{M} - f_{i,k}^{m}, \qquad i \in R \tag{3.38}$$

and the $n \times n$ diagonal matrix

$$B(k) = \text{diag}[B_{i,k}, i \in R] \tag{3.39}$$

Furthermore, define the $n \times n$ lower triangular matrix M by

$$m_{ii} = -1, \qquad i \in R$$
$$m_{(j+1)j} = 1, \qquad j = 1, \ldots, n-1 \tag{3.40}$$

Then the modified cost functional in equation (3.29) becomes

$$J_0(u(k), x(k-1)) = E\left[H^T x(K) + \sum_{k=1}^{K} \{u^T(k)B(k)x(k-1) \right.$$
$$+ (\lambda(k) + \mu(k))^T x(k-1) - \lambda^T(k)x(k)$$
$$+ (A(k) + M^T\lambda(k)$$
$$\left. + M^T\mu(k) + \psi(k))^T u(k)\} \right] \tag{3.41}$$

We will need the following identity:

$$\sum_{k=1}^{K} \lambda^T(k)x(k) = -\lambda^T(0)x(0) + \lambda^T(K)x(K)$$
$$+ \sum_{k=1}^{K} \lambda^T(k-1)x(k-1) \tag{3.42}$$

Then we can write the cost functional (3.41) as

$$J_0[u(k), x(k-1)] = E[(H - \lambda(K))^T x(K) + \lambda^T(0)x(0)$$
$$+ \sum_{k=1}^{K} \{u^T(k)B(k)x(k-1)$$
$$+ (\lambda(k) - \lambda(k-1) + \mu(k))^T x(k-1)$$
$$+ (A(k) + M^T\lambda(k) + M^T\mu(k) + \psi(k))^T u(k)\}] \tag{3.43}$$

Define the $2n \times 1$ column vectors

$$X(k) = \text{col}[x(k-1), u(k)] \tag{3.44}$$
$$R(k) = \text{col}[(\lambda(k) - \lambda(k-1) + \mu(k)),$$
$$(A(k)M^T\lambda(k) + M^T\mu(k) + \psi(k))] \tag{3.45}$$

and also the $2n \times 2n$ matrix $L(k)$ as

$$L(k) = \begin{bmatrix} 0 & \frac{1}{2}B(k) \\ \frac{1}{2}B(k) & 0 \end{bmatrix} \tag{3.46}$$

Using these definitions, (3.43) becomes

$$J_0[x(K), X(k)] = E[(H - \lambda(K))^T x(K) + \lambda^T(0)x(0)$$

$$+ \sum_{k=1}^{K} \{X^T(k)L(k)X(k) + R^T(k)X(k)\}] \quad (3.47)$$

Equation (3.47) is composed of a boundary term and a discrete integral part, which are independent of each other. To maximize J_0 in equation (3.47), one maximizes each term separately:

$$\max J_0[x(K), X(k)] = \max_{x(K)} E[(H - \lambda(K))^T x(K) + \lambda^T(0)x(0)$$

$$+ \max_{X(k)} E\left[\sum_{k=1}^{K} \{X^T(k)L(k)X(k) + R^T(k)X(k)\}\right]$$

$$(3.48)$$

3.3.2.3. The Optimal Solution

There is exactly one optimal solution to the problem formulated in (3.48). The boundary part in (3.48) is optimized when

$$E[\lambda(K) - H] = [0] \quad (3.49)$$

because $\delta x(K)$ is arbitrary and $x(0)$ is constant.

We define the $2n \times 1$ vector $V(k)$ as

$$V(k) = L^{-1}(k)R(k) \quad (3.50)$$

Now, the discrete integral part of equation (3.48) can be written as

$$\max_2[X(k)] = \max_{X(k)} E\left[\sum_{k=1}^{K} \{(X(k) + \tfrac{1}{2}V(k))^T L(Lk)(X(k) + \tfrac{1}{2}V(k))\right.$$

$$\left. - \tfrac{1}{4}V^T(k)L(k)V(k)\}\right] \quad (3.51)$$

The last term in equation (3.51) does not depend explicitly on $X(k)$, so that it is necessary only to consider

$$\max J_2[X(k)] = \max_{X(k)} E\left[\sum_{k=1}^{K} \{(X(k) + \tfrac{1}{2}V(k))^T L(k)(X(k) + \tfrac{1}{2}V(k))\}\right]$$

$$(3.52)$$

Equation (3.52) defines a norm. This norm is considered to be an element of a Hilbert space because $X(k)$ is always positive. Equation (3.52) can be written as

$$\max J_2[X(k)] = \max_{X(k)} E\|X(k) + \tfrac{1}{2}V(k)\|_{L(k)} \quad (3.53)$$

The maximization of $J_2[X(k)]$ is mathematically equivalent to the minimiz-ation of the norm of equation (3.53). The minimum of the norm in equation (3.53) is clearly achieved when

$$E\|X(k) + \tfrac{1}{2}V(k)\| = [0] \tag{3.54}$$

Substituting from equation (3.50) into equation (3.54) for $V(k)$, one finds the optimal solution is given by

$$E[R(k) + 2L(k)X(k)] = [0] \tag{3.55}$$

Writing equation (3.55) explicitly and adding the reservoir dynamic equation, one obtains the long-term optimal equations as

$$E[-x(k) + x(k-1) + I(k) + Mu(k) + Ms(k)] = [0] \tag{3.56}$$

$$E[\lambda(k) - \lambda(k-1) + \mu(k) + B(k)u(k)] = [0] \tag{3.57}$$

$$E[A(k) + M^T\lambda(k) + M^T\mu(k) + \psi(k) + B(k)x(k-1)] = [0] \tag{3.58}$$

We can now state the optimal solution of equations (3.56)–(3.58) in component form as

$$E[-x_{i,k} + x_{i,k-1} + I_{i,k} + u_{(i-1),k} - u_{i,k} + s_{(i-1),k} - s_{i,k}] = 0$$

$$i = 1, \ldots, n, \qquad k = 1, \ldots, K \tag{3.59}$$

$$E[\lambda_{i,k} - \lambda_{i,k-1} + \mu_{i,k} + c_k b_i u_{i,k}] = 0, \qquad i = 1, \ldots, n, \qquad k = 1, \ldots, K \tag{3.60}$$

$$E[c_k a_i + \lambda_{(i+1),k} - \lambda_{i,k} + \mu_{(i+1),k} - \mu_{i,k} + \psi_{i,k} + c_k b_i x_{i,k-1}] = 0$$

$$i = 1, \ldots, n, \qquad k = 1, \ldots, K \tag{3.61}$$

Besides the above equations, one has the Kuhn–Tucker exclusion equations, which must be satisfied at the optimum as

$$e_{i,k}(x_i^m - x_{i,k}) = 0, \qquad e_{i,k}^1(x_{i,k} - x_i^M) = 0 \tag{3.62}$$

$$f_{i,k}(u_{i,k}^m - u_{i,k}) = 0, \qquad f_{i,k}^1(u_{i,k} - u_{i,k}^M) = 0 \tag{3.63}$$

also we have the following limits on the variables:

$$\begin{aligned} &\text{if } x_{i,k} < x_i^m, &&\text{then we put } x_{i,k} = x_i^m \\ &\textit{if } x_{i,k} > x_i^M, &&\text{then we put } x_{i,k} = x_i^M \\ &\text{if } u_{i,k} < u_{i,k}^m, &&\text{then we put } u_{i,k} = u_{i,k}^m \\ &\text{if } u_{i,k} > u_{i,k}^M, &&\text{then we put } u_{i,k} = u_{i,k}^M \end{aligned} \tag{3.64}$$

Equations (3.59)–(3.64) completely specify the optimal solution.

Table 3.7. Characteristics of the Installations

Site	Minimum capacity x_i^m (Mm³)	Maximum capacity x_i^M (Mm³)	Maximum effective discharge (m³/sec)	Minimum effective discharge (m³/sec)	Average monthly productibility (MWh/Mm³)	Reservoir's constants a_i (MWh/Mm³)	b_i [MWh/(Mm³)²]
1	0	9628	400	0	18.31	11.8	1.3×10^{-3}
2	0	570	547	0	234.36	231.5	0.532×10^{-3}
3	0	50	594	0	216.14	215.82	12.667×10^{-3}
4	0	3420	1180	0	453.44	437.00	11.173×10^{-3}

3.3.2.4. Practical Application (see Table 3.7)

A computer program was written to solve equations (3.57)–(3.62) iteratively using the steepest descent method. Figure 3.4 shows the main outline of this program. This program was applied to the system of Section 3.3.1.2. The reservoir constants a_i and b_i were obtained by least-squares curve fitting to the data given in Table 3.2, where m is the ratio between the storage x_i and the capacity of the reservoir x_i^M.

In Table 3.8 we give the optimal discharges and the profits realized during the year of high flow, year 1. Also, in Table 3.9 we report the results obtained for the same system during the year of low flow, year 2.

The computing time to obtain the optimal solution for a period of a year was 1.2 sec in CPU units on the University of Alberta Amdahl 470V/6 computer.

3.3.2.5. Comments

We presented in this section the application of the minimum norm theorem to the optimization of the total benefits from a multireservoir power

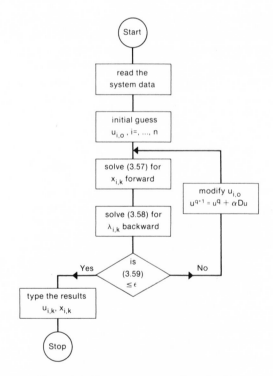

Figure 3.4. Computer program flow chart.

Table 3.8. Optimal Monthly Releases from the Reservoirs and the Profits Realized in Year 1

Month k	$u_{1,k}$ (Mm^3)	$u_{2,k}$ (Mm^3)	$u_{3,k}$ (Mm^3)	$u_{4,k}$ (Mm^3)	Profits ($)
1	0	368	528	2253	990,816
2	1037	1418	1499	2700	1,828,432
3	853	1076	1118	1928	1,473,793
4	1071	1465	1547	2244	2,300,938
5	968	1130	1098	2855	2,656,644
6	1071	1336	1400	3161	2,785,338
7	1037	1127	1133	1580	1,475,190
8	0	0	229	1310	609,669
9	0	446	627	2326	1,051,100
10	0	506	711	2330	979,250
11	0	491	637	2117	873,319
12	0	363	495	1794	788,087

Value of water remaining in the reservoirs at the end of the year		10,414,598
Total profits		28,227,174

system connected in series on a river. It was found that this algorithm can deal with a large-scale power system with stochastic inflows. In this section new optimal equations are derived; if these equations are solved forward and backward in time, one can easily obtain the optimal long-term scheduling for maximum total benefits from any number of series reservoirs.

Table 3.9. Monthly Releases from the Reservoirs and the Profits Realized in Year 2

Month k	$u_{1,k}$ (Mm^3)	$u_{2,k}$ (Mm^3)	$u_{3,k}$ (Mm^3)	$u_{4,k}$ (Mm^3)	Profits ($)
1	0	215	343	878	405,647
2	1037	1262	1412	2242	1,573,185
3	881	1052	1074	1624	1,308,419
4	1071	1465	1537	3107	2,829,756
5	968	1319	1341	2855	2,759,020
6	1071	1260	1299	2628	2,380,707
7	1037	1118	1131	1418	1,383,345
8	0	0	209	1470	675,513
9	0	0	64	1454	541,237
10	0	115	166	91	75,287
11	0	169	191	114	93,025
12	0	229	304	154	144,909

Value of water remaining in the reservoirs at the end of the year		7,165,203
Total benefits		21,335,253

We have compared the minimum norm approach with the dynamic programming and decomposition approach, which is currently used by many utility companies. For the same system we obtained increased benefits using a smaller computing time as indicated above.

3.3.3. A Nonlinear Model: Minimum Norm Approach

In the previous two sections, a constant water conversion factor (MWh/Mm^3) is used. For power systems in which the water heads vary by a small amount, this assumption is adequate, but for power systems in which this variation is large, using this assumption is not adequate. On the other hand, the linear storage-elevation curve used in modeling the reservoirs is adequate only for small-capacity reservoirs of rectangular cross sections. In practice, most reservoirs are nearly trapezoidal in cross section.

This section discusses the solution of the long-term optimal operating problem of multireservoir power systems having a variable water conversion factor and a nonlinear storage-elevation curve. The optimal solution is obtained using the minimum norm formulation in the framework of functional analysis optimization technique.

3.3.3.1. Formulation

Again, the problem of the power system of Figure 3.1 is to determine the optimal discharge $u_{i,k}$, $i = 1, \ldots, n$, $k = 1, \ldots, K$ subject to satisfying the following constraints:

(a) The total expected benefits from the system (benefits from the energy generated by hydropower systems over the planning period plus the expected future benefits from the water left in storage at the end of that period) is a maximum.

(b) The MWh generated per Mm^3 ($1 Mm^3 = 10^6 m^3$) discharge (the water conversion factor) as a function of the storage is adequately described by

$$h_{i,k} = \alpha_i + \beta_i x_{i,k} + \gamma_i (x_{i,k})^2, \qquad i = 1, \ldots, n, \qquad k = 1, \ldots, K \quad (3.65)$$

where α_i, β_i, and γ_i are constants, These were obtained by least-squares curve fitting to typical plant data available.

(c) The reservoir dynamics is given by equation (3.23) and the operational constraints are given by equations (3.24) and (3.25).

(d) The MWh generated by a hydropower plant during a period k is given by

$$G_i[u_{i,k}, \tfrac{1}{2}(x_{i,k} + x_{i,k-1})] = \alpha_i u_{i,k} + \tfrac{1}{2}\beta_i u_{i,k}(x_{i,k} + x_{i,k-1}) + \tfrac{1}{4}\gamma_i u_{i,k}(x_{i,k} + x_{i,k-1})^2,$$

$$i = 1, 2, \ldots, n \quad (3.66)$$

Substituting for $x_{i,k}$ from equation (3.23) into equation (3.66) yields

$$G_i[u_{i,k}, \tfrac{1}{2}(x_{i,k} + x_{i,k-1})] = b_{i,k}u_{i,k} + d_{i,k}u_{i,k}x_{i,k-1} + f_{i,k}(u_{i-1,k} - u_{i,k})u_{i,k}$$
$$+ \gamma_i u_{i,k}y_{i,k-1} + \tfrac{1}{4}\gamma_i u_{i,k}(z_{i,k} + z_{i-1,k})$$
$$+ \gamma_i r_{i,k-1}(u_{i-1,k} - u_{i,k}) - \tfrac{1}{2}\gamma_i z_{i,k}u_{i-1,k} \qquad (3.67)$$

where

$$q_{i,k} = I_{i,k} + s_{i-1,k} - s_{i,k}, \qquad i = 1, 2, \ldots, n \qquad (3.68a)$$
$$b_{i,k} = \alpha_i + \tfrac{1}{2}\beta_i q_{i,k} + \tfrac{1}{4}\gamma_i (q_{i,k})^2, \qquad i = 1, 2, \ldots, n \qquad (3.68b)$$
$$d_{i,k} = \beta_i + \gamma_i q_{i,k}, \qquad i = 1, 2, \ldots, n \qquad (3.68c)$$
$$f_{i,k} = \tfrac{1}{2}d_{i,k}, \qquad i = 1, 2, \ldots, n \qquad (3.68d)$$

and the following are pseudostate variables:

$$y_{i,k} = (x_{i,k})^2, \qquad i = 1, 2, \ldots, n, \qquad k = 1, 2, \ldots, K \qquad (3.69)$$
$$z_{i,k} = (u_{i,k})^2, \qquad i = 1, 2, \ldots, n, \qquad k = 1, 2, \ldots, K \qquad (3.70)$$
$$r_{i,k-1} = u_{i,k}x_{i,k-1}, \qquad i = 1, 2, \ldots, n, \qquad k = 1, 2, \ldots, K \qquad (3.71)$$

In mathematical terms, the object of the optimizing computation is to find the discharge $u_{i,k}$ that maximizes

$$J = E\left[\sum_{i=1}^{n} \sum_{j=1}^{n} x_{i,K}(\alpha_j + \beta_j x_{j,K} + \gamma_j y_{j,K}) \right.$$
$$+ \sum_{k=1}^{K} \{c_k b_{i,k}u_{i,k} + c_k d_{i,k}u_{i,k}x_{i,k-1} + c_k f_{i,k}u_{i,k}(u_{i-1,k} - u_{i,k})$$
$$+ c_k \gamma_i u_{i,k}y_{i,k-1} + \tfrac{1}{4}c_k \gamma_i u_{i,k}(z_{i,k} + z_{i-1,k})$$
$$\left. + c_k \gamma_i r_{i,k-1}(u_{i-1,k} - u_{i,k}) - \tfrac{1}{2}c_k \gamma_i z_{i,k}u_{i-1,k}\} \right] \qquad \text{in \$} \qquad (3.72)$$

Subject to satisfying the following constraints:

1. Equality constraints given by equation (3.23) and equations (3.69)–(3.71);
2. Inequality constraints given by equations (3.24) and (3.25).

The cost functional in equation (3.72) can be written in vector form as

$$J = E\left[A^T x(K) + x^T(K)\beta x(K) + x^T(K)\gamma y(K) + \sum_{k=1}^{K} \{B^T(k)u(k) \right.$$
$$+ u^T(k)d(k)x(k-1) + u^T(k)f(k)Mu(k) + u^T(k)\gamma(k)y(k-1)$$
$$\left. + \tfrac{1}{4}u^T(k)\gamma(k)Nz(k) + r^T(k-1)\gamma(k)Mu(k) - \tfrac{1}{2}z^T(k)\gamma(k)Lu(k)\} \right]$$
$$(3.73)$$

and the equality and inequality constraints can be written in vector form as

(1) Equality constraints given by

$$x(k) = x(k-1) + q(k) + Mu(k) \tag{3.74}$$

$$y(k) = x^T(k)\mathbf{H}x(k) \tag{3.75}$$

$$z(k) = u^T(k)\mathbf{H}u(k) \tag{3.76}$$

$$r(k-1) = u^T(k)\mathbf{H}x(k-1) \tag{3.77}$$

(2) Inequality constraints given by

$$x^m \le x(k) \le x^M \tag{3.78}$$

$$u^m(k) \le u(k) \le u^M(k) \tag{3.79}$$

where $x(k)$, $u(k)$, $y(k)$, $z(k)$, $r(k-1)$ are n-dimensional vectors at the end of period k; their components are $x_{i,k}$, $u_{i,k}$, $y_{i,k}$, $z_{i,k}$, and $r_{i,k-1}$, respectively; $d(k)$, $f(k)$, $\gamma(k)$ are $n \times n$ diagonal matrices whose elements are $d_{ii,k} = c_k d_{i,k}$, $f_{ii,k} = c_k f_{i,k}$ and $\gamma_{ii,k} = c_k \gamma_i$, $i = 1, \ldots, n$; M is a lower triangular matrix whose elements are $m_{ii} = -1$, $i = 1, \ldots, n$; $m_{(j+1)j} = 1$, $j = 1, \ldots, n-1$; N is a lower triangular matrix whose elements are $n_{ii} = 1$, $i = 1, \ldots, n$, $n_{(j+1)j} = 1$, $j = 1, \ldots, n-1$; β is an $n \times n$ matrix whose elements are

$$\beta_{ii} = \beta_i, \qquad i = 1, \ldots, n, \qquad \beta_{(j+1)i} = \beta_{i(j+1)} = \tfrac{1}{2}\beta_{(j+1)},$$

$$j = 1, \ldots, n-1$$

γ is a $n \times n$ upper triangular matrix whose elements are $\gamma_{ii} = \gamma_i$; $\gamma_{i(j+1)} = \gamma_{(j+1)}$; $i = 1, \ldots, n, j = 1, \ldots, n-1$; $\mathbf{H}$ is a vector matrix in which the vector index varies from 1 to n while the matrix dimension of H is $n \times n$; A, $B(k)$ are n-dimensional vectors whose components are given by

$$A_i = \sum_{j=i}^{n} \alpha_j, \qquad \beta_{i,k} = c_k b_{i,k}, \qquad i = 1, \ldots, n$$

The augmented cost functional is obtained by adjoining to the cost function in equation (3.73), the equality constraints via Lagrange multipliers and the inequality constraints via Kuhn–Tucker multipliers as

$$\tilde{J} = E\bigg[A^T x(K) + x^T(K)\beta x(K) + x^T(K)\gamma y(K) + \sum_{k=1}^{K} \{B^T(k)u(k)$$

$$+ u^T(k)d(k)x(k-1) + u^T(k)f(k)Mu(k) + u^T(k)\gamma(k)y(k-1)$$

$$+ \tfrac{1}{4}u^T(k)\gamma(k)Nz(k) + r^T(k-1)\gamma(k)Mu(k) - \tfrac{1}{2}z^T(k)\gamma(k)Lu(k)$$

$$+ \lambda^T(k)(-x(k) + x(k-1) + q(k) + Mu(k)) + \phi^T(k)(-y(k)$$

$$+ x^T(k)\mathbf{H}x(k)) + \mu^T(k)(-z(k) + u^T(k)\mathbf{H}u(k))$$

$$+ \psi^T(k)(-r(k-1) + u^T(k)\mathbf{H}x(k-1))$$

$$+ e^{mT}(k)(x^m - x(k)) + e^{MT}(k)(x(k) - x^M)$$

$$+ g^{mY}(k)(u^m(k) - u(k)) + g^{MT}(k)(u(k) - u^M(k))\Big] \tag{3.80}$$

where $\lambda(k)$, $\phi(k)$, $\mu(k)$, and $\psi(k)$ are Lagrange multipliers in \$/Mm^3. These are obtained so that the corresponding equality constraints are satisfied, and $e^m(k)$, $e^M(k)$, $g^m(k)$, and $g^M(k)$ are Kuhn-Tucker multipliers; these are equal to zero if the constraints are not violated and greater than zero if the constraints are violated.

Employing the discrete version of integration by parts, and dropping constant terms we obtain

$$\tilde{J} = E\Big[x^T(K)(\beta + \phi^T(K)\mathbf{H})x(K) + (A - \lambda(K))^Tx(K) - \phi^T(K)y(K)$$

$$+ x^T(K)\gamma y(K) + x^T(0)\phi^T(0)\mathbf{H}x(0) + \lambda^T(0)x(0) + \phi^T(0)y(0)$$

$$+ \sum_{k=1}^{K} \{x^T(k-1)\phi^T(k-1)\mathbf{H}x(k-1) + u^T(k)d(k)x(k-1)$$

$$+ u^T(k)(f(k)M + \mu^T(k)\mathbf{H})u(k) + u^T(k)\gamma(k)y(k-1)$$

$$+ \tfrac{1}{4}u^T(k)\gamma(k)Nz(k) + r^T(k-1)\gamma(k)Mu(k) - \tfrac{1}{2}z^T(k)\gamma(k)luy(k)$$

$$+ u^T(k)\psi^T(k)\mathbf{H}x(k-1) + (\lambda(k) - \lambda(k-1) + \nu(k))^Tx(k-1)$$

$$+ (M^T\nu(k) + B(K) + M^T\lambda(k) + \sigma(k))^Tu(k) - \phi^T(k-1)y(k-1)$$

$$- \mu^T(k)z(k) - \psi^T(k)r(k-1)\} \Big] \tag{3.81}$$

where

$$\nu(k) = e^M(k) - e^m(k) \tag{3.82a}$$

$$\sigma(k) = g^M(k) - g^m(k) \tag{3.82b}$$

We define

$$Z^T(K) = [x^T(K), y^T(K)] \tag{3.83}$$

$$X^T(k) = [x^T(k-1), y^T(k-1), u^T(k), z^T(k), r^T(k-1)] \tag{3.84}$$

$$N(K) = \begin{bmatrix} \beta + \phi^T(K)\mathbf{H} & \tfrac{1}{2}\gamma \\ \tfrac{1}{2}\gamma^T & 0 \end{bmatrix} \tag{3.85}$$

$$L(k) = \begin{bmatrix} \phi^T(k-1)\mathbf{H} & 0 & \frac{1}{2}[\psi^T(k)\mathbf{H} + d(k)] & 0 & 0 \\ 0 & 0 & \frac{1}{2}\gamma(k) & 0 & 0 \\ \frac{1}{2}[\psi^T(k)\mathbf{H} + d(k)] & \frac{1}{2}\gamma(k) & f(k)M + \mu^T(k)\mathbf{H} & \frac{1}{8}\gamma(k)N - \frac{1}{4}L^T\gamma(k) & \frac{1}{2}M^T\gamma(k) \\ 0 & 0 & (\frac{1}{8}N^T\gamma(k) - \frac{1}{4}\gamma(k)L & 0 & 0 \\ 0 & 0 & \frac{1}{2}\gamma(k)M & 0 & 0 \end{bmatrix}$$

$$\text{(3.86)}$$

Furthermore, if we define the following vector such that

$$Q(K) = N^{-1}(K)W(K) \tag{3.87}$$

$$V(k) = L^{-1}(k)R(k) \tag{3.88}$$

then the augmented cost functional in equation (3.81) can be written as

$$\tilde{J} = E[(Z(K) + \tfrac{1}{2}Q(K))^T N(K)(Z(K) + \tfrac{1}{2}Q(K)) - \tfrac{1}{4}Q^T(K)N(K)Q(K)$$

$$+ x^T(0)\phi^T(0)\mathbf{H}x(0) + \lambda^T(0)x(0) + \phi^T(0)y(0)]$$

$$+ E\left[\sum_{k=1}^{K} \{(X(k) + \tfrac{1}{2}V(k))^T L(k)(X(k) + \tfrac{1}{2}V(K)) \right.$$

$$\left. - \tfrac{1}{4}V^T(k)L(k)V(k)\} \right] \tag{3.89}$$

Since it is desired to maximize $\tilde{J}$ with respect to $Z(K)$ and $X(k)$, the problem is equivalent to

$$\max_{Z(K),X(k)} \tilde{J} = \max E[(Z(K) + \tfrac{1}{2}Q(K))^T N(K)(Z(K) + \tfrac{1}{2}Q(K))]$$

$$+ \max E\left[\sum_{k=1}^{K} \{(X(k) + \tfrac{1}{2}V(K))^T L(k)(X(k) + \tfrac{1}{2}V(k))\} \right] \tag{3.90}$$

because $Q(K)$ and $V(k)$ are independent of $Z(K)$ and $X(k)$, respectively, and $x(0)$ and $y(0)$ are constants.

It will be noticed that $\tilde{J}$ in equation (3.90) is composed of a boundary part and a discrete integral part, which are independent of each other. To maximize $\tilde{J}$ in equation (3.90), one maximizes each term separately.

The boundary part in equation (3.90) defines a norm. Hence we can write this part as

$$\max_{Z(K)} J_1 = \max_{Z(K)} E\|Z(K) + \tfrac{1}{2}Q(K)\|_{N(K)} \tag{3.91}$$

Also, the discrete integral part in equation (3.90) defines a norm, we can write this part as

$$\max_{X(k)} J_2 = \max_{X(k)} E\|X(k) + \tfrac{1}{2}V(k)\|_{L(k)} \tag{3.92}$$

3.3.3.2. The Optimal Solutions

There is only one optimal solution to the problems formulated in equations (3.91) and (3.92). The maximization of J_1 is mathematically equivalent to the minimization of the norm in equation (3.91). The minimum of the norm in equation (3.91) is clearly achieved when

$$E[Z(K) + \tfrac{1}{2}Q(K)] = [0] \tag{3.93}$$

Substituting from equations (3.83), (3.84), and (3.87) into equation (3.93), we obtain

$$E[A - \lambda(K) + 2(\beta + \phi^T(K)\mathbf{H})x(K) + \gamma y(K)] = [0] \tag{3.94}$$

$$E[-\phi(K) + \gamma^T x(K)] = [0] \tag{3.95}$$

Equations (3.94) and (3.95) give the values of Lagrange multipliers at last period studied.

Also, the maximum of J_2 in equation (3.92) is achieved when

$$E[X(k) + \tfrac{1}{2}V(k)] = [0] \tag{3.96}$$

Substituting from equations (3.84), (3.86), and (3.88) into equation (3.96), and adding the equality constraints (3.74)-(3.77) we obtain

$$E[-x(k) + x(k-1) + q(k) + Mu(k)] = [0] \tag{3.97}$$

$$E[-y(k) + x^T(k)\mathbf{H}(k)] = [0] \tag{3.98}$$

$$E[-z(k) + u^T(k)\mathbf{H}u(k)] = [0] \tag{3.99}$$

$$E[-r(k-1) + u^T(k)\mathbf{H}x(k-1)] = [0] \tag{3.100}$$

$$E[2\phi^T(k-1)\mathbf{H}x(k-1)$$
$$+ (\psi^T(k)\mathbf{H} + d(k))u(k) + \lambda(k-1) + \nu(k)] = [0] \tag{3.101}$$

$$E[\gamma(k)u(k) - \phi(k-1)] = [0] \tag{3.102}$$

$$E[(\psi^T(k)\mathbf{H} + d(k))x(k-1) + \gamma(k)y(k-1) + 2(f(k)M + \mu^T(k)\mathbf{H})u(k)$$
$$+ (\tfrac{1}{4}\gamma(k)N - \tfrac{1}{2}L^T\gamma(k))z(k) + M^T\gamma(k)rz(k-1) + B(k)$$
$$+ M^T\lambda(k) + M^T\nu(k) + \sigma(k)] = [0] \tag{3.103}$$

$$E[(\tfrac{1}{4}N^T\gamma(k) - \tfrac{1}{2}\gamma(k)L)u(k) - \mu(k)] = [0] \tag{3.104}$$

$$E[\gamma(k)Mu(k) - \psi(k)] = [0] \tag{3.105}$$

Besides the above equations, we have Kuhn–Tucker exclusion equations which must be satisfied at the optimum

$$e_{i,k}^m(x_i^m - x_{i,k}) = 0, \qquad e_{i,k}^M(x_{i,k} - x_i^M) = 0 \tag{3.106}$$

$$g_{i,k}^m(u_{i,k}^m - u_{i,k}) = 0, \qquad g_{i,k}^M(u_{i,k} - u_{i,k}^M) = 0 \tag{3.107}$$

Table 3.10. Characteristics of the Installations

Site	x_i^M (Mm^3)	Maximum effective discharge (m^3/sec)	Reservoir's constants		
			α_i	β_i	γ_i
1	9628	400	11.41	0.15226×10^{-2}	-0.19131×10^{-7}
2	570	547	231.53	0.10282×10^{-1}	-0.13212×10^{-5}
3	50	594	215.82	0.12586×10^{-1}	0.2979×10^{-5}
4	3420	1180	432.20	-0.12972×10^{-1}	-0.12972×10^{-5}

Table 3.11. Monthly Inflows and the Associated Probabilities for One of the Reservoirs in the System

Month k	$I_{1,k}$ (Mm^3)	P	Month k	$I_{1,k}$ (Mm^3)	P
October	437.2	0.0668	April	46.8	0.0668
	667.8	0.2417		118.6	0.2417
	886.2	0.3830		186.6	0.3830
	1104.6	0.2417		254.6	0.2417
	1335.3	0.0668		326.4	0.0668
November	376.5	0.0668	May	120.2	0.0668
	550.3	0.2417		425.8	0.2417
	714.9	0.3830		715.2	0.3830
	879.4	0.2417		1004.6	0.2417
	1053.2	0.0668		1310.2	0.3830
December	323.9	0.0668	June	491.5	0.0668
	414.1	0.2417		857.8	0.2417
	499.6	0.3830		1204.6	0.3830
	585.0	0.2417		1551.5	0.2417
	675.2	0.0668		1917.8	0.0668
January	212.4	0.0668	July	448.1	0.0668
	269.3	0.2417		739.1	0.2417
	323.1	0.3830		1014.7	0.3830
	377.0	0.2417		1290.2	0.2417
	757.0	0.0668		1581.3	0.0668
February	130.3	0.0668	August	255.1	0.0668
	168.8	0.2417		547.7	0.2417
	205.2	0.3830		824.7	0.3830
	241.6	0.2417		1101.8	0.2417
	280.0	0.0668		1394.4	0.0668
March	109.0	0.0668	September	263.4	0.0668
	147.5	0.2417		551.5	0.2417
	183.9	0.3830		824.8	0.3830
	220.3	0.2417		1097.2	0.2417
	258.7	0.0668		1385.4	0.0668

one also has the following limits on the variables:

$$\text{If } x_{i,k} < x_i^m, \qquad \text{then we put } x_{i,k} = x_i^m$$

$$\text{If } x_{i,k} > x_i^M, \qquad \text{then we put } x_{i,k} = x_i^M$$

$$\text{If } u_{i,k} < u_{i,k}^m, \qquad \text{then we put } u_{i,k} = u_{i,k}^m$$

$$\text{If } u_{i,k} < u_{i,k}^M, \qquad \text{then we put } u_{i,k} = u_{i,k}^M$$

Equations (3.97)-(3.108) with equations (3.94) and (3.95) completely specify the optimal long-term operation of a series multireservoir power system.

3.3.3.3. Practical Application (see Table 3.10)

A computer program was written to solve these equations according to the main flow chart given in Figure 3.4, for the same system mentioned in the previous section. In Table 3.11 we give the inflows in Mm^3 to the first reservoir and the associated probabilities during each month. In Table 3.12, we give the monthly release from each reservoir and the profits realized during the optimization interval. In Table 3.13, we give the optimal monthly reservoir storage during the same year.

The computing time to get the optimal solution for a period of a year was 0.9 sec in CPU units on the Amdahl 470V/6 computer.

Table 3.12. Optimal Monthly Releases from the Four Reservoirs and the Profits Realized

Month k	$u_{1,k}$ (Mm^3)	$u_{2,k}$ (Mm^3)	$u_{3,k}$ (Mm^3)	$u_{4,k}$ (Mm^3)	Profits ($)
1	0	315	474	1867	817,745
2	1025	1289	1378	2270	1,566,862
3	1071	1256	1305	2029	1,596,784
4	1071	1367	1420	1894	1,997,250
5	953	1035	1058	2459	2,151,664
6	1071	1035	1058	2459	2,151,664
7	1024	999	1022	1039	1,137,618
8	0	234	273	950	533,420
9	0	483	660	2194	1,042,055
10	337	675	773	1370	703,741
11	0	305	407	1060	457,890
12	0	305	407	1695	723,348

Value of water remaining in the reservoirs at the end of the year	9,380,718
Total profits	24,537,888

Table 3.13. Optimal Monthly Reservoir Storage

Month	$x_{1,k}$ (Mm^3)	$x_{2,k}$ (Mm^3)	$x_{3,k}$ (Mm^3)	$x_{4,k}$ (Mm^3)
1	7575	570	0	3254
2	7265	570	0	3409
3	6693	570	13	3417
4	5945	393	0	3417
5	5196	342	0	2235
6	4308	446	0	1104
7	3470	540	0	1360
8	4185	570	50	1731
9	5390	533	22	1962
10	6067	570	50	2853
11	6892	570	50	3408
12	7717	570	50	3328

3.3.3.4. Comments

In this section, the problem of the long-term optimal operation of a series of reservoirs with a nonlinear model is discussed. The nonlinear model used was a quadratic function of the average storage to avoid underestimation in the hydroelectric production for rising water levels and overestimation for falling water levels.

The section dealt with a nonlinear cost functional and explained how to deal with such cost functionals; we introduced a set of pseudostate variables to cast the problem into a quadratic form that can be solved by the minimum norm formulation (Refs. 3.11–3.14).

References

3.1. ARVANITIDIES, N. V., and ROSING, J., "Composite Representation of a Multireservoir Hydroelectric Power System," *IEEE Transactions on Power Apparatus and Systems* **PAS-89**(2), 319–326 (1970).

3.2. ARVANITIDIES, N. V., and ROSING, J., "Optimal Operation of Multireservoir Systems Using a Composite Representation," *IEEE Transactions on Power Apparatus and Systems* **PAS-89**(2), 327–335 (1970).

3.3. DURAN, H., *et al.*, "Optimal Operation of Multireservoir Systems Using an Aggregation-Decomposition Approach," *IEEE Transactions on Power Apparatus and Systems*, **PAS-104**(8), 2086–2092 (1985).

3.4. TURGEON, A., "A Decomposition/Projection Method for the Multireservoir Operating Problem," Paper presented at the Joint National TIMS/ORSA Meeting, Los Angeles, California November, 1978.

3.5. TURGEON, A., "Optimal Operation of Multireservoir Power System with Stochastic Inflows," *Water Resources Research* **16**(6), 275–283 (1980).

3.6. TURGEON, A., "A Decomposition Method for the Long-Term Scheduling of Reservoirs in Series," *Water Resources Research* **17**(6), 1565–1570 (1981).

3.7. OLCER, S., *et al.*, "Application of Linear and Dynamic Programming to the Optimization of the Production of Hydroelectric Power," *Optimal Control Application and Methods* **6**, 43–56 (1985).

3.8. GRYGIER, J. C., and STEDINGER, J. R., "Algorithms for Optimizing Hydropower System Operation," *Water Resources Research* **21**(1), 1–10 (1985).

3.9. HALLIBURTON, T. S., and SIRISENA, H. R., "Development of a Stochastic Optimization for Multireservoir Scheduling," *IEEE Transactions on Automatic Control* **AC-29**, 82–84 (1984).

3.10. MARINO, M. A., and LOAICIGA, H. A., "Quadratic Model for Reservoir Management: Applications to the Central Valley Project," *Water Resources Research* **21**(5), 631–641 (1985).

3.11. SAGE, A. P., *Optimal Systems Control*, Prentice-Hall, Englewood Cliffs, New Jersey, 1968.

3.12. EL-HAWARY, M. A., and CHRISTENSEN, G. S., *Optimal Economic Operation of Electric Power System*, Academic Press, New York, 1979.

3.13. SHAMALY, A., CHRISTENSEN, G. S., and EL-HAWARY, M. A., "A Transformation for Necessary Optimality Conditions for Systems with Polynomial Nonlinearities," *IEEE Transactions on Automatic Control* **AC-24**(6), 983–985 (1979).

3.14. SHAMALY, A., CHRISTENSEN, G. S., and EL-HAWARY, M. A., "Optimal Control of Large Turboalternator," *Journal of Optimization Theory and Applications* **34**(1), 83–97 (1981).

3.15. SOLIMAN, S. A., and CHRISTENSEN, G. S., "Discrete Stochastic Optimal Long-Term Scheduling of Series Reservoir," 14th IASTED International Conference, Vancouver, June 4–6, 1986.

3.16. SOLIMAN, S. A., CHRISTENSEN, G. S., and ABDEL-HALIM, M. A., "Optimal Operation of Multireservoir Power System Using Functional Analysis," *Journal of Optimization Theory and Applications* **49**(3), 449–461 (1986).

4

Long-Term Operation of Multichain Power Systems

4.1. Introduction (Refs. 4.1–4.23)

The problem of determining the optimal operating policy of a multi-reservoir power system is a difficult problem for the following reasons:

- It has a nonlinear objective function of the discharge and the head which itself is a function of the storage.
- The production-energy function of the hydroplant is a nonseparable function of the discharge and the head.
- There are linear constraints on both the state (storage or the head) and decision (release) variables.
- It is a stochastic problem with respect to the river flows and demand for electricity.

The intent of this chapter is to solve the long-term optimal operating problem of a multichain power system. The rivers in this chapter may or may not be independent of each other. In this chapter, we shall discuss different approaches used to solve the problem. Section 4.2 is devoted to the general formulation of the problem. In Section 4.3.2 we discuss the stochastic dynamic programming approach used to solve the problem. Section 4.3.3 is devoted to the aggregation–decomposition approach, we shall compare this approach with another well-known approach called "The One-at-a-Time Method." Section 4.4 is devoted to solving the problem by the discrete maximum principle. Finally, Section 4.5 explains the solution of the optimal long-term problem by the minimum norm formulation of functional analysis developed by the authors.

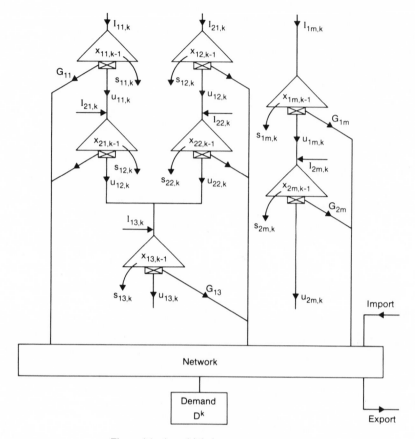

Figure 4.1. A multichain power system.

4.2. Problem Formulation

4.2.1. The System under Study

The system under consideration consists of m rivers; the rivers may or may not be independent of each other, with one or several reservoirs and power plants in series on each, and interconnection lines to neighboring systems through which energy may be exchanged (Figure 4.1).

4.2.2. The Objective Function (Ref. 4.24)

The long-term stochastic optimization problem determines the discharge $u_{ij,k}$; $i = 1, \ldots, n_j$; $j = 1, \ldots, m$; $k = 1, \ldots, K$ such that:

(1) The expected total benefits from the system (benefits from energy generated by a hydro power system over the planning period plus the

expected future returns from the water left in storage at the end of that period) is a maximum.

(2) The water conservation equation for each reservoir may be adequately described by the continuity-type equation

$$x_{ij,k} = x_{ij,k-1} + I_{ij,k} + z_{ij,k} - u_{ij,k} - s_{ij,k} \qquad (4.1)$$

where $z_{ij,k}$ is the dependent water inflow connecting hydroelectric plants over the same hydraulic valley and is given by

$$z_{ij,k} = \sum_{\substack{v \in R_u \\ j \in R_r}} (u_{vj,k} + s_{vj,k}) \qquad (4.2)$$

where R_u is the set of plants immediately upstream from plant i on river j and R_r is the set of upstream rivers and

$$s_{ij,k} = \begin{cases} (x_{ij,k-1} + I_{ij,k} + z_{ij,k} - x_{ij,k}) - u^M_{ij,k}, \\ \quad \text{if } (x_{ij,k-1} + I_{ij,k} + z_{ij,k} - x_{ij,k}) > u^M_{ij,k} \quad \text{and} \quad x_{ij,k} = x^M_{ij} \\ 0 \quad \text{otherwise} \end{cases} \qquad (4.3)$$

Note that if the rivers are independent of each other, then we have

$$z_{ij,k} = u_{(i-1)j,k} + s_{(i-1)j,k} \qquad (4.4)$$

(3) In order to be realizable and also to satisfy multipurpose stream use requirements, such as flood control, irrigation, navigation fishing, and other purposes, if any, the following upper and lower limits on the variables should be satisfied:

a. Upper and lower bounds on the storage

$$x^m_{ij} \le x_{ij,k} \le x^M_{ij} \qquad (4.5)$$

b. Upper and lower bounds on the discharge

$$u^m_{ij,k} \le u_{ij,k} \le u^M_{ij,k} \qquad (4.6)$$

(4) The energy generated by a hydropower plant during a period k is a function of the discharge and the average storage, to avoid overestimation of production for falling water level and underestimation for rising water levels. In a power system where the heads vary with a small amount, the energy generated can be considered as a constant times the discharge from the reservoir, and this constant is the water conversion factor, MWh/Mm^3.

4.3. The Aggregation Approach (Turgeon Approach)
(Refs. 4.26–4.31)

The system used in this approach has m independent rivers; the following assumptions are used:

(1) The optimal operating policy of river i is such that spillage will not occur in period k, or will occur at every reservoir of the river and a

shortage of water will not occur in period k or will occur in every reservoir of the river.

(2) The amount of electrical energy generated by any power plant in period k is a constant times the discharge.

Following these hypotheses, Turgeon built a composite model of each river (Figure 4.2). It suffices to convert the water stored at each plant into its at-site and downstream generating capability and sum over all plants.

The resulting composite model of river j has the following characteristics: $I_{j,k}$, $x_{j,k}$, $u_{j,k}$, and $s_{j,k}$, where $I_{j,k}$ is the inflow of potential energy to composite reservoir j in period k:

$$I_{j,k} = \sum_i \sum_{p \geq i} h_{pj} I_{ij,k} \tag{4.7}$$

where h_{pj} is the water conversion factor, MWh/Mm^3, of water at site p of river j; $x_{j,k}$ is the potential energy content of composite reservoir j at the end of peroid k:

$$x_{j,k} = \sum_i \sum_{p \geq i} h_{pj} x_{ij,k}, \qquad x_j^m \leq x_{j,k} \leq x_j^M \tag{4.8}$$

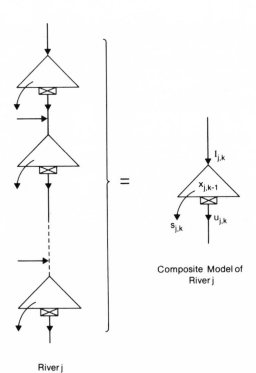

River j

Figure 4.2. Composite model of river j.

where

$$x_j^m = \sum_i \sum_{p \geq i} h_{pj} x_{ij}^m \tag{4.9}$$

and

$$x_j^M = \sum_i \sum_{p \geq i} h_{pj} x_{ij}^M \tag{4.10}$$

$u_{j,k}$ is the discharge from the composite hydroplant j in period k:

$$u_{j,k} = \sum_i h_{ij} u_{ij,k}, \qquad u_{j,k}^m \leq u_{j,k} \leq u_{j,k}^M \tag{4.11}$$

where

$$u_{j,k}^m = \sum_i h_{ij} u_{ij,k}^m \tag{4.12}$$

$$u_{j,k}^M = \sum_i h_{ij} u_{ij,k}^M \tag{4.13}$$

and $s_{j,k}$ is the amount of potential energy spilt from composite reservoir j in period k:

$$s_{j,k} = \sum_i h_{ij} s_{ij,k} \tag{4.14}$$

4.3.1. The Objective

If we replace every river in Figure 4.1 by its composite model found in the preceding section, we obtain a network of m reservoir hydroplant complexes in parallel, shown in Figure 4.3.

In this section we shall find the optimal operating policy of this network. More precisely, we shall find the discharges $u_{1,k}, u_{2,k}, \ldots, u_{m,k}; k = 1, \ldots, K$ that minimize the expected production cost given by

$$J = E\left[\sum_{k=1}^{K} F_k\left(D_k - \sum_{j=1}^{m} u_{j,k} \right) \right] \tag{4.15}$$

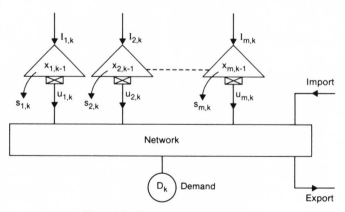

Figure 4.3. The aggregate network.

subject to the following constraints:

$$x_{j,k} = x_{j,k} + I_{j,k} - u_{j,k} - s_{j,k}, \qquad j = 1, \ldots, m, \qquad k = 1, \ldots, K \qquad (4.16)$$

$$x_j^m \le x_{j,k} \le x_j^M, \qquad j = 1, \ldots, m, \qquad k = 1, \ldots, K \qquad (4.17)$$

$$u_{j,k}^m \le u_{j,k} \le u_{j,k}^M, \qquad j = 1, \ldots, m, \qquad k = 1, \ldots, K \qquad (4.18)$$

In equation (4.15) the symbol E stands for the expected value. The expected value is taken with respect to the variable $I_{j,k}, j = 1, \ldots, m$, and the function $F_k(\cdot)$ represents the production cost in period k (Figure 4.4).

When $D_k = \sum_{j=1}^m u_{j,k}$, the production cost is zero, for the hydroelectric power plants cost nothing to operate. When $D_k > \sum_{j=1}^m u_{j,k}$, energy is imported from neighboring producers or produced at a thermal or nuclear plant at the cost shown in Figure 4.4. When $D_k < \sum_{j=1}^m u_{j,k}$, the producer may choose to increase his revenues by producing surplus energy rather than spilling water and selling it to neighboring producers. These revenues correspond to the negative cost in Figure 4.4.

For the purpose of clarity the demand for electricity is assumed to be known hereafter. One may verify, however, that no additional difficulty is created by supposing it known in probability only.

4.3.2. The Solution by Dynamic Programming

In this section we discuss how to use the stochastic dynamic programming to solve the problem formulated in equations (4.15)-(4.18).

Let u_k be the set of $\{u_{1,k}, u_{2,k}, \ldots, u_{m,k}\}$ that satisfies constraints (4.16)-(4.18) and let the variable $u_{j,k}$ be a function of the variables $x_{1,k-1}$, $x_{2,k-1}, \ldots, x_{m,k-1}, I_{1,k}, \ldots, I_{m,k}$. If m, number of reservoir-hydroplant complexes in the aggregate network, is small ($m \le 4$), problems (4.15)-(4.18) can be solved by dynamic programming. This consists in solving the following functional equation recursively, starting in period K with

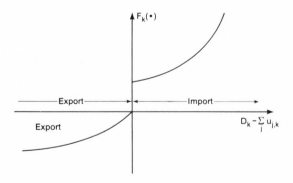

Figure 4.4. The production cost.

$Z_{0,K+1}(x_{1,K-1}, \ldots, X_{m,K+1})$ for all values of $x_{1,K}, \ldots, x_{m,K}$ and going backward in time:

$$Z_{0,k}(x_{1,k-1}, \ldots, x_{m,k-1}) = E[\min_{\{u_{1,k}, \ldots, u_{m,k}\} \in U_k} F_k(D_k - u_{1,k}, \ldots, u_{m,k}$$

$$+ Z_{0,k+1}(x_{1,k}, \ldots, x_{m,k})\}] \qquad (4.19)$$

However, if the number of reservoir–hydroplant complexes is greater than four, solving (4.19) requires a capacity of storage that is much beyond that of known computers. For these large networks, (4.19) may be manipulated before being programmed for the computer. In the following two sections, two possible manipulations are presented. The first consists in solving (4.19) with respect to one variable at a time. The second consists in breaking up this m-dimensional problem into m problems of two dimensions.

4.3.3. The One-at-a-Time Method

This method consists in breaking up the original m-state variable problem into a series of one-state variable subproblems in such a manner that the sequence of optimizations over the subproblems converges to the solution to the original problem. To apply this idea to our stochastic problem, however, we must adopt the approximation in which each control law $u_{j,k}$ is a function of the variables $x_{j,k-1}$ and $I_{j,k}$ only or in other words, is a "local feedback." Then we assign values to $u_{2,k}(x_{2,k-1}, I_{2,k})$, $u_{3,k}(x_{3,k-1}, I_{3,k}), \ldots, u_{m,k}(x_{m,k-1}, I_{m,m-1})$, $k = 1, \ldots, K$, and proceed with determining the corresponding optimal value of $u_{1,k}(x_{1,k-1}, I_{1,k})$, $k = 1, \ldots, K$ by solving the following one-state Bellman's functional equation recursively as (4.19):

$$Z_{1,k}(x_{1,k-1}) = E[\min_{I_{1,k} u_{1,k} \in U_k} [F_{1,k}(u_{1,k}) + Z_{1,k+1}(x_{1,k})\}] \qquad (4.20)$$

where

$$F_{1,k}(u_{1,k}) = E\left[F_k\left\{ D_k - u_{1,k} - \sum_{j=2}^{m} u_{j,k}(x_{j,k-1}, I_{j,k}) \right\} \right] \qquad (4.21)$$

In (4.20) the expectation is taken with respect to the random variable $I_{1,k}$; in (4.21) it is taken with respect to the variables $x_{j,k-1}$ and $I_{j,k}, j = 2, \ldots m$.

The probability distribution of the variable $x_{j,k-1}, j = 2, \ldots, n$ can be determined by solving recursively, starting in period 1 and going forward in time, the following Markov process:

$$\text{Prob}(x_{j,k-1} = \alpha) = \sum_{\xi} \text{Prob}(x_{j,k-1} = \xi) \cdot \text{Prob}(I_{j,k-1} - u_{j,k-1}(\xi, I_{j,k-1})$$

$$- s_{j,k-1}(\xi, I_{j,k-1}) = \alpha - \xi) \qquad (4.22)$$

where

$$s_{j,k-1}(\xi, I_{j,k-1}) = \max[0, \xi + I_{j,k-1} - u_{j,k-1}(\xi, I_{j,k-1}) - x_j^M] \quad (4.23)$$

and where "Prob" stands for the probability.

With the solution to (4.20) and the nominal solutions for $u_{3,k}(x_{3,k-1}, I_{3,k})$, $u_{4,k}(x_{4,k-1}, I_{4,k}), \ldots, u_{m,k}(x_{m,k-1}, I_{m,k})$, $k = 1, \ldots, K$ an optimization of $u_{2,k}(x_{2,k-1}, I_{2,k})$ can be performed analogously. Similarly, $u_{3,k}(x_{3,k-1}, I_{3,k})$, $u_{4,k}(x_{4,k-1}, I_{4,k}), \ldots, u_{m,k}(x_{m,k-1}, I_{m,k})$ can be optimized one at a time by applying the same procedure repetitively. For the optimum, this sequence of one-dimensional optimizations must be repeated until no change occurs in any of the values of the control laws.

4.3.3.1. Algorithm of Solution

In this section, we explain the algorithm of solution of the previous approach. In this algorithm all variables are discretized.

Step 1. Set $x_{j,0} = \alpha_j \forall j$. Then let $\mathrm{Prob}(x_{j,0} = B)$ be equal to 1 for $B = \alpha_j$ and to 0 otherwise, $j = 2, \ldots, m$. Set $u_{j,k}(x_{j,k-1}, I_{j,k-1}) = \min[I_{j,k}, u_{j,k}^M]\forall j, k$ and $Z_{j,k}(x_{j,K}) = 0$ for all values of $x_{j,K}$; $j = 1, \ldots, m$. Let $N = 1, j = 2$, and go to step 2.

Step 2. Compute $\mathrm{Prob}(x_{j,k-1} = \alpha)$, $k = 2, \ldots, K$ using (4.22). If $N = 1$, go to step 3. Otherwise, go to step 4.

Step 3. Set $P_{j,k-1}(\alpha) = \mathrm{Prob}(x_{j,k-1} = \alpha)\forall \alpha$, k, and $\mathrm{Prob}(x_{j,0} = \alpha) = \mathrm{Prob}(x_{j,k} = \alpha)\forall \alpha$. Then let $N = 12$ and go back to step 2.

Step 4. If $\mathrm{Prob}(x_{j,k-1} = \alpha) = P_{j,k-1}(\alpha) \pm \varepsilon \forall \alpha$, k, where ε is an infinitesimal value, go to step 5. Otherwise go to step 3.

Step 5. Set $j = j + 1$. If $j > m$, go to step 6. Otherwise, set $N = 1$, and go to step 2.

Step 6. Let

$$A_{1,k} = \sum_{j \neq 1} u_{j,k}^*(x_{j,k-1}, I_{j,k})$$

and compute

$$\mathrm{Prob}(A_{1,k} = \alpha) = \mathrm{Prob}\left(\sum_{j \neq 1} u_{j,k}^*(x_{j,k-1}, I_{j,k}) = \alpha\right)$$

for all α, k. Set $j = 1$, $k = K$, $N = 0$ and go to step 7.

Step 7. Let $k = k - 1$. If $k = 0$, go to step 9. Otherwise, compute

$$F_{j,k}(u_{j,k}) = \sum_\alpha F_k(D_k - u_{j,k} - \alpha) \cdot \text{Prob}(A_{j,k} = \alpha)$$

for all possible values of $u_{j,k}$, and go to step 8.

Step 8. Determine the release policy $u_{j,k}(x_{j,k-1}, I_{j,k})$ that satisfies the following equation:

$$Z_{j,k}(x_{j,k-1}) = E\{\min_{u_{j,k} \in U_k} [F_{j,k}(u_{j,k}) + Z_{j,k+1}(x_{j,k})]\}$$

and go back to step 7.

Step 9. If $u_{j,k}(x_{j,k-1}, I_{j,k}) = u_{j,k}^*(x_{j,k-1}, I_{j,k}) \forall x_{j,k-1}, I_{j,k}$, we set $N = N + 1$, and go to step 10. Otherwise set

$$u_{j,k}^*(x_{j,k-1}, I_{j,k}) = u_{j,k}(x_{j,k-1}, I_{j,k}) \forall x_{j,k-1}, I_{j,k}, k$$

and compute $\text{Prob}(x_{j,k-1} = \alpha) \forall \alpha, k$ by proceeding as in steps 2, 3, and 4. Then go to step 10.

Step 10. Set $j = j + 1$. If $j > m$, go to step 11. Otherwise, compute

$$\text{Prob}(A_{j,k} = \alpha) = \text{Prob}(A_{j-1,k} + u_{j-1,k}^*(x_{j-1,k-1}, I_{j-1,k}) - u_{j,k}^*(x_{j,k-1}, I_{j,k})$$
$$= \alpha)$$

for all α, k. Set $k = K$, and go to step 7.

Step 11. If $N = m$, stop. Otherwise, go to step 6.

4.3.3.2. Practical Example

In this section, the one-at-a-time method is applied to a network of six reservoir–hydroplant complexes in parallel; the characteristics of the installations are given in Table 4.1. The optimization is done in a weekly

Table 4.1. Characteristics of the Installations

Complex j	x_j^M GWh^a	$u_{j,k}^M$ GWh^a/week
1	232	7.4
2	162	3.9
3	24	2.9
4	317	4.2
5	150	6.0
6	25	4.5

^{a} GWh = 10^9 watt hours.

Table 4.2. Mean and Standard Deviation of the Inflows to Reservoir 1

Week	Mean (GWh)	Standard deviation (GWh)	Week	Mean (GWh)	Standard deviation (GWh)
1	4.59	1.62	27	1.05	0.20
2	4.56	1.49	28	1.12	0.34
3	4.80	1.72	29	1.22	0.37
4	5.05	2.08	30	1.96	1.42
5	4.65	1.57	31	3.05	2.36
6	4.46	1.57	32	5.1	5.1
7	3.86	1.13	33	7.79	5.57
8	3.42	0.74	34	10.86	5.34
9	3.03	0.71	35	12.21	4.92
10	2.77	0.61	36	14.04	5.49
11	2.58	0.76	37	13.57	5.46
12	2.53	0.89	38	12.18	4.7
13	2.35	0.93	39	11.44	4.92
14	2.12	0.77	40	9.42	3.89
15	1.88	0.48	41	8.1	3.5
16	1.81	0.54	42	7.06	2.95
17	1.68	0.52	43	6.18	2.77
18	1.51	0.38	44	5.62	2.36
19	1.42	0.31	45	5.70	2.45
20	1.35	0.28	46	5.12	1.68
21	1.28	0.27	47	4.76	1.78
22	1.22	0.26	48	4.41	1.48
23	1.16	0.24	49	4.04	1.25
24	1.11	0.24	50	4.48	1.54
25	1.06	0.21	51	4.72	2.04
26	1.05	0.20	52	4.38	1.53

time basis for a period of 52 weeks. The weekly inflows to the reservoirs are assumed normally distributed. For instance, the means and standard deviations of inflows for reservoir 1 are given in Table 4.2. The long-term average weekly inflows to each of the six reservoirs are 4.56, 3.53, 1.05, 2.3, 3.92, and 1.64 GWh, respectively. The yearly demand for electrical energy is set to 887 GWh and distributed among the 52 weeks of the year as given in Table 4.3. The production cost in week k, F, is assumed to vary with $z = D_k - \sum_j u_{j,k}$ as follows:

$$F(z) = 6z + \tfrac{1}{9}z^2, \qquad 0 \le z \le 3$$

$$F(z) = 25 + 10z + \tfrac{1}{9}z^2, \qquad z > 3$$

$$F(z) = 4z - \tfrac{1}{36}z^2, \qquad z < 0$$

Table 4.3. Weekly Demand for Electrical Energy

Week	Demand (GWh)	Week	Demand (GWh)	Week	Demand (GWh)	Week	Demand (GWh)
1	16.94	14	18.27	27	17.12	40	16.05
2	16.94	15	18.18	28	17.03	41	16.05
3	16.94	16	18.09	29	16.85	42	16.14
4	16.94	17	17.92	30	16.76	43	16.23
5	17.03	18	17.83	31	16.68	44	16.41
6	17.21	19	17.83	32	16.68	45	16.50
7	17.21	20	17.74	33	16.59	46	16.59
8	17.21	21	17.74	34	16.50	47	16.59
9	17.56	22	17.65	35	16.41	48	16.68
10	18.09	23	17.65	36	16.23	49	16.68
11	18.18	24	17.47	37	16.05	50	16.76
12	18.18	25	17.47	38	15.97	51	16.76
13	18.27	26	17.30	39	15.97	52	16.85

The computer program that executes the one-at-a-time method has 425 statements and requires 134 K bytes of computer memory. One hundred fifty minutes of CPU time on an IBM/370 model 168 are required to execute the algorithm for the network of six reservoir–hydroplant complexes described above where the variables $x_{j,k}$ and $I_{j,k}$ are allowed to take 35 and 9 possible values, respectively, for all j,k.

4.3.4. The Aggregation–Decomposition Method

In this section, we discuss the aggregation–decomposition approach to obtain a closed-loop solution, since a local feedback and an open solution have already been found above. So let us aggregate all the reservoir–hydroplant complexes shown in Figure 4.3 except complex j to obtain the network shown in Figure 4.5. Here

$$I_{jc,k} = \sum_{\sigma \neq j} I_{\sigma,k}, \qquad x_{jc,k-1} = \sum_{\sigma \neq j} x_{\sigma,k-1}, \qquad x_{jc}^M = \sum_{\sigma \neq j} x_{\sigma}^M$$

$$u_{jc,k} = \sum_{\sigma \neq j} u_{\sigma,k}, \qquad u_{jc,k}^M = \sum_{\sigma \neq j} u_{\sigma,k}^M, \qquad S_{jc,k} = \sum_{\sigma \neq j} S_{\sigma,k}$$

Now, the optimal operating policy of this new network is to minimize

$$J = E\left[\sum_{k=1}^{K} F_k(D_k - u_{j,k} - u_{jc,k}) \right] \tag{4.24}$$

Subject to satisfying the following constraints:

$$x_{i,k} = x_{i,k-1} + I_{i,k} - u_{i,k} - s_{i,k}, \qquad i = j, jc, k = 1, \ldots, K \tag{4.25}$$

$$x_i^m \leq x_{i,k} \leq x_i^M, \qquad i = j, jc, k = 1, \ldots, K \tag{4.26}$$

$$u_{i,k}^m \leq u_{i,k} \leq u_{i,k}^M, \qquad i = k, jc, k = 1, \ldots, K \tag{4.27}$$

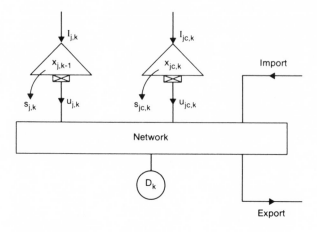

Figure 4.5. The network with complex j and its complement.

The above problem has two state variables only, which can be solved by dynamic programming by solving the following recursive equation:

$$Z_{j,k}(x_{j,k-1}, x_{jc,k-1}) = \{ \min_{u_{j,k}, u_{jc,k} \in U_{j,k}} [F_k(D_k - u_{j,k} - u_{jc,k})$$

$$+ Z_{j,k+1}(x_{j,k}, x_{jc,k})]\} \qquad (4.28)$$

where $U_{j,k}$ is the set of points $\{u_{j,k}, u_{jc,k}\}$ that satisfy constraints (4.25)–(4.27).

Let us solve (4.28) for $j = 1, 2, \ldots, m$, that is, for the m reservoir-hydroplant complexes, to obtain the cost functions $Z_{1,k}(x_{1,k-1}, x_{1c,k}, Z_{2,k}(x_{2,k-1}, x_{2c,k-1}), \ldots, Z_{m,k}(x_{m,k-1}, x_{mc,k-1})$, $k = 1, \ldots, K$. Then choose the operating policy for the m reservoir-hydroplant complex in period k as follows:

$$\min_{\{u_{1,k},\ldots,u_{m,k}\} \in U_k} \left[F_k(D_k - u_{1,k}, \ldots, u_{m,k}) + \frac{1}{m} \sum_{j=1}^{m} Z_{j,k+1}(x_{j,k}, x_{jc,k}) \right] \qquad (4.29)$$

In other words, we substitute $(1/m) \sum_j Z_{j,k+1}(x_{j,k}, x_{jc,k})$ for $Z_{0,k+1}(x_{1,k}, \ldots, x_{m,k})$ in the right-hand side of equation (4.19).

It was concluded that a closed loop operating policy for the network of m reservoir-hydroplant complexes in parallel can be obtained by first solving the m dynamic programming problems (4.28) and then by adjusting $u_{1,k}, u_{2,k}, \ldots u_{m,k}$ so as to satisfy (4.29). In the next section we discuss the algorithm of solution for the aggregation–decomposition approach.

4.3.4.1. Algorithm of Solution

In this algorithm all variables are discretized. The following steps explain the aggregation–decomposition algorithm written by Turgeon in his well-known work:

Step 1. Set $Z_{j,K}(x_{j,K}, x_{jc,K}) = 0 \; \forall x_{j,K}, x_{jc,K}, j$. Then let $j = 0$, and go to step 2.

Step 2. Let $j = j + 1$. If $j > m$, stop. Otherwise, set

$$N = 1, x_{jc}^M = \sum_{\sigma \neq j} x_\sigma^M, u_{jc,k}^M = \sum_{\sigma \neq j} u_{\sigma,k}^M$$

and

$$\text{Prob}(I_{jc,k} = \alpha) = \text{Prob}\left(\sum_{\sigma \neq j} I_{\sigma,k} = \alpha \right)$$

for all α, k, and go to step 3.

Step 3. Determine $u_{j,k}(w_{j,k}, w_{jc,k})$, $k = 1, \ldots, K$, where $w_{j,k} = x_{j,k-1} + I_{j,k}$ and $w_{jc,k} = x_{jc,k} + I_{jc,k}$, by solving recursively, starting in period K and going backward in time, functional equation (4.28). If $N = 1$, go to step 4. Otherwise, go to step 5.

Step 4. Set $u_{j,k}^*(w_{j,k}, w_{jc,k}) = u_{j,k}(w_{j,k}, w_{jc,k}) \forall w_{j,k}, w_{jc,k}, k$; $Z_{j,K}(x_{j,K-1}, x_{jc,K-1}) = Z_{j,1}(x_{j,K-1}, x_{jc,K-1}) \forall x_{j,K-1}, x_{jc}$; $N = N + 1$; and go to step 3.

Step 5. If $u_{j,k}(w_{j,k}, w_{jc,k}) = u_{j,k}^*(w_{j,k}, w_{jc,k}) \pm \varepsilon \forall w_{j,k}, w_{jc,k}$, where ε is an infinitesimal value, go to step 2. Otherwise go to step 4.

4.3.4.2. Practical Example

In this section, the aggregation–decomposition algorithm of the above section is applied to the same example mentioned in Section 4.3.3.2, which consists of six reservoir-hydroplant complexes in parallel, for a period of 52 weeks.

The computer program that executes the aggregation–decomposition method has 224 statements and requires 80 Kbytes of computer memory. The computing time to obtain the optimal operating policy for the network of six complexes was about 150 minutes of CPU time on an IBM/370 model 168 when the variables $x_{j,k}$, $x_{jc,k}$ and $I_{j,k}$ are assigned, respectively, 25, 35, and 9 different values, or about the same computer time as the one-at-a-time algorithm.

4.3.5. Comments

To compare the operating policies obtained with the one-at-a-time and aggregation–decomposition methods, Turgeon simulated the operation of the six-reservoir-hydroplant complex with these two policies. After 10 years of simulation Turgeon found that, with the local feedback operating policy, the total operating cost was \$3167.13. After the same 10 years of simulation with the closed-loop operating policy, the cost was \$899.04; hence a difference of \$2268.09 was obtained between the two policies. In the 10 years of simulation 229.9 and 163.33 GWh of secondary energy were generated with the closed-loop and local feedback operating policies for total revenues of \$906.23 and \$551.31, respectively, or an average revenue of \$3.94 per GWh with the closed-loop policy as compared with \$3.42 with the local feedback.

In the simulation, 213.75 GWh were imported with the closed-loop policy at a cost of \$1805.27 and 367.71 GWh were imported with the local feedback policy at a cost of \$3722.8 for an average cost of \$6.45 per GWh with the closed-loop policy and \$10.12 per GWh with the local feedback.

It is clear from the above simulation that the aggregation–decomposition method gives a better operating policy than the known successive-approximation–dynamic-programming method, and this with the same processing time and computer memory. Whether this is generally true cannot be stated with certainty. It is true that 150 min of processing time on an IBM 370/168 is enormous, which is the computing time taken by both methods for a system of six complex reservoirs. It has been found that this time can easily be reduced by assigning better values to the water remaining in the reservoir at the end of the planning horizon. In the algorithm given in Section 4.3.4.1 $Z_{j,K}(x_{j,K}, x_{jc,K})$ is initialized to zero for all values of $x_{j,K}$ and $x_{jc,K}$, which is very far from reality. Because of that, over 10 years of iteration were required to solve the problem. For a 5-year-period the computing time will be reduced to 75 min rather than 150 min.

In addition to the fact that it gives better results, the aggregation-decomposition method has another advantage over the method of successive-approximation–dynamic-programming, which is that the computational effort increases only linearly with the number of reservoirs. More precisely, for each new reservoir added to the system, only one additional dynamic programming problem of two-state variables has to be solved.

4.4. Discrete Maximum Principle
(Refs. 4.26 and 4.32–4.40)

The discrete maximum principle has not been a popular approach to multireservoir network control. This is because of numerical difficulties in solving the resulting two-point boundary value problem (TPBVP), especially when state variable constraints are included in the problem formulation. In this section we discuss the application of the discrete maximum principle to the optimization of multireservoir power systems.

4.4.1. General Problem Formulation

The basic problem is to find the control variable $u(k)$ that minimizes (maximizes) a cost functional

$$J = E[\phi(x(K)) + \sum_{k=0}^{K-1} F(x(k), u(k), k)] \qquad (4.30)$$

subject to satisfying the state equation given by

$$x(k+1) = f(x(k), u(k), k), \qquad k = 0, \ldots, K-1 \qquad (4.31)$$

$$x(0) = x_0 \qquad (4.32)$$

and control constraints

$$u^m(k) \le u(k) \le u^M(k), \qquad k = 0, \ldots, K-1 \qquad (4.33)$$

and state variable constraints of the form

$$x^m \le x(k) \le x^M, \qquad k = 1, \ldots, K-1 \qquad (4.34)$$

where $x \in R^n$ and $u \in R^m$ are the state and control vector, respectively; ϕ, F, and f are nonlinear differentiable functions; and K is the fixed optimization horizon.

The above formulation can be extended to consider a broader class of optimal control problems. Some of these extensions are as follows: First, the nonlinear functionals ϕ, F, and f may contain in their arguments past values of the control variables; for example,

$$x(k+1) = f(x(k), u(k), u(k-1), \ldots, u(k-\tau), k) \qquad (4.35)$$

can be considered instead of equation (4.31). Another extension would be to consider nonseparable objective functionals. In this case, sufficient initial values must be provided for the retarded variables. For example, if equation (4.35) is used instead of equation (4.31), we need the values of $u(-\tau), \ldots, u(-1)$ in order to solve the dynamic optimization problem for a time horizon $k = 0, \ldots, K$.

The inequality constraints (4.34) may be included in an augmented cost functional $\tilde{J}$ by use of appropriate penalty functions as

$$\tilde{J} = J + \tfrac{1}{2} \sum_{k=0}^{K-1} [\|\psi(x(k) - x^m)\|_Q^2 + \|\psi(x(k) - x^M)\|_Q^2] \qquad (4.36)$$

where $\|x(k) - x^m\|_Q^2 = [x(k) - x^m]^T Q(x(k) - x^m)$, $Q > 0$ is a diagonal weighting matrix, and $\psi(\eta)$ is a vector penalty fundtion with components

$$\psi_i(\eta) = \min(0, \eta_i) \qquad (4.37)$$

note that η is equal to either $[x(k) - x^m]$ or $[x(k) - x^M]$.

Similar arguments apply to the case of fixed terminal conditions

$$x(K) = x_E \tag{4.38}$$

In this case ϕ should be augmented by a penalty term to become

$$\phi^1 = \phi + \tfrac{1}{2}\|x(K) - x_E\|_P^2 \tag{4.39}$$

where $P > 0$ is a diagonal weighting matrix.

It is well known that higher values of the weighting matrix Q lead to the constraints (4.34) being exceeded for a shorter time and to a lower convergence rate. Usually, one specifies Q so as to guarantee only slight violation of (4.34).

4.4.2. Solution Algorithm

Papageorgiou suggested in his work (Ref. 4.26) the following algorithm for solving the problem just described above: For a given admissible trajectory $u(k)$, $k = 0, \ldots, K - 1$ the resulting state trajectory $x(k)$ can be found by solving forward in stages the state equation (4.31) with the initial condition (4.32), and hence, the cost functional can be regarded as depending on the control variables only, i.e., $J = J(u)$. The gradient of J with respect to $u(k)$ on the equality constraints surface is given by

$$g(k) = \frac{\partial J}{\partial u(k)} + \sum_{z=k}^{K} \left[\frac{\partial f(z)}{\partial u(k)}\right]^T \lambda(z + 1) \tag{4.40}$$

where the costate variables $\lambda \in R^n$ satisfy

$$\lambda(k) = \frac{\partial J}{\partial x(k)} + \frac{\partial f^T}{\partial x(k)}\lambda(k + 1), \qquad k = 0, \ldots, K - 1 \tag{4.41}$$

and

$$\lambda(K) = \frac{\partial \phi}{\partial x(K)} \tag{4.42}$$

The gradient $g(k)$ defined in (4.39) corresponds to the Hamiltonian of the optimal problem if no retarded variables are included in the cost functional J and in the state functional f. The costate variable λ corresponds to the Lagrange multipliers of static optimization problems.

The ith component of the projected gradient with respect to the admissible control $u(k)$, $\gamma_i(k)$, is given by

$$\gamma_i(k) = g_i(k) \qquad \text{for } u_i^m(k) \le u_i(k) \le u_i^M(k)$$
$$\gamma_i(k) = 0 \qquad \text{otherwise} \tag{4.43a}$$

One of the necessary conditions of optimality besides equations (4.39)-(4.41) is that the gradient given by equation (4.43) is zero; i.e.,

$$\gamma_i(k) = 0 \qquad \forall i, \qquad \forall k \tag{4.43b}$$

The necessary conditions for optimality constitute a two-point boundary value problem. The following steps are used to solve this problem.

Step 1. Assume an initial guess for the admissible control trajectory $u(k)$, $k = 0, \ldots, K - 1$ such that

$$u^m(k) \le u^i(k) \le u^M(k), \qquad \text{set } i = 0$$

Step 2. Solve equation (4.31) forward in stages with equation (4.32) as initial condition to obtain $x^i(k + 1)$.

Step 3. Solve equation (4.41) backward in stages with equation (4.42) as terminal condition.

Step 4. Calculate the gradient $g^i(k)$ from equation (4.40). If $g^i(k) \le \varepsilon$, where is a specified parameter for terminating the iteration, terminate the iteration. Otherwise go to step 5.

Step 5. Calculate a new set of the admissible control variables from

$$u^{i+1}(k) = u^i(k) + \alpha^i g^i(k)$$

where α^i is the scalar step length. Several techniques for specifying α_i can be applied including steepest descent conjugate gradient and variable metric.

The algorithm described is known to be insensitive with respect to poor initial guess trajectories $u^0(k)$.

4.4.3. Practical Example

The above algorithm is applied to solve the problem of optimal operation of multireservoir networks of the ten-reservoir network of Figure 4.6. The objective function for this system is given by

$$J = \tfrac{1}{2} \sum_{k=0}^{K-1} \left[D(k) - \sum_{j=1}^{m} \sum_{i=1}^{n_j} \nu_{ij} q_{ij,k} \right]^2 \tag{4.44}$$

where $D(k)$ represents the energy demand, $\nu_{ij} q_{ij,k}$ is the energy generated by reservoir i on river j at time k, m is the number of rivers, which is equal to 9 in this example, and n_j is the number of series reservoirs on river j. More realistic performance functionals including storage can be considered by the method if required.

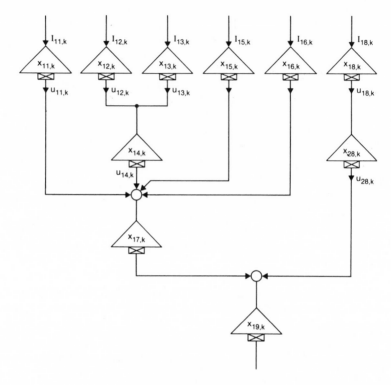

Figure 4.6. Example network.

Table 4.4. Characteristics of the Installations

River	Reservoir	x_{ij}^m $(\text{Mm}^3)^a$	x_{ij}^M $(\text{Mm}^3)^a$	Maximum effective discharge (m^3/sec)	Minimum effective discharge (m^3/sec)	Reservoir constants a_{ij}	b_{ij}
1	R_{11}	9949	24763	1119	85	268.672	7.62×10^{-3}
	R_{21}	3734	5304	1538	85	148.467	3.30×10^{-2}
2	R_{12}	33197	74225	1877	283	274.100	1.91×10^{-3}
	R_{22}	0	0	1930	283	100.735	0
3	R_{13}	24467	45264	1632	283	224.061	5.621×10^{-3}
	R_{23}	8886	9175	1876	283	116.500	1.892×10^{-2}

^{a} 1 Mm^3 = 10^6 m^3.

The objective functional of equation (4.44) (1) minimizes the energy deficit and (2) distributes deficit, if any, uniformly over the individual time intervals.

The natural inflows to the sites and energy demand over the optimization horizon are given in Table 4.4. The optimization is done on a monthly time basis for a period of 20 months. We assume an initial storage

$$x(0) = [6\ 6\ 3\ 8\ 8\ 7\ 15\ 6\ 5\ 15]^T$$

The maximum and minimum releases from the reservoir i on river j are given by

$$u_{ij,k}^M = 0.0864d^k \qquad \text{(maximum effective discharge in m}^3/\text{sec)}$$

$$u_{ij,k}^m = 0.0864d^k \qquad \text{(minimum effective discharge in m}^3/\text{sec)}$$

where the maximum and minimum effective discharges are given by

$$q^M = [4\ 4.5\ 2.12\ 7.0\ 6.43\ 4.21\ 17.1\ 3.1\ 4.2\ 1.9]$$

$$q^m = 10^{-3}[5\ 5\ 5\ 5\ 6\ 6\ 10\ 8\ 8\ 10]^T$$

and d^k is the number of days in period k. The parameters v_{ij} of the objective functional (4.44) were taken to be

$$v = [1.1\ 1.4\ 1\ 1.1\ 1\ 1.4\ 2.6\ 1\ 1\ 2.7]^T$$

The weighting matrix Q for the penalty terms was set to $Q = \text{diag}(100, \ldots, 100)$.

The computing time to obtain the optimal operation was 13 CPU sec in the rather slow DEC2060 digital computer. The final value for the objective functional was zero after 40 iterations; this indicates that the energy demand has been completely satisfied and that there was no significant violation of the state variable constraints. Now, assume that we desire the state variables to reach the terminal condition $x(K) = x_E = x_0$, and we set the weighting matrix $P = \text{diag}(10, \ldots, 10)$. The terminal state found by solution of the optimal control problem in this case was

$$x(K) = [5.42\ 5.29\ 2.32\ 7.41\ 7.42\ 6.39\ 14.52\ 5.57\ 4.66\ 14.75]$$

The energy deficit was almost identical for all time intervals

$$D(k) - \sum_{j=1}^{m} \sum_{i=1}^{n_j} v_{ij}q_{ij,k} = 0.9, \qquad k = 0, \ldots, K-1$$

and the computing time was 24 CPU sec. If the weighting matrix increases to $P = \text{diag}[100, \ldots, 100]$ we get

$$x(K) = [5.86\ 5.83\ 2.84\ 7.86\ 7.87\ 6.86\ 14.89\ 5.90\ 4.92\ 14]$$

and a uniformly distributed energy deficit of 2.15. Optimal results have been achieved in the same computer after 25 CPU sec.

Finally, if the weighting matrix increased to $P = \text{diag}[1000, \ldots, 1000]$ the optimal solution was reached after 31 CPU sec, and the terminal state was

$$x(K) = [5.98 \quad 5.98 \quad 2.98 \quad 7.98 \quad 7.98 \quad 6.98 \quad 14.99 \quad 5.99 \quad 4.99 \quad 14.99]^T$$

and the energy deficit of each time interval increased to 2.5. Figures 4.7–4.9 depict the solution trajectories for this last optimization run.

4.4.4. Comments

In this section, we discuss the application of the discrete maximum principle to multireservoir control. The main advantage of the algorithm presented is in the moderate computational effort as well as the fact that

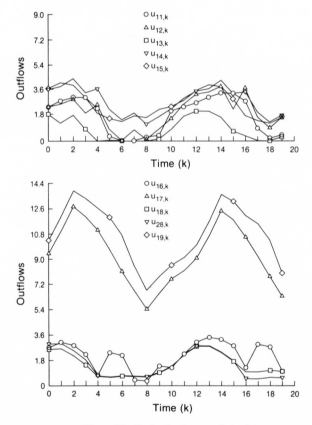

Figure 4.7. Optimal release trajectories.

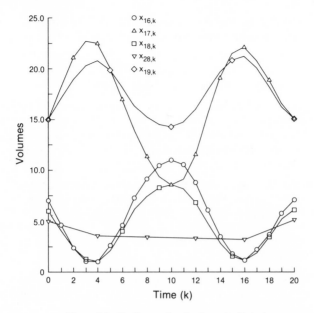

Figure 4.8. Further optimal release trajectories.

both storage space and CPU time increase roughly linearly with the problem dimension permitting solution of large-scale problems. Flow times between subsequent reservoirs can be explicitly considered in the problem formulation. Nonlinear objectives and nonlinear state equations can be treated. The algorithm avoids operations with high-dimensional matrices and does not require discretization of state variables. State variable inequality constraints are handled by use of appropriate penalty functions. In view of the

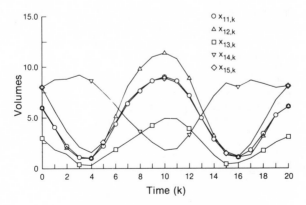

Figure 4.9. Optimal storage trajectories.

short computation times, on-line application of the algorithm for repeated optimization with updated inflow predictions appears to be a feasible task.

4.5. A Minimum Norm Approach, Linear Model

In this section, we explain a new approach used for solving the optimal long-term scheduling problem for a multireservoir hydroelectric power system. The problem is formulated as a minimum norm problem in the framework of functional analytic optimization technique. The algorithm developed here attempts to maximize the value of energy generated by hydropower systems over the planning period plus the expected future returns from water left in storage at the end of that period.

4.5.1. The Objective Function

The long-term stochastic optimization problem determines the discharge $u_{ij,k}$, $i = 1, \ldots, n_j$, $j = 1, \ldots, m$, $k = 1, \ldots, K$ under the following constraints:

(1) The total expected benefits from the system (benefits from energy generated by a hydro power system over the planning period plus the expected future returns from the water left in storage at the end of that period).

(2) The MWh generated per Mm^3 of discharge (the water conversion factor) as a function of the storage is adequately described by the following equation:

$$h_{ij,k} = a_{ij} + b_{ij}x_{ij,k}, \qquad i = 1, \ldots, n_j, \qquad j = 1, \ldots, m, \qquad k - 1, \ldots, K$$

$$(4.45)$$

where a_{ij} and b_{ij} are constants. These were obtained by least-squares curve fitting to typical plant data available. In equation (4.45) $h_{ij,k}$ stands for the water conversion factor of reservoir i on river j during period k and $x_{ij,k}$ is the storage of reservoir i on river j at the end of period k in Mm^3 ($1 \ Mm^3 = 10^6 \ m^3$).

(3) The water conservation equation for each reservoir may adequately be described by the continuity-type equation

$$x_{ij,k} = x_{ij,k-1} + z_{ij,k} + I_{ij,k} - u_{ij,k} - s_{ij,k} \tag{4.46}$$

where $z_{ij,k}$ is the dependent water inflow connecting hydroelectronics plants over the same hydraulic valley, and is expressed by

$$z_{ij,k} = \sum_{\substack{\sigma \in R_u \\ j \in R_r}} (u_{\sigma j,k} + s_{\sigma j,k}) \tag{4.47}$$

where R_u is the set of plants immediately upstream from plant i on river j and R_r is the set of upstream rivers and

$$
S_{ij,k} = \begin{cases} (x_{ij,k-1} + z_{ij,k} + I_{ij,k} - x_{ij,k}) - u_{ij,k}^M \\ \quad \text{if } (x_{ij,k-1} + z_{ij,k} + I_{ij,k} - x_{ij,k}) > u_{ij,k}^M \\ \quad \text{and } x_{ij,k} = x_{ij}^M \\ 0, \quad \text{otherwise} \end{cases} \tag{4.48}
$$

If the rivers are independent of each other, then we have

$$
z_{ij,k} = u_{(i-1)j,k} + s_{(i-1)j,k} \tag{4.49}
$$

(4) To satisfy multipurpose stream use requirements, such as flood control, irrigation, fishing, nagivation, and other purposes, if any, the following upper and lower limits on the variables should be satisfied:

a. Upper and lower bounds on the storage

$$
x_{ij}^m \le x_{ij,k} \le x_{ij}^M \tag{4.50}
$$

where x_{ij}^m and x_{ij}^M are the minimum and maximum storages, respectively;

b. Upper and lower bounds on the discharge

$$
u_{ij,k}^m \le u_{ij,k} \le u_{ij,k}^M \tag{4.51}
$$

where $u_{ij,k}^m$ and $u_{ij,k}^M$ are the minimum and the maximum discharges.

(5) The energy generated by a hydropower plant at a month k is given by

$$
G_{ij,k}(u_{ij,k}, \tfrac{1}{2}(x_{ij,k} + x_{ij,k-1})) = a_{ij}u_{ij,k} + \tfrac{1}{2}b_{ij}u_{ij,k}(x_{ij,k} + x_{ij,k-1}) \tag{4.52}
$$

In the above equation an average of begin and end-of-time step storage has been used to avoid overestimation of production for falling water levels and underestimation for rising water levels.

In mathematical terms, the object of the optimizing computation is to find the discharge $u_{ij,k}$ that maximizes

$$
J = E\left[\sum_{j=1}^m \sum_{i=1}^{n_j} \sum_{\sigma=i}^{n_j} (a_{\sigma j} + b_{\sigma j}x_{\sigma j,K})x_{ij,K} \right.
$$

$$
\left. + \sum_{j=1}^m \sum_{i=1}^{n_j} \sum_{k=1}^K \{c_{j,k}a_{ij}u_{ij,k} + c_{j,k}b_{ij}u_{ij,k}(x_{ij,k} + x_{ij,k-1})\} \right] \tag{4.53}
$$

Subject to satisfying the equality constraints (4.46)–(4.49) and the inequality constraints (4.50) and (4.51). The first term in equation (4.53) presents the amount of energy left in storage at the end of period K in dollars and the second term gives the amount of energy generated during the optimization interval in dollars. $c_{j,k}$ is the value in dollars of one MWh generated anywhere on the river. The symbol E in equation (4.53) stands for the expected value. The expectation is taken with respect to the variable $I_{ij,k}$.

The monthly natural inflows into the reservoirs form a random sequence that depends on natural phenomena such as rainfall and snowfall. The statistical parameters of the inflow can be determined from historical records. In reality, there is a statistical correlation between inflows of successive months. However, it is assumed here that the inflow in each month is statistically independent of previous inflows.

4.5.2. A Minimum Norm Formulation

The cost functional in equation (4.53) can be written in the following vector form:

$$J = E\left[A^T x(K) + x^T(K)\beta x(K) \right.$$

$$\left. + \sum_{k=1}^{K} \{\alpha^T(k)u(k) + u^T(k)\beta(k)(x(k) + x(k-1))\} \right] \quad (4.54)$$

subject to satisfying the following constraints in vector form:

$$x(k) = x(k-1) + I(k) + Mu(k) + Ms(k), \qquad k = 1, \ldots, K \quad (4.55)$$

$$x^m \leq x(k) \leq x^M, \qquad k = 1, \ldots, K \quad (4.56)$$

$$u^m(k) \leq u(k) \leq u^M(k), \qquad k = 1, \ldots, K \quad (4.57)$$

In the above equations, the following n, $n = \sum_{j=1}^{n_j} n_j$, vectors are defined: $u(k)$ is an n-dimensional decision vector of penstock releases at the end of month k; its components are $u_{ij,k}$; $i = 1, \ldots, n_j, j = 1, \ldots, m$. $x(k)$ is an n-dimensional state vector of storage at the end of month k; its components are $x_{ij,k}$; $i = 1, \ldots, n_j$; $j = 1, \ldots, m$. $s(k)$ is an n-dimensional vector of spillage at the end of month k; its components are $s_{ij,k}$; $i = 1, \ldots, n_j$, $j = 1, \ldots, m$. $I(k)$ is an n-dimensional vector of forecast inflow during a month k; its components are $I_{ij,k}$; $i = 1, \ldots, n_j$, $j = 1, \ldots, m$. A is an n-dimensional vector; its components are $A_{ij} = \sum_{\sigma=i}^{n_j} a_{\sigma j}$; $i = 1, \ldots, n_j, j = 1, \ldots, m$. $\alpha(k)$ is an n-dimensional vector; its components are $\alpha_{ij,k} = a_{ij}c_{j,k}$, $i = 1, \ldots, n_j$; $j = 1, \ldots, m$.

Furthermore, the following $(n \times n)$-dimensional vectors are defined: $\beta(k)$ is an $n \times n$ diagonal matrix; its component elements are $\beta_{ij,k} = \frac{1}{2}b_{ij}c_{j,k}$. M is an $n \times n$ lower triangular matrix that contains the topological arrangement of the system. β is an $n \times n$ diagonal matrix; its component is β_j; $j = 1, \ldots, m$ and the elements of the matrix β_j are $\beta_{ij} = b_{ij}$, $i = 1, \ldots, n_j$, $j = 1, \ldots, m$ and $\beta_{(\sigma+1)\sigma} = \beta_{\sigma(\sigma+1)} = \frac{1}{2}b_{(\sigma+1)j}$, $\sigma = 1, \ldots, n_{j-1}$.

We can now form an augmented cost functional, $\tilde{J}$, by adjoining to the cost functional in equation (4.54) the equality constraint (4.55) via Lagrange multipliers $\lambda(k)$, and the inequality constraints (4.56) and (4.57) via Kuhn–Tucker multipliers. One thus obtains

$$\tilde{J} = E\left[A^T x(k) + x^T(K)\beta x(K) + \sum_{k=1}^{K} \{\alpha^T(k)u(k) + u^T(k)\beta(k)x(k-1)\right.$$

$$+ u^T(k)\beta(k)Mu(k) + x^T(k-1)\beta(k)u(k) + \lambda^T(k)(-x(k) + x(K-1)$$

$$+ Mu(k)) + e^{mT}(k)(x^m - x(k)) + e^{MT}(k)(x(k) - x^M)$$

$$\left. + g^{mT}(k)(u^m(k) - u(k)) + g^{MT}(k)(u(k) - u^M(k))\} \right] \qquad (4.58)$$

In the above equations, constant terms are dropped and $e^m(k)$, $e^M(k)$, $g^m(k)$ and $g^M(k)$ are $(n \times 1)$-dimensional Kuhn-Tucker multipliers, these are equal to zero if the constraints are not violated, and greater than zero if the constraints are violated.

Employing the discrete version of integration by parts, substituting for $x(k)$, and dropping terms explicitly independent of $x(k-1)$ and $u(k)$ one obtains

$$\tilde{J} = E\left[x^T(K)\beta x(K) + (A - \lambda(K))x(K) + \lambda^T(0)x(0) \right.$$

$$+ \sum_{k=1}^{K} \{u^T(k)\beta(k)Mu(k) + x^T(k-1)\beta(k)u(k) + u^T(k)\beta(k)x(k-1)$$

$$+ (\lambda(k) - \lambda(k-1) + \mu(k))^T x(k-1)$$

$$\left. + (\alpha(k) + M^T\lambda(k) + M_\mu^T(k) + \psi(k))^T u(k)\} \right] \qquad (4.59)$$

where

$$\mu(k) = e^M(k) - e^m(k) \qquad (4.60a)$$

$$\psi(k) = g^M(k) - g^m(k) \qquad (4.60b)$$

If one defines the following vectors

$$N(K) = A - \lambda(K) \qquad (4.61)$$

$$W(K) = \beta^{-1}N(K) \qquad (4.62)$$

$$X^T(k) = [x^T(k-1), u^T(k)] \qquad (4.63)$$

$$L(k) = \begin{bmatrix} 0 & \beta(k) \\ \beta(k) & \beta(k)M \end{bmatrix} \quad (2n \times 2n) \qquad (4.64)$$

$$R^T(k) = [(\lambda(k) - \lambda(k-1) + \mu(k))^T, (\alpha(k) + M^T\lambda(k)$$

$$+ M^T\mu(k) + \psi(k))] \qquad (4.65)$$

and

$$V(k) = L^{-1}(k)R(k) \tag{4.66}$$

then, the cost functional in equation (4.59) can be written in the following form by a process similar to completing the squares:

$$
\begin{aligned}
\tilde{J} = E\Bigg[&\{(x(K) + \tfrac{1}{2}W(K))^T\beta(x(K) + \tfrac{1}{2}W(K)) - \tfrac{1}{4}W^T(K)\beta W(K) \\
&+ \lambda^T(0)x(0)\} + \sum_{k=1}^{K} \{(X(k) + \tfrac{1}{2}V(k))^T L(k)(X(k) + \tfrac{1}{2}V(k)) \\
&- \tfrac{1}{4}V^T(k)L(k)V(k)\} \Bigg]
\end{aligned} \tag{4.67}
$$

Since it is desired to maximize $\tilde{J}$ with respect to $x(K)$ and $X(k)$ the problem is equivalent to

$$
\begin{aligned}
\max_{x(K),X(k)} \tilde{J} = &\max_{x(K)} E[(x(K) + \tfrac{1}{2}W(K))^T\beta(x(K) + \tfrac{1}{2}W(K))] \\
&+ \max_{X(k)} E\Bigg[\sum_{k=1}^{K} (X(k) + \tfrac{1}{2}V(k))^T L(k)(X(k) + \tfrac{1}{2}V(k)) \Bigg]
\end{aligned} \tag{4.68}
$$

because $W(K)$ and $V(k)$ are, respectively, independent of $x(K)$ and $X(k)$ and $x(0)$ is constant. It will be noticed that $\tilde{J}$ in equation (4.68) is composed of a boundary part and a discrete integral part which are independent of each other. To maximize $\tilde{J}$ in equation (4.68), one maximizes each term separately.

The boundary term in equation (4.68), the first term, defines a norm; hence, this term can be written as

$$\max_{x(K)} J_1 = \max_{x(K)} E \|x(K) + \tfrac{1}{2}W(K)\|_\beta \tag{4.69}$$

The discrete integral part in equation (4.68), the second term, also defines a norm; hence, it can be written as

$$\max_{X(k)} J_2 = \max_{X(k)} E \|X(k) + \tfrac{1}{2}V(k)\|_{L(k)} \tag{4.70}$$

4.5.3. The Optimal Solution

There is exactly one optimal solution to the problem formulated in equations (4.69) and (4.70), namely, the optimal vector. The maximum of J_1 in equation (4.69) is achieved when the norm of that equation is equal to zero:

$$E[x(K) + \tfrac{1}{2}W(K)] = [0] \tag{4.71}$$

provided that the matrix β is negative definite. Substituting from equations (4.61) and (4.62) into equation (4.71), one obtains

$$E[\lambda(K)] = E[A + 2\beta x(K)] \tag{4.72}$$

Equation (4.72) gives the value of Lagrange's multiplier at the last period studied, K.

The maximum of J_2 is also achieved, when the norm of equation (4.70) is equal to zero

$$E[X(k) + \tfrac{1}{2}V(k)] = [0] \tag{4.73}$$

provided that the matrix $L(k)$ is negative definite. Substituting from equations (4.63), (4.64), and (4.65) into equation (4.73) and adding the continuity equation (4.55), one obtains the· following set of optimal equations:

$$E[x(k)] = E[x(k-1) + I(k) + Mu(k) + Ms(k)] \tag{4.74}$$

$$E[\lambda(k-1)] = E[\lambda(k) + \mu(k) + 2\beta(k)u(k)] \tag{4.75}$$

$$E[\alpha(k) + M^T\lambda(k) + M^T\mu(k) + \psi(k) + 2\beta(k)x(k-1) + 2\beta(k)Mu(k)] \tag{4.76}$$

Besides the above equations, one has the following limits on the variables:

$$
\begin{aligned}
&\text{If } x(k) < x^m, &&\text{then we put } x(k) = x^m \\
&\text{If } x(k) > x^M, &&\text{then we put } x(k) = x^M \\
&\text{If } u(k) < u^m(k), &&\text{then we put } u(k) = u^m(k) \\
&\text{If } u(k) > u^M(k), &&\text{then we put } u(k) = u^M(k)
\end{aligned} \tag{4.77}
$$

Finally, the following Kuhn-Tucker exclusion equations must be satisfied at the optimum:

$$e^m_{ij,k}(x^m_{ij} - x_{ij,k}) = 0 \tag{4.78}$$

$$e^M_{ij,k}(x_{ij,k} - x^M_{ij}) = 0 \tag{4.79}$$

$$g^m_{ij,k}(u^m_{ij,k} - u_{ij,k}) = 0 \tag{4.80}$$

$$g^M_{ij,k}(u_{ij,k} - u^M_{ij,k}) = 0 \tag{4.81}$$

Equations (4.74)-(4.81) with equation (4.72) completely specify the optimal solution.

4.5.4. Algorithm of Solution

Given the number of rivers (m), the number of reservoirs on each river (n_l), the initial storage $x(0)$, the expected natural inflows to the reservoirs, $E[I(k)]$, and the cost of energy on each river, $c_{j,k}$ in \$/MWh.

Step 1. Assume initial guess for the admissible control variables $u(k)$, such that

$$u^m(k) \le u(k) \le u^M(k)$$

Step 2. Assume first that $s(k)$ is equal to zero; solve equation (4.74) forward in stages with $x(0)$ given.

Step 3. Check the limits on $x(k)$. If $x(k) < x^m$, put $x(k) = x^m$, and if $x(k) > x^M$, put $x(k) = x^M$ and go to step 4.

Step 4. Calculate the new discharge $u(k)$ by solving

$$E[u(k)] = E[[M]^{-1}(x(k) - x(k-1) - I(k))]$$

forward in stages from $k = 1$ to K, and go to step 5.

Step 5. Check the limits on $u(k)$. If $u(k) < u^m(k)$, then put $u(k) = u^m(k)$, and if $u(k) > u^M(k)$, then put $u(k) = u^M(k)$ and go to step 6.

Step 6. Calculate the spill at month k from the following equation:

$$E[s(k)] = E[[M]^{-1}(x(k) - x(k-1) - I(k)) - u^M(k)]$$

If $s(k) < 0$, put $s(k) = 0$; otherwise go to step 7.

Step 7. Calculate the discharge from the following equation:

$$E[u(k)] = E[[M]^{-1}(x(k) - x(k-1) - I(k) - Ms(k))]$$

and use the value of $s(k)$ obtained from step 6 and go to step 8.

Step 8. Solve equation (4.74) forward in stages with $x(0)$ given. If $x(k)$ satisfies the inequality

$$x^m < x(k) < x^M$$

go to step 9. Otherwise put $x(k)$ to its limits and go to step 4.

Step 9. Solve equation (4.75) with $\mu(k) = 0$ backward in stages with equation (4.72) as terminal conditions.

Step 10. Calculate Kuhn-Tucker multipliers for $u(k)$, $\psi(k)$ from the following equation:

$$E[\psi(k)] = E[2M^T\beta(k)u(k) - M^T\lambda(k-1) - \alpha(k) - 2\beta(k)x(k) - 1)$$
$$- 2\beta(k)Mu(k)]$$

If $u(k)$ satisfies the inequality

$$u^m(k) < u(k) < u^M(k)$$

put $\psi(k) = 0$.

Step 11. Determine a new control iterate from the following:

$$u^{i+1}(k) = u^i(k) + \alpha Du^i(k)$$

where

$$E[Du(k)] = E[\alpha(k) + M^T\lambda(k) + \psi(k)2\beta(k)x(k-1) + 2\beta(k)Mu(k)]$$

and α is a positive scalar which is chosen with consideration to such factors as convergence.

Step 12. Check the limits on $u^{i+1}(k)$. If $u^{i+1}(k)$ satisfies the inequality

$$u^m(k) < u^{i+1}(k) < u^M(k)$$

go to step 13; otherwise put $u^{i+1}(k)$ to its limits and go to step 2.

Step 13. Solve the following equation forward in stages:

$$E[\lambda(k-1)] = E\{[M^T]^{-1}[2M^T\beta(k)u(k) - \alpha(k) - 2\beta(k)x(k-1)$$
$$- 2\beta(k)Mu(k)\}$$

Step 14. Determine the Kuhn-Tucker multiplier for $x(k)$, $\mu(k)$ from the following:

$$E[\mu(k)] = -[M^T]^{-1}[\alpha(k) + M^T\lambda(k) + 2\beta(k)x(k-1) + 2\beta(k)Mu(k)]$$

If $x(k)$ satisfies the inequality

$$x^m < x(k) < x^M$$

put $\mu(k) = 0$.

Step 15. Determine a new state iterate from the following equation:

$$x^{i+1}(k) = x^i(k) + \alpha Dx^i(k)$$

where

$$E[Dx(k)] = E[\lambda(k) - \lambda(k-1) + \mu(k) + 2\beta(k)u(k)]$$

Step 16. Repeat the calculation starting from step 3. Continue until the state $x(k)$ and the control $u(k)$ do not change significantly from iteration to iteration and J in equation (4.53) is a maximum.

4.5.5. Practical Example

The algorithm of the last section has been used to determine the optimal monthly operation of a real system in operation. The system consists of three rivers; each river has two series reservoirs. The characteristics of the installation are given in Table 4.4.

If we let d^k denote the number of days in month k, then the maximum and minimum releases during a period k are given by

$$u_{ij,k}^m = 0.086d^k \text{ (minimum effective discharge in } \text{m}^3/\text{sec)} \text{ Mm}^3$$

$$u_{ij,k}^M = 0.0864d^k \text{ (maximum effective discharge in } \text{m}^3/\text{sec)} \text{ Mm}^3$$

where the minimum and maximum effective discharges are given in Table 4.4.

The optimization is done on a monthly time basis for a period of a year. The times of water travel from upstream to downstream reservoirs are assumed shorter than a month. These times are neglected; the transmission line losses are also neglected. The expected monthly natural inflows to the sites in the year of high flow, which we call year 1, and the cost of energy, are given in Table 4.5. In Table 4.6 we give the optimal monthly release from each reservoir, and the profits realized in year 1. In Table 4.7, we give the optimal storage for each reservoir during year 1.

We have simulated the monthly operation of the system for widely different water conditions. The expected monthly natural inflow to the sites in year 2 which is the year of low flow are given in Table 4.8. In Table 4.9

Table 4.5. Expected Monthly Inflows to the Sites in Year 1 and the Cost of Energy

Month[a] k	$I_{11,k}$ (Mm^3)	$I_{21,k}$ (Mm^3)	$I_{12,k}$ (Mm^3)	$I_{22,k}$ (Mm^3)	$I_{13,k}$ (Mm^3)	$I_{23,k}$ (Mm^3)	$c_{j,k}$[b] ($\$/\text{MWh}$)
1	948	326	2526	30	2799	318	1.4
2	482	189	1226	14	1632	193	1.4
3	350	148	1001	15	1380	221	1.4
4	300	113	849	8	1035	234	0.8
5	238	83	724	7	825	164	0.8
6	225	78	644	8	767	169	0.8
7	385	160	462	7	794	229	0.8
8	1388	910	4558	53	2017	373	0.8
9	4492	2143	17322	147	15509	1920	0.8
10	5028	2026	7660	76	7453	999	0.8
11	2685	963	5195	69	4953	712	1.1
12	1402	580	2349	29	3376	414	1.1

[a] The water year starts with September.

[b] The average of $c_{j,k}$ during the year is one $\$/\text{MWh}$.

Table 4.6. Optimal Releases from the Reservoirs and the Profits Realized during Year 1

Month k	$u_{11,k}$ (Mm^3)	$u_{21,k}$ (Mm^3)	$u_{12,k}$ (Mm^3)	$u_{22,k}$ (Mm^3)	$u_{13,k}$ (Mm^3)	$u_{23,k}$ (Mm^3)	Profits ($)
1	2396	3285	4483	4514	4371	4689	15,441,752
2	2220	1875	4391	4405	4230	4422	13,816,672
3	1924	2117	4550	4565	4178	4399	13,337,588
4	831	1355	3348	3356	3054	3535	5,366,083
5	621	1523	2959	2966	2744	2903	4,506,654
6	688	1058	3348	3357	3079	3249	4,741,587
7	710	870	3219	3226	2983	3212	4,445,930
8	1002	1912	3361	3414	3152	3526	4,943,712
9	973	3116	3294	3441	3179	4863	6,453,121
10	2479	3897	4761	4837	4371	5025	9,092,486
11	2997	2997	4928	4996	4371	5025	12,942,779
12	2900	3480	3532	3562	423	4643	11,800,057

Value of water remaining in the reservoirs at the end of the year   93,236,784
Total profits   200,125,205

we give the optimal monthly release from each reservoir and the profits realized in year 2. In Table 4.10, we give the optimal storage for each reservoir during year 2. We start both years with the reservoirs full.

The processing time required to determine the optimal monthly releases for a period of a year for the system just described was 1.6 sec in CPU units on the Amdahl 470/6 computer, University of Alberta.

Table 4.7. Optimal Reservoirs Storage during Year 1

Month k	$x_{11,k}$ (Mm^3)	$x_{21,k}$ (Mm^3)	$x_{12,k}$ (Mm^3)	$x_{13,k}$ (Mm^3)	$x_{23,k}$ (Mm^3)
1	23315	4771	72297	44100	9132
2	21577	5304	69131	41502	9132
3	20003	5258	65582	38703	9132
4	19472	4848	63083	36683	8886
5	19080	4026	60822	34735	8886
6	18617	3734	58117	32422	8886
7	18292	3734	55858	30232	8886
8	18677	3734	57056	29097	8886
9	22196	3734	71083	41426	9122
10	24744	4342	73982	44508	9132
11	24432	5304	74248	45091	9132
12	22933	5304	73064	44236	9132

Table 4.8. Expected Monthly Inflows to the Sites in Year 2 (Dry Year)

Month k	$I_{11,k}$ (Mm^3)	$I_{21,k}$ (Mm^3)	$I_{12,k}$ (Mm^3)	$I_{22,k}$ (Mm^3)	$I_{13,k}$ (Mm^3)	$I_{23,k}$ (Mm^3)
1	551	200	1987	23	2260	531
2	361	285	1226	14	991	169
3	316	298	1054	15	849	84
4	273	129	744	10	720	54
5	245	10	491	6	599	72
6	265	92	563	7	579	90
7	608	251	1182	14	598	96
8	1593	925	3898	54	2336	432
9	3851	2065	7487	73	5872	668
10	4566	1630	2556	30	4179	675
11	2701	1137	1828	22	3299	865
12	2348	1080	1813	22	2664	852

4.6. A Minimum Norm Approach, Nonlinear Model

The model used in Section 4.5 for the water conversion factor, MWh/Mm^3, is a linear model. For power systems in which the water heads vary by a small amount, this assumption is true, but for power systems in which water heads vary by a considerable amount this assumption is not adequate and may cause overestimation and underestimation in the production of energy. In this section, we offer a formulation to the problem

Table 4.9. Optimal Releases from the Reservoirs in Year 2 and the Profits Realized

Month k	$u_{11,k}$ (Mm^3)	$u_{21,k}$ (Mm^3)	$u_{12,k}$ (Mm^3)	$u_{22,k}$ (Mm^3)	$u_{13,k}$ (Mm^3)	$u_{23,k}$ (Mm^3)	Profits ($)
1	2997	3197	4420	4443	4371	4902	16,013,177
2	2485	2769	4369	4684	4230	4398	14,196,242
3	1296	2884	4567	4582	3448	3532	11,565,489
4	227	636	1873	1883	975	1275	2,148,487
5	205	205	1485	1491	710	780	1,464,047
6	227	227	1873	1879	1039	1129	1,965,371
7	220	220	1747	1761	1007	1102	1,873,458
8	529	1140	1903	1956	1268	1700	2,747,102
9	2900	4061	1843	1916	1667	2335	4,961,431
10	2997	4239	2003	2064	2852	3281	6,446,289
11	2997	4134	3405	3428	4195	5025	12,075,924
12	2900	3980	3317	3339	4230	4863	11,599,919

Value of water remaining in the reservoirs at the end of the year   84,799,056
Total profits   171,855,992

Table 4.10. Optimal Reservoirs Storage during Year 2

Month k	$x_{11,k}$ (Mm^3)	$x_{21,k}$ (Mm^3)	$x_{12,k}$ (Mm^3)	$x_{13,k}$ (Mm^3)	$x_{23,k}$ (Mm^3)
1	22317	5304	71821	43561	9132
2	20192	5304	68679	40322	9132
3	19212	4014	65164	37722	9132
4	19258	4737	64035	37466	8886
5	19289	3744	63023	37334	8886
6	19327	3836	61713	36874	8886
7	19714	4086	61148	36464	8886
8	20778	4400	63143	37532	8886
9	21728	5304	68786	41736	8886
10	23297	5304	69308	43063	9132
11	23000	5304	67730	42167	9132
12	22480	5304	66226	40601	8132

mentioned in Section 4.5 based on the minimum norm approach, but the model used for the water conversion factor is a nonlinear model of the storage; we use a quadratic model.

The object of the optimizing computation is to find the discharge $u_{ij,k}$ that maximizes the total expected benefits from the system. In mathematical terms the problem is to find $u_{ij,k}$ that maximizes

$$J = E\left[\sum_{j=1}^{m} \sum_{i=1}^{n_j} \left\{ V_{ij}(x_{ij,K}) + \sum_{k=1}^{K} c_{j,k} G_{ij,k}(u_{ij,k}, \tfrac{1}{2}(x_{ij,k} + x_{ij,k-1})) \right\} \right] \quad (4.82)$$

Subject to satisfying the equality constraints given by equations (4.46)–(4.48) and the inequality constraints given by equations (4.50) and (4.51).

4.6.1. Modeling of the System

The value in dollars of the amount of water left in storage at the end of the planning period $V_{ij}(x_{ij,K})$ is given by

$$V_{ij}(x_{ij,K}) = \sum_{\sigma=i}^{n_j} x_{ij,K}[\alpha_{\sigma j} + \beta_{\sigma j} x_{\sigma j,K} + \gamma_{\sigma j}(x_{\sigma j,K})^2] \quad (4.83)$$

The above equation is obtained by multiplying the amount of water left in storage by the water conversion factor of at-a-site and downstream hydroplants; i.e., we convert this amount of water to the equivalent electrical energy MWh. Since no one knows when this energy will be used in the future, we assume the value of this energy is one dollar per MWh, the average cost during a year.

The generation of a hydroelectric plant is a nonlinear function of the water discharge $u_{ij,k}$ and the reservoir head, which itself is a function of the storage. To avoid underestimation of production for rising water levels and overestimation for falling water levels, an average of begin and end of time step (month) storage is used. We may choose the following for the function $G_{ij}(u_{ij,k}, \frac{1}{2}(x_{ij,k} + x_{ij,k-1}))$:

$$G_{ij}(u_{ij,k}, \tfrac{1}{2}(x_{ij,k} + x_{ij,k-1})) = \alpha_{ij} u_{ij,k} + \tfrac{1}{2}\beta_{ij} u_{ij,k}(x_{ij,k} + x_{j,k-1})$$
$$+ \tfrac{1}{4}\gamma_{ij} u_{ij,k}(x_{ij,k} + x_{ij,k-1})^2 \qquad (4.84)$$

where α_{ij}, β_{ij}, γ_{ij} are constants. These were obtained by least squares curve fitting to typical plant data available.

In one assumes that the rivers are independent of each other, and substitutes from equations (4.46) and (4.49) into equation (4.84), we obtain

$$G_{ij}(u_{ij,k}, \tfrac{1}{2}(x_{ij,k} + x_{ij,k-1}))$$
$$= b_{ij,k} u_{ij,k} + u_{ij,k} d_{ij,k} x_{ik,k-1}$$
$$+ u_{ij,k} f_{ij,k}(u_{(i-1)j,k} - u_{ij,k}) + \gamma_{ij} u_{ij,k}(x_{ij,k-1})^2$$
$$+ \tfrac{1}{4} u_{ij,k} \gamma_{ij}[(u_{ij,k})^2 + (u_{(i-1)j,k})^2]$$
$$+ \gamma_{ij} u_{ij,k} x_{ij,k-1}[u_{(i-1)j,k} - u_{ij,k}] - \tfrac{1}{2}\gamma_{ij} u_{(i-1)j}(u_{ij,k}) \qquad (4.85)$$

where

$$q_{ij,k} = I_{ij,k} + s_{(i-1)j} - s_{ij,k} \qquad (4.86)$$
$$b_{ij,k} = \alpha_{ij} + \tfrac{1}{2}\beta_{ij} q_{ij,k} + \tfrac{1}{4}\gamma_{ij}(q_{ij,k})^2 \qquad (4.87)$$
$$d_{ij,k} = \beta_{ij} + \gamma_{ij} q_{ij,k} \qquad (4.88)$$
$$f_{ij,k} = \tfrac{1}{2}\beta_{ij} + \tfrac{1}{2}\gamma_{ij} q_{ij,k} \qquad (4.89)$$

To cast the generating function in equation (4.85) into a quadratic function, we may define the following pseudostate variables as:

$$y_{ij,k} = (x_{ij,k})^2, \qquad i = 1, \ldots, n_j, \quad j = 1, \ldots, m, \quad k = 1, \ldots, K \qquad (4.90)$$
$$z_{ij,k} = (u_{ij,k})^2, \qquad i = 1, \ldots, n, \quad j = 1, \ldots, m, \quad k = 1, \ldots, K \qquad (4.91)$$
$$r_{ij,k} = u_{ij,k} x_{ij,k-1}, \qquad i = 1, \ldots, n, \quad j = 1, \ldots, m, \quad k = 1, \ldots, K \qquad (4.92)$$

Then, the cost functional in equation (4.85) becomes

$$G_{ij}(u_{ij,k}, \tfrac{1}{2}(x_{ij,k} + x_{ij,k-1})) = b_{ij,k} u_{ij,k} + u_{ij,k} d_{ij,k} x_{ij,k-1}$$
$$+ u_{ij,k} f_{ij,k}(u_{(i-1)j,k} - u_{ij,k})$$
$$+ \gamma_{ij} u_{ij,k} y_{ij,k-1}$$
$$+ \tfrac{1}{4} u_{ij,k} \gamma_{ij}(z_{(i-1)j,k} + z_{ij,k})$$
$$+ \gamma_{ij} r_{ij,k-1}(u_{l(i-1)j,k} - u_{ij,k})$$
$$- \tfrac{1}{2}\gamma_{ij} z_{ij,k} u_{(i-1)j,k} \qquad (4.93)$$

Now, the cost functional in equation (4.83) becomes

$$J = E\left[\sum_{j=1}^{m} \sum_{i=1}^{n_j} \sum_{\sigma=i}^{n_j} x_{ij,K}(\alpha_{\sigma j} + \beta_{\sigma j}x_{\sigma j,K} + \gamma_{\sigma j}y_{\sigma j,K}) \right.$$

$$+ \sum_{j=1}^{m} \sum_{i=1}^{n_j} \sum_{k=1}^{K} \{c_{j,k}b_{ij,k}u_{ij,k} + c_{j,k}u_{ij,k}d_{ij,k}x_{ij,k-1}$$

$$+ c_{j,k}u_{ij,k}f_{ij,k}[u_{(i-1)j,k} - u_{ij,k}]$$

$$+ c_{j,k}\gamma_{ij}u_{ij,k}y_{ij,k-1} + \tfrac{1}{4}c_{j,k}u_{ij,k}\gamma_{ij}[z_{(i-1)j,k} + z_{ij,k}]$$

$$\left. + c_{j,k}\gamma_{ij}r_{ij,k-1}[u_{(i-1)j,k} - u_{ij,k}] - \tfrac{1}{2}c_{j,k}\gamma_{ij}z_{ij,k}u_{(i-1)j,k}\} \right] \tag{4.94}$$

Subject to satisfying the following constraints:
1. Equality constraints given by

$$y_{ij,k} = (x_{ij,k})^2 \tag{4.95}$$

$$z_{ij,k} = (u_{ij,k})^2 \tag{4.96}$$

$$r_{ij,k-1} = u_{ij,k}x_{ij,k-1} \tag{4.97}$$

$$x_{ij,k} = x_{ij,k-1} + q_{ij,k} + u_{(i-1)j,k} - u_{ij,k} \tag{4.98}$$

2. Inequality constraints given by

$$x_{ij}^m \leq x_{ij,k} \leq x_{ij}^M \tag{4.99}$$

$$u_{ij,k}^m \leq u_{ij,k} \leq u_{ij,k}^M \tag{4.100}$$

Now, the problem is that of maximizing (4.94) subject to satisfying the constraints (4.95)–(4.100). We can form an augmented cost functional, $\tilde{J}$, by adjoining to the cost function in equation (4.94) the equality constraints (4.95)–(4.100) via Lagrange multipliers, and the inequality constraints (4.99) and (4.100) via Kuhn-Tucker multipliers. One thus obtains

$$\tilde{J} = E\left[\sum_{j=1}^{m} \sum_{i=1}^{n_j} \sum_{\sigma=i}^{n_j} x_{ij,K}(\alpha_{\sigma j} + \beta_{\sigma j}x_{\sigma j,K} + \gamma_{\sigma j}y_{\sigma j,K}) \right.$$

$$+ \sum_{j=1}^{m} \sum_{i=1}^{n_j} \sum_{k=1}^{K} \{c_{j,k}b_{ij,k}u_{ij,k} + c_{j,k}u_{ij,k}d_{ij,k}x_{ij,k-1}$$

$$+ c_{j,k}u_{ij,k}f_{ij,k}(u_{(i-1)j,k} - u_{ij,k}) + c_{j,k}\gamma_{ij}u_{ij,k}y_{ij,k-1}$$

$$+ \tfrac{1}{4}c_{j,k}u_{ij,k}\gamma_{ij}(z_{(i-1)j,k} + z_{ij,k})$$

$$+ c_{j,k}\gamma_{ij}r_{ij,k-1}(u_{(i-1)j,k} - u_{ij,k}) - \tfrac{1}{2}c_{j,k}\gamma_{ij}z_{ij,k}u_{(i-1)j,k}$$

$$+ \mu_{ij,k}(-y_{ij,k} + (x_{ij,k})^2)$$

$$+ \phi_{ij,k}(-z_{ij,k} + (u_{ij,k})^2) + \psi_{ij,k}(-r_{ij,k-1} + u_{ij,k}x_{ij,k-1})$$

$$+ \lambda_{ij,k}(-x_{ij,k} + x_{ij,k-1} + q_{ij,k} + u_{(-1)j,k} - u_{ij,k}) + e_{ij,k}(x_{ij}^m - x_{ij,k})$$

$$\left. + e_{ij,k}^1(x_{ij,k} - x_{ij}^M) + g_{ij,k}(u_{ij,k}^m - u_{ij,k}) + g_{ij,k}^1(u_{ij,k} - u_{ij,k}^M)\} \right] \tag{4.101}$$

In the above equation $\mu_{ij,k}$, $\phi_{ij,k}$, $\psi_{ij,k}$, and $\lambda_{ij,k}$ are Lagrange multipliers. They are to be determined so that the corresponding equality constraints are satisfied and $e_{ij,k}$, $e^1_{ij,k}$, $g_{ij,k}$, and $g^1_{ij,k}$ are Kuhn–Tucker multipliers; these are equal to zero if the constraints are not violated, and greater than zero if the constraints are violated.

If R_e and R_v are the sets of the series reservoirs and rivers, respectively, and if one defines the following vectors:

$$A = \text{col}(A_{ij}, i \in R_e, j \in R_v) \tag{4.102}$$

$$b(k) = \text{col}(c_{j,k}b_{ij,k}, i \in R_e, j \in R_v) \tag{4.103}$$

$$x(k) = \text{col}(x_{ij,k}, i \in R_e, j \in R_v) \tag{4.104}$$

$$y(k) = \text{col}(y_{ij,k}, i \in R_e, j \in R_v) \tag{4.105}$$

$$u(k) = \text{col}(u_{ij,k}, i \in R_e, j \in R_v) \tag{4.106}$$

$$z(k) = \text{col}(z_{ij,k}, i \in R_e, j \in R_v) \tag{4.107}$$

$$r(k) = \text{col}(r_{ij,k}, i \in R_e, j \in R_v) \tag{4.108}$$

$$\mu(k) = \text{col}(\mu_{ij,k}, i \in R_e, j \in R_v) \tag{4.109}$$

$$\phi(k) = \text{col}(\phi_{ij,k}, i \in R_e, j \in R_v) \tag{4.110}$$

$$\psi(k) = \text{col}(\psi_{ij,k}, i \in R_e, j \in R_v) \tag{4.111}$$

$$\lambda(k) = \text{col}(\lambda_{ij,k}, i \in R_e, j \in R_v) \tag{4.112}$$

$$q(k) = \text{col}(q_{ij,k}, i \in \cdot R_e, j \in R_v) \tag{4.113}$$

$$s(k) = \text{col}(s_{ij,k}, i \in R_e, j \in R_v) \tag{4.114}$$

$$\nu(k) = \text{col}(\nu_{ij,k}, i \in R_e, j \in R_v) \tag{4.115}$$

$$\nu_{ij,k} = e^1_{ij,k} - e_{ij,k} \tag{4.116}$$

$$\sigma(k) = \text{col}(\sigma_{ij,k}, i \in R_e, j \in R_v) \tag{4.117}$$

$$\sigma_{ij,k} = g^1_{ij,k} - g_{ij,k} \tag{4.118}$$

Furthermore, if one defines the following matrices:

$$B = \text{diag}(B_1, \ldots, B_m) \tag{4.119}$$

B_m is an $n_m \times n_m$ matrix with

$$b_{ii} = \beta_{im}, \qquad i \in R_e$$

$$b_{(\sigma+1)i} = b_{i(\sigma+1)} = \tfrac{1}{2}\beta_{(\sigma+1)m}, \qquad \sigma = 1, \ldots, n_m - 1, \qquad i \in R_e$$

$$C = \text{diag}(C_1, \ldots, C_m) \tag{4.120}$$

C_m is an $n_m \times n_m$ upper triangular matrix with

$$c_{ii} = \gamma_{im}, \qquad i \in R_e$$

$$c_{i(\sigma+1)} = \gamma_{(\sigma+1)i}, \qquad \sigma = 1, \ldots, n_m - 1, \qquad i \in R_e$$

$$d(k) = \text{diag}(c_{j,k}d_{ij,k}, i \in R_e, j \in R_v) \qquad (4.121)$$

$$f(k) = \text{diag}(c_{j,k}f_{ij,k}; i \in R_e, j \in R_v) \qquad (4.122)$$

$$C(k) = \text{diag}(c_{j,k}\gamma_{ij}, i \in R_e, j \in R_v) \qquad (4.123)$$

$$L = \text{diag}(L_1, \ldots, L_m) \qquad (4.124)$$

L_m is an $n_m \times n_m$ lower triangular matrix depending on the topological arrangement of the reservoirs. For nonindependent rivers, the elements of the matrix L_m are given by

$$l_{(\sigma+1)\sigma} = 1, \qquad \sigma = 1, \ldots, n_j - 1, \qquad j \in R_v$$

Then, the augmented cost functional in equation (4.101) becomes

$$\tilde{J} = E[A^T x(K) + x^T(K)Bx(K) + x^T(K)Cy(K) + \sum_{k=1}^{K} \{b^T(k)u(k)$$

$$+ u^T(k)d(k)x(k-1) + u^T(k)f(k)Mu(k) + u^T(k)C(k)y(k-1)$$

$$+ \tfrac{1}{4}u^T(k)C(k)Nz(k) + r^T(k-1)C(k)Mu(k)$$

$$- \tfrac{1}{2}z^T(k)C(k)Lu(k) + \mu^T(k)(-y(k) + x^T(k)\mathbf{H}x(k))$$

$$+ \phi^T(k)(-z(k) + u^T(k)\mathbf{H}u(k))$$

$$+ \psi^T(k)(-r(k-1) + u^T(k)\mathbf{H}x(k-1))$$

$$+ \lambda^T(k)(-x(k) + x(k-1) + q(k) + Mu(k))$$

$$+ \nu^T(k)(x(k-1) + q(k) + Mu(k)) + \sigma^T(k)u(k)\}] \qquad (4.125)$$

In the above equation M is a diagonal matrix whose elements depend on the topological arrangement of the reservoirs. For independent rivers M is given by

$$M = \text{diag}(M_1, \ldots, M_m) \qquad (4.126)$$

where any matrix $M_1, \ldots, M_m$ is a lower triangular matrix, whose elements are given by

1. $m_{ii} = -1, \qquad i = 1, \ldots, n_j, \qquad j = 1, \ldots, m$
2. $m_{(\sigma+1)\sigma} = 1, \qquad \sigma = 1, \ldots, n_j - 1$

and N is also a diagonal matrix, the elements of which depend on the topological arrangement of the reservoirs. For independent river systems N is given by

$$N = \text{diag}(N_1, \ldots, N_m) \qquad (4.127)$$

where any matrix $N_1, \ldots, N_m$ is a lower triangular matrix whose elements are given by

    1. $n_{ii} = 1$, $i = 1, \ldots, n_j$

    2. $n_{(\sigma+1)\sigma} = 1$, $\sigma = 1, \ldots, n_j - 1$

Furthermore $\mathbf{H}$ is a vector matrix in which the vector index varies from 1 to $\sum_{j=1}^{m} n_j$, while the matrix dimension of $\mathbf{H}$ is $\sum_{j=1}^{m} n_j \times \sum_{j=1}^{m} n_j$.

We shall need the following identity:

$$\sum_{k=1}^{K} \lambda^T(k)x(k) = -\lambda^T(0)x(0) + \lambda^T(K)x(K) + \sum_{k=1}^{K} \lambda^T(k-1)x(k-1)$$

$$(4.128)$$

Then, we can write the cost functional in equation (4.125) as

$$
\begin{aligned}
\tilde{J}[u(k), x(k-1), x(K)] = E\Big[& x^T(K)(B + \mu^T(K)\mathbf{H})x(K) \\
& + x^T(K)Cy(K) \\
& + (A - \lambda(K))^T x(K) - \mu^T(K)y(K) \\
& + x^T(0)\mu^T(0)\mathbf{H}x(0) \\
& + \lambda^T(0)x(0) + \mu^T(0)y(0) \\
& + \sum_{k=1}^{K} \{x^T(k-1)\mu^T(k-1)\mathbf{H}x(k-1) \\
& + u^T(k)d(k)x(k-1) + u^T(k)f(k)Mu(k) \\
& + u^T(k)C(k)y(k-1) + \tfrac{1}{4}u^T(k)C(k)Nz(k) \\
& + r^T(k-1)C(k)Mu(k) - \tfrac{1}{2}z^T(k)C(k)Lu(k) \\
& - \mu^T(k-1)y(k-1) - \phi^T(k)z(k) \\
& + u^T(k)\phi^T(k)\mathbf{H}u(k) \\
& - \psi^T(k)r(k-1) + u^T(k)\psi^T(k)\mathbf{H}x(k-1) \\
& + (\lambda(k) - \lambda(k-1) + \nu(k))^T x(k-1) \\
& + (b(k) + M^T\lambda(k) + M^T\nu(k) + \sigma(k))^T u(k)\}\Big]
\end{aligned}
$$

$$(4.129)$$

If one defines the following vectors such that

$$Z^T(K) = [x^T(K), y^T(K)] \qquad (4.130)$$

$$W(K) = \begin{bmatrix} B + \mu^T(K)\mathbf{H} & \tfrac{1}{2}C \\ \tfrac{1}{2}C^T & 0 \end{bmatrix} \qquad (4.131)$$

$$Q^T(K) = [(A - \lambda(K))^T, \ -\mu^T(K)] \qquad (4.132)$$

$$X^T(k) = [x^T(k-1), y^T(k-1), z^T(k), r^T(k-1), u^T(k)] \qquad (4.133)$$

$$
L(k) = \begin{bmatrix}
\mu^T(k-1)\mathbf{H} & 0 & 0 & 0 & \frac{1}{2}[d(k) + \psi^T(k)\mathbf{H}] \\
0 & 0 & 0 & 0 & \frac{1}{2}C(k) \\
0 & 0 & 0 & 0 & \frac{1}{8}N^TC(k) - \frac{1}{4}C(k)L \\
0 & 0 & 0 & 0 & \frac{1}{2}C(k)M \\
\frac{1}{2}[d(k) + \psi^T(k)\mathbf{H}] & \frac{1}{2}C(k) & \frac{1}{8}C(k)N^T - \frac{1}{4}L^TC(k) & \frac{1}{2}M^TC(k) & f(k)M + \phi^T(k)\mathbf{H}
\end{bmatrix}
$$

$$(4.134)$$

and

$$
R^T(k) = [(\lambda(k) - \lambda(k-1) + \nu(k))^T, -\mu^T(k-1), -\phi^T(k), -\psi^T(k),
$$

$$
(b(k) + M^T\lambda(k) + M^T\nu(k) + \sigma(k))^T] \tag{4.135}
$$

then, the augmented cost functional in equation (4.129) can be written as

$$
\tilde{J}[Z(K), X(k)] = E\Big[Z^T(K)W(K)Z(K) + Q^T(K)Z(K)
$$

$$
+ x^T(0)\mu^T(0)\mathbf{H}x(0) + \lambda^T(0)x(0) + \mu^T(0)y(0)
$$

$$
+ \sum_{k=1}^{K} \{X^T(k)L(k)X(k) + R^T(k)X(k)\} \Big] \tag{4.136}
$$

Furthermore, if one defines the following vectors such that

$$
N(K) = W^{-1}(K)Q(K) \tag{4.137}
$$

and

$$
V(k) = L^{-1}(k)R(k) \tag{4.138}
$$

then, by using the process of completing the squares, equation (4.136) can be written as

$$
\tilde{J}[Z(K), X(k)] = E\Big[\{(Z(K) + \tfrac{1}{2}N(K))^TW(K)(Z(K) + \tfrac{1}{2}N(K))
$$

$$
+ \tfrac{1}{4}N^T(K)W(K)N(K) + x^T(0)\mu^T(0)\mathbf{H}x(0)
$$

$$
+ \lambda^T(0)x(0) + \mu^T(0)y(0)\}
$$

$$
+ \sum_{k=1}^{K} \{(X(k) + \tfrac{1}{2}V(k))^TL(k)(X(k) + \tfrac{1}{2}V(k))
$$

$$
- \tfrac{1}{4}V^T(k)L(k)V(k)\} \Big] \tag{4.139}
$$

Because $x(0)$ and $y(0)$ are constants and $N(K)$ and $V(k)$ are independent of $Z(K)$ and $X(k)$, respectively, then equation (4.139) can be written as

$$
\tilde{J}[Z(K), X(k)] = E\Big[\{(Z(K) + \tfrac{1}{2}N(K))^TW(K)(Z(K) + \tfrac{1}{2}N(K))\}
$$

$$
+ \sum_{k=1}^{K} \{(X(k) + \tfrac{1}{2}V(k))^TL(k)(X(k) + \tfrac{1}{2}V(k))\} \Big] \tag{4.140}
$$

It will be noticed that $\tilde{J}$ in equation (4.140) is composed of a boundary part and a discrete integral part, which are independent of each other. $\tilde{J}$ in equation (4.140) can be written as

$$\tilde{J}[Z(K), X(k)] = J_1[Z(K)] + J_2[X(k)] \tag{4.141}$$

with

$$J_1[Z(K)] = E[(Z(K) + \tfrac{1}{2}N(K))^T W(K)(Z(K) + \tfrac{1}{2}N(K))] \tag{4.142}$$

and

$$J_2[X(k)] = E[(X(k) + \tfrac{1}{2}V(k))^T L(k)(X(k) + \tfrac{1}{2}V(k))] \tag{4.143}$$

To maximize $\tilde{J}$ in equation (4.140), one maximizes the boundary part and the discrete part separately:

$$\max_{Z(K), X(k)} \tilde{J}[Z(K), X(k)] = \max_{Z(K)} J_1[Z(K)] + \max_{X(k)} J_2[X(k)] \tag{4.144}$$

J_1 in equation (4.142) defines a norm; hence one can write equation (4.142) as

$$J_1[Z(K)] = E\|Z(K) + \tfrac{1}{2}N(K)\|_{W(K)} \tag{4.145}$$

Also J_2 in equation (4.143) defines a norm, hence J_2 can be written as

$$J_2[X(k)] = E\|X(k) + \tfrac{1}{2}V(k)\|_{L(k)} \tag{4.146}$$

4.6.2. The Optimal Solution

The boundary term in equation (4.145) is optimized, when the norm in that equation is equal to zero

$$E[Z(K) + \tfrac{1}{2}N(K)] = [0] \tag{4.147}$$

Substituting from equation (4.137) into equation (4.147) for $N(K)$, we obtain

$$E[2W(K)Z(K) + Q(K)] = [0] \tag{4.148}$$

Writing equation (4.148) explicitly by substituting from equations (4.130)–(4.132) into equation (4.148) one obtains

$$E[\lambda(K)] = E[A + 2Bx(K) + 2\mu^T(K)Hx(K) + Cy(K)] \tag{4.149}$$

$$E[\mu(K)] = E[C^T x(K)] \tag{4.150}$$

Equations (4.149) and (4.150) give the values of Lagrange multipliers, $\lambda(K)$ and $\mu(K)$, at the end of the last period studied.

The discrete integral part in equation (4.143) is optimized when the norm in that equation is equal to zero, i.e.,

$$E[X(k) + \tfrac{1}{2}V(k)] = [0] \qquad (4.151)$$

Substituting from equation (4.138) into equation (4.151), one obtains the optimality equation

$$E[R(k) + 2L(k)X(k)] = [0] \qquad (4.152)$$

Writing equation (4.152) explicitly, by substituting from equations (4.133)-(4.135) and adding the equality constraints given by equations (4.95)-(4.98), one obtains

$$E[x(k)] = E[x(k-1) + q(k) + Mu(k)] \qquad (4.153a)$$

where

$$q(k) = I(k) + Ms(k) \qquad (4.153b)$$

$$E[-y(k) + x^T(k)\mathbf{H}x(k)] = [0] \qquad (4.154)$$

$$E[-z(k) + u^T(k)\mathbf{H}u(k)] = [0] \qquad (4.155)$$

$$E[r(k-1) + u^T(k)\mathbf{H}x(k-1)] = [0] \qquad (4.156)$$

$$E[\lambda(k) - \lambda(k-1) + v(k) + 2\mu^T(k-1)\mathbf{H}x(k-1) + d(k)u(k)$$
$$+ \psi^T(k)\mathbf{H}u(k)] = [0] \qquad (4.157)$$

$$E[-\mu(k-1) + C(k)u(k)] = [0] \qquad (4.158)$$

$$E[-\phi(k) + \tfrac{1}{4}N^TC(k)u(k) - \tfrac{1}{2}C(k)u(k)] = [0] \qquad (4.159)$$

$$E[-\psi(k) + C(k)Mu(k)] = [0] \qquad (4.160)$$

$$E[b(k) + M^T\lambda(k) + M^Tv(k) + \sigma(k) + d(k)x(k-1) + \psi^T(k)\mathbf{H}x(k-1)$$
$$+ C(k)y(k-1) + \tfrac{1}{4}C(k)N^Tz(k) - \tfrac{1}{4}L^TC(k)z(k) + M^TC(k)r(k-1)$$
$$+ 2f(k)Mu(k) + 2\phi^T(k)\mathbf{H}u(k)] = [0] \qquad (4.161)$$

Besides the above equations, one has the following limits on the variables:

$$\text{If } x(k) < x^m, \quad \text{then we put } x(k) = x^m \qquad (4.162a)$$

$$\text{If } x(k) > x^M, \qquad \text{then we put } x(k) = x^M \qquad (4.162\text{b})$$

$$\text{If } u(k) < u^m(k), \qquad \text{then we put } u(k) = u^m(k) \qquad (4.163\text{a})$$

$$\text{If } u(k) > u^M(k), \qquad \text{then we put } u(k) = u^M(k) \qquad (4.163\text{b})$$

One also has the following exclusion equation, which must be satisfied at the optimum:

$$e_{ij,k}(x_{ij}^m - x_{ij,k}) = 0 \qquad (4.164\text{a})$$

$$e_{ij,k}^1(x_{ij,k} - x_{ij}^M) = 0 \qquad (4.164\text{b})$$

$$g_{ij,k}(u_{ij,k}^m - u_{ij,k}) = 0 \qquad (4.165\text{a})$$

$$g_{ij,k}^1(u_{ij,k} - u_{ij,k}^M) = 0 \qquad (4.165\text{b})$$

Equations (4.153)–(4.165) with equations (4.149) and (4.150) completely specify the optimal long-term operation of the system.

4.6.3. Algorithm of Solution

Assume given the number of rivers (m), the number of reservoirs on each river (n_j), the expected natural inflows to each site, and the cost of energy on each valley $c_{j,k}$ in \$/MWh. The following steps are used to solve the optimal equations of the above section.

Step 1. Assume initial guess for the control variables $u(k)$, admissible region for $u(k)$, such that

$$u^m(k) \le u^i(k) \le u^M(k), \ i = \text{iteration number}, \qquad i = 0$$

Step 2. Assume first that $s(k)$ is equal to zero; solve equations (4.153)–(4.156) forward in stages with $x(0)$ given.

Step 3. Check the limits on $x(k)$. If $x(k)$ satisfies the inequality

$$x^m < x(k) < x^M$$

go to step 10; otherwise put $x(k)$ to its limits and go to step 4.

Step 4. Calculate the new discharge from the following equation:

$$E[u(k)] = E[[M]^{-1}(x(k) - x(k-1) - I(k))]$$

Step 5. Check the limits on $u(k)$. If $u(k)$ satisfies the inequality

$$u^m(k) < u(k) < u^M(k)$$

go to step 14; otherwise put $u(k)$ to its limits and go to step 6.

Step 6. Calculate the spill $s(k)$ at month k from the following equation:

$$E[s(k)] = E[(M)^{-1}(x(k) - x(k-1) - I(k)) - u^M(k)]$$

Note that water is spilled only when $u(k) > u^M(k)$ and the reservoir is filled to capacity.

Step 7. Solve again equations (4.153)-(4.156) with equations (4.158)-(4.160) forward in stages with $x(0)$ given, with the values of $s(k)$ obtained from step 6.

Step 8. Check the limits on $x(k)$. If $x(k)$ satisfies thè inequality

$$x^m < x(k) < x^M$$

go to step 9; otherwise put $x(k)$ to its limits and go to step 4.

Step 9. With $\nu(k) = 0$, solve equation (4.157) backward in stages with equation (4.149) as the terminal conditions.

Step 10. Calculate Kuhn-Tucker multipliers for $u(k)$, $\sigma(k)$, from

$$\begin{aligned}
E[\sigma(k)] = E[(M^T)(2\mu^T(k-1)\mathbf{H}x(k-1) + d(k)u(k) + \psi^T(k)\mathbf{H}u(k) \\
- \lambda(k-1) - C(k)r(k-1) - b(k) - d(k)x(k-1) \\
- \psi^T(k)\mathbf{H}x(k-1) - C(k)y(k-1) - \tfrac{1}{4}C(k)N^Tz(k) \\
+ \tfrac{1}{4}L^TC(k)z(k) - 2f(k)Mu(k) - 2\phi^T(k)\mathbf{H}u(k)]
\end{aligned}$$

Note that the above equation is obtained by multiplying equation (4.157) by M^T and subtracting the resulting equation from equation (4.161). The above equation gives explicitly the value of Kuhn-Tucker multipliers for $u(k)$. If $u(k)$ satisfies the inequality

$$u^m(k) < u(k) < u^M(k)$$

then, $\sigma(k) = 0$.

Step 11. Determine a new control iterate $u^{i+1}(k)$ as

$$E[u^{i+1}(k)] = E[u^i(k) + \alpha Du^i(k)]$$

where

$$
\begin{aligned}
E[Du(k)] = E[b(k) &+ M^T\lambda(k) + \sigma(k) + d(k)x(k-1) \\
&+ \psi^T(k)Hx(k-1) + C(k)y(k-1) \\
&+ \tfrac{1}{4}C(k)N^Tz(k) - \tfrac{1}{4}L^TC(k)z(k) \\
&+ M^TC(k)r(k-1) + 2f(k)Mu(k) + 2\phi^T(k)Hu(k)]
\end{aligned}
$$

and α is a positive scalar that is chosen with consideration to such factors as convergence.

Step 12. Check the limits on $u^{i+1}(k)$. If $u^{i+1}(k)$ satisfies the inequality

$$u^m(k) < u^{i+1}(k) < u^M(k)$$

go to step 13; otherwise put $u^{i+1}(k)$ to its limits and go to step 6.

Step 13. Solve the following equation forward in stages:

$$
\begin{aligned}
E[\lambda(k-1)] = E[2\mu^T(k-1)Hx(k-1) &+ d(k)u(k) + \psi^T(k)Hu(k) \\
- [M^T]^{-1}(b(k) &+ d(k)x(k-1) + \psi^T(k)Hx(k-1) \\
&+ C(k)y(k-1) + \tfrac{1}{4}C(k)N^Tz(k) - \tfrac{1}{4}L^TC(k)z(k) \\
&+ C(k)r(k-1) + 2f(k)Mu(k) + 2\phi^T(k)Hu(k))]
\end{aligned}
$$

Step 14. Determine Kuhn–Tucker multipliers for $x(k)$, $\nu(k)$, from the following equation:

$$
\begin{aligned}
E[\nu(k)] = [M^T]^{-1}E[-b(k) &- M^T\lambda(k) - d(k)x(k-1) - \psi^T(k)Hx(k-1) \\
-C(k)y(k-1) &- \tfrac{1}{4}C(k)N^Tz(k) + \tfrac{1}{4}L^TC(k)z(k) - C(k)r(k-1) \\
-2f(k)Mu(k) &- 2\phi^T(k)Hu(k)]
\end{aligned}
$$

If $x(k)$ satisfies the inequality

$$x^m < x(k) < x^M$$

then we put $\nu(k) = 0$.

Step 15. Determine a new state iterate from the following equation:

$$E[px^{i+1}(k)] = E[x^i(k) + \alpha Dx^i(k)]$$

where

$$
\begin{aligned}
E[Dx(k)] = E[\lambda(k) &- \lambda(k-1) + \nu(k) + 2\mu^T(k-1)Hx(k-1) \\
&+ d(k)u(k) + \psi^T(k)Hu(k)]
\end{aligned}
$$

Step 16. Repeat the calculation starting from step 3. Continue until the state $x(k)$ and the control $u(k)$ do not change significantly from iteration to iteration and the cost functional in equation (4.94) is a maximum.

Table 4.11. Characteristics of the Installations

| River | Site name | x_{ij}^M (Mm³) | x_{ij}^M (Mm³) | Maximum effective discharge (m³/sec) | Minimum effective discharge (m³/sec) | Reservoir constants | | |
						α_{ij}	β_{ij}	γ_{ij}
1	R_{11}	24763	9949	1119	85	212.11	146.96×10^{-4}	$-20503142.65 \times 10^{-14}$
	R_{21}	5304	3734	1583	85	117.20	569.71×10^{-4}	$-368119890.49 \times 10^{-14}$
2	R_{12}	74255	33196	1877	283	232.46	359.45×10^{-4}	$-1603544.32 \times 10^{-14}$
	R_{22}	0	0	1930	283	100.74	0	0
3	R_{13}	45672	24467	1632	283	176.28	105.626×10^{-4}	$-10022665.72 \times 10^{-14}$
	R_{23}	9132	8886	1876	283	131.44	200.897×10^{-4}	$-34741725.6 \times 10^{-14}$

Table 4.12. Optimal Releases from the Reservoirs and the Profits Realized in Year 1

Month k	$u_{11,k}$ (Mm³)	$u_{21,k}$ (Mm³)	$u_{12,k}$ (Mm³)	$u_{22,k}$ (Mm³)	$u_{13,k}$ (Mm³)	$u_{23,k}$ (Mm³)	Profits ($)
1	1649	3451	3821	3851	3641	4010	9,898,372
2	1589	1453	3703	3718	3538	3730	8,685,940
3	1552	1578	3839	3855	3601	3871	8,970,313
4	1318	401	3690	3689	3294	3725	4,554,491
5	1177	1257	3304	3311	2931	3089	4,214,638
6	1322	1400	3695	3703	3236	3406	4,658,819
7	1293	1453	3567	3574	3092	3321	4,491,206
8	1387	2298	3699	3752	3185	3559	4,886,224
9	1318	3461	3853	3730	3150	4824	5,598,097
10	1637	3662	4540	4616	4371	5025	6,627,686
11	2266	3229	5027	5169	4371	5025	9,571,348
12	2706	3286	3416	3445	4029	4442	8,345,312

Value of water left in storage at the end of the year	99,334,960
Total benefits from the system	179,837,406

4.6.4. Practical Example

The algorithm of the last section has been used to solve the same example mentioned earlier in Section 4.5.5 using the model of Section 4.6.1, quadratic model. The characteristics of the installations are given in Table 4.11.

The expected natural inflows to the sites for a year of high flow, which we call year 1, and the cost of energy in $/MWh for that year are given in

Table 4.13. Optimal Reservoirs Storage during Year 1 (Wet Year)

Month k	$x_{11,k}$ (Mm³)	$x_{21,k}$ (Mm³)	$x_{12,k}$ (Mm³)	$x_{13,k}$ (Mm³)	$x_{23,k}$ (Mm³)
1	24062	3828	72959	44779	9132
2	22955	4152	70481	42873	9132
3	21752	4274	67642	40652	9083
4	20734	5304	64801	38393	8886
5	19786	5304	62916	36258	8886
6	18689	5304	59145	33788	8886
7	17780	5304	56539	31489	8886
8	17780	5304	57397	30320	8886
9	20953	5304	71135	42678	9132
10	24345	5304	74255	45672	9132
11	24763	5304	74255	45672	9132
12	23458	5304	73187	45018	9132

Table 4.14. Optimal Release from the Turbines and the Profits Realized in Year 2 (Dry Year)

Month k	$u_{11,k}$ (Mm^3)	$u_{21,k}$ (Mm^3)	$u_{12,k}$ (Mm^3)	$u_{22,k}$ (Mm^3)	$u_{13,k}$ (Mm^3)	$u_{23,k}$ (Mm^3)	Profits ($)
1	2997	3197	2426	2449	2260	2037	8,143,159
2	2900	3185	2335	2350	1711	1880	7,073,434
3	2997	3576	2491	2506	1957	2041	7,588,778
4	2300	2148	1888	1898	922	975	2,764,157
5	2214	2223	1508	1515	685	754	2,399,639
6	2310	2401	1902	1909	1026	1117	2,843,036
7	1704	1955	1779	1794	968	1064	2,439,814
8	1592	2518	1922	1975	1389	1821	2,978,143
9	1423	3488	2358	2432	3861	4283	4,949,586
10	2174	3805	2555	2586	4178	4854	5,618,122
11	2997	4134	1827	1850	4371	5025	7,907,461
12	2900	3980	2115	2137	4230	4863	7,929,455
Value of water left in storage at the end of the year							90,212,080
Total benefits							152,846,864

Table 4.5. In Table 4.12, we give the optimal monthly releases from each reservoir and the profits realized at the end of year 1 for the optimal global-feedback solution. In Table 4.13, we give the optimal storage schedule for each reservoir during year 1.

We applied our algorithm to the system when the natural inflows are very low, a dry year, which we call year 2. The monthly expected natural inflows to the sites during that year are given in Table 4.8. In Table 4.14 we give the optimal monthly releases from each reservoir and the profits at

Table 4.15. Optimal Reservoirs Storage during Year 2

Month k	$x_{11,k}$ (Mm^3)	$x_{21,k}$ (Mm^3)	$x_{12,k}$ (Mm^3)	$x_{13,k}$ (Mm^3)	$x_{23,k}$ (Mm^3)
1	22316	5304	73816	45672	8886
2	19777	5304	72706	44951	8886
3	17096	5023	71269	43843	8886
4	15068	5304	70124	43641	8886
5	13091	5304	69089	43534	8886
6	11045	5304	67749	43085	8886
7	9949	5304	67151	42715	8886
8	9949	5304	69127	43661	8886
9	12376	5304	74255	45672	9132
10	14768	5304	74255	45672	9132
11	14472	5304	74255	44600	9132
12	13918	5304	73953	42034	9132

the end of year 2. In Table 4.15 we give the optimal storage for each reservoir during year 2. We started both years with the reservoirs full.

The computing time to obtain the optimal solution for a period of a year was 3.5 sec, in CPU units on Amdahl 470V/6 of the University of Alberta Computing Services.

4.6.5. Comments

In Section 4.6, we discussed the optimal long-term operating problem of a multichain power system. The model used for this system is a highly nonlinear model. By using this accurate model, it is clear from the results obtained that the total benefits increased by a considerable amount. Although the computing time is increased, it is still very small compared to what has been done so far using other approaches. In this chapter, we assumed that the times of water travel from upstream reservoirs to downstream reservoirs are shorter than a month, for which reason those times are neglected; also we assumed a constant tail-water elevation.

References

4.1. ACKOFF, R. L., *Progress in Operations Research*, Vol. 1, John Wiley, New York, 1961.
4.2. AMIR, R., "Optimal Operation of a Multi-reservoir Water Supply System," Report No. EEP-24, Progress in Engineering and Economic Planning, Stanford University, Stanford, California, 1967.
4.3. ARUNKUMAR, S., "Characterization of Optimal Operating Policies for Finite Dams, *Journal of Mathematical Analysis and Applications* **49**(2), 267–274 (1975).
4.4. ARUNKUMAR, S., "Optimal Regulation Policies for a Multipurpose Reservoir with Seasonal Input and Return Function," *Journal of Optimization Theory and Applications* **21**(3), 319–328 (1977).
4.5. ARUNKUMAR, S., and CHON, K., "On Optimal Regulation for Certain Multireservoir Systems," *Operations Research* **26**(4), 551–562 (1978).
4.6. ARUNKUMAR, S., and YEH, W. W-G., "Probabilistic Models in the Design and Operation of a Multi-purpose Reservoir System," Contribution No. 144, California Water Resources Center, University of Calif., Davis, December 1973.
4.7. ASKEW, A. J., "Optimum Reservoir Operating Policies and the Imposition of a Reliability Constraint," *Water Resources Research* **10**(1), 51–56 (1974).
4.8. ASKEW, A. J., "Chance-Constrained Dynamic Programming and the Optimization of Water Resource Systems," *Water Resources Research* **10**(6), 1099–1106 (1974).
4.9. ASKEW, A. J., "Use of Risk Premiums in Chance-Constrained Dynamic Programming," *Water Resources Research* **11**(6), 862–866 (1975).
4.10. ASKEW, A. J., YEH, W. W-G., and HALL, W. A., "Use of Monte Carlo Techniques in the Design and Operation of a Multi-purpose Reservoir System," *Water Resources Research* **6**(4), 819–826 (1971).
4.11. BATHER, J. A., "Optimal Regulation Policies for Finite Dams," *Journal of the Society for Industrial and Applied Mathematics* **10**(3), 395–423 (1962).

4.12. BEARD, L. B., "Hydrologic Simulation in Water-Yield Analysis," Technical Paper No. 10, Hydrolics Engineering Center, U.S. Army Corps of Engineers, Davis, California, 1964.

4.13. BEARD, L. R., "Status of Water Resources Systems Analysis," paper presented at Seminar on Hydrological Aspects of Project Planning, Hydrolics Engineering Center, U.S. Army Corps of Engineers, Davis, California, March 1972,

4.14. BEARD, L. R., WEISS, A. C., and AUSTIN, T. A., "Alternative Approaches to Water Resource System Simulation," Technical Paper 32, Hydrolics Engineering Center, U.S. Army Corps of Engineers, Davis, California, 1972.

4.15. BECHARD, D., *et al.*, "The Ottawa River Regulation Modeling System (ORRMS)," in Proceedings of International Symposium on Real-Time Operation of Hydrosystems, Vol. 1, pp. 179-198, University of Waterloo, Ontario, 1981.

4.16. BECKER, L., and YEH, W. W-G., "Optimization of Real Time Operation of Multiple-Reservoir System," *Water Resources Research* 10(6), 1107-1112 (1974).

4.17. BECKER, L., YEH, W. W-G., FULTS, D., and SPARKS, D., "Operations Models for the Central Valley Project," *Journal of Water Resource Planning and Management Division of the American Society of Civil Engineers* 102(WR1), 101-115 (1976).

4.18. BELLMAN, R. E., *Dynamic Programming*, Princeton University Press, Princeton, New Jersey, 1957.

4.19. BELLMAN, R., and DREYFUS, S., *Applied Dynamic Programming*, Princeton University Press, Princeton, New Jersey, 1962.

4.20. BERHOLTZ, B., and GRAHAM, L. J., "Hydrothermal Economic Scheduling, I, Solution by Incremental Dynamic Programming," *IEEE Transactions on Power Apparatus and Systems* 50, 921-932 (1960).

4.21. BHASKAR, N. R., and WHITLATCH, E. E., Jr., "Derivation of Monthly Reservoir Release Policies," *Water Resources Research* 16(6), 983-987 (1980).

4.22. BODIN, L. D., and ROEFS, T. G., "A Decomposition Approach to Nonlinear Programs as Applied to Reservoir Systems," *Networks* 1(1), 59-73 (1971).

4.23. BOWER, T. A., HUFSCHMIDT, M. M., and REEDY, W. W., "Operating procedures: Their Role in Design of Water Resources Systems by Simulation Analysis," in *Design of Water Resources Systems*, edited by A. Maass, pp. 444-460, Harvard University Press, Cambridge, Massachusetts, 1962.

4.24. SOLIMAN, S. A., and CHRISTENSEN, G. S., "Application of Functional Analysis to Optimization of a Variable Head Multireservoir Power System for Long-Term Regulation," *Water Resources Research* 22(6), 852-858 (1986).

4.25. CHRISTENSEN, G. S., and SOLIMAN, S. A., "On the Application of Functional Analysis to the Optimization of the Production of Hydroelectric Power," IEEE/PES, /JPGC 660-6, Portland, Oregon, October 19-23, 1986.

4.26. PAPAGEORGIOU, M., "Optimal Multireservoir Network Control by the Discrete Maximum Principle," *Water Resources Research* 21(12), 1824-1830 (1985).

4.27. TURGEON, A., "Optimal Operation of a Multireservoir Hydro-Steam Power System," in *Large Scale Engineering Applications*, edited by M. Singh and A. Titli, pp. 540-555, North-Holland, Amsterdam, 1979.

4.28. TURGEON, A., "Optimal Operation of Multireservoir Power Systems with Stochastic Inflows," *Water Resources Research* 16(2), 275-283 (1980).

4.29. TURGEON, A., "Optimal Short-Term Scheduling of Hydroplants in Series—A Review," paper presented at Proceedings, International Symposium on Real-Time Operation of Hydrosystems, University of Waterloo, Waterloo, Ontario, June 1981.

4.30. TURGEON, A., "Optimal Short-Term Hydro Scheduling from the Principle of Progressive Optimality," *Water Resources Research* 17(3), 481-486 (1981a).

4.31. TURGEON, A., "A Decomposition Method for the Long-Term Scheduling of Reservoirs in Series," *Water Resources Research* 17(6), 1565-1570 (1981b).

4.32. TURGEON, A., "Incremental Dynamic Programming May Yield Nonoptimal Solutions," *Water Resources Research* **18**(6), 1599–1604 (1982).

4.33. CHARA, A. M. and PANT, A. K., "Optimal Discharge Policy of Multireservoir Linked System using Successive Variations," *International Journal of Systems Science* **15**(1), 31–52 (1984).

4.34. CHU, W. S., YEH, W. G., "A Nonlinear Programming Algorithm for Real-Time Hourly Reservoir Operations," *Water Resources Bulletin* **14**, 1048–1063 (1978).

4.35. FLETCHER, R., and REEVES, C. M., "Function Minimization by Conjugate Gradients," *Computer Journal* **7**, 149–154 (1964).

4.36. GAGNON, C. R., HICKS, R. H., JACOBY, S. L. S., and KOWALIK, J. S., "A Nonlinear Programming Approach to a Very Large Hydroelectric System Optimization," *Mathematical Programming* **6**, 28–41 (1974).

4.37. LEE, F. S., and WAZIRUDDIN, S., "Applying Gradient Projection and Conjugate Gradient to the Optimum Operation of Reservoirs," *Water Resources Bulletin* **6**(5), 713–724 (1970).

4.38. PAPAGEORGIOU, M., "Automatic Control Strategies for Combined Sewer Systems," *Journal of Environmental Engineering Division of the American Society of Civil Engineers* **109**(6), 1385–1402 (1983).

4.39. PAPAGEORGIOU, M., and MAYR, R., "Optimal Real-Time Control of Combined Sewer Networks," in *IFAC Conference on System Analysis Applied to Water and Related Land Resources*, pp. 17–22, Pergamon, New York, 1985.

4.40. PEARSON, J. R., Jr., and SRIDHAR, R., "A Discrete Optimal Control Problem," *IEEE Transactions on Automatic Control* **AC-11**, 171–174 (1966).

4.41. SAGE, A. P., *Optimum System Control*, Prentice-Hall, Englewood Cliffs, New Jersey, 1968.

5

Modeling and Optimization of a Multireservoir Power System for Critical Water Conditions

5.1. Introduction (Refs. 5.10–5.15)

The period in which reservoirs are drawn down from full to empty is referred to as the "critical period," and the stream flows that occur during the critical period are called "critical period stream flow" because they are the lowest on record. The duration of the critical period is determined by the amount of reservoir storage in the hydroelectric system and on the amount of energy support available from thermal, gas turbine plants, and possible purchase, and it depends on how these resources are committed to support the hydroelectric system.

The basic requirement for the critical period is that the generation during this period should be uniform during each year of the critical period and should supply the required load on the system, and at the same time the reservoirs are drawn down to empty at the end of the critical period; we mean by empty the minimum storage.

Over the past several years a number of methods have been developed to solve the problem. These methods are limited to either dynamic programming, linear programming, or a combination of them, but they suffer from major problems when they are applied to multidimensional systems, including excessive demands on computing time and storage requirements (Refs. 5.1–5.4).

In this chapter, we maximize the production of hydroelectric power for a multireservoir power system. Section 5.2 is devoted to the maximization of the generation from the system using a linear model of the discharge and the average storage; the generation during each year of the critical period is uniform. In Section 5.3, we maximize the generation from the same system, but the model used in this case, is a quadratic model of the discharge and the average storage; the storage-elevation curve is a quadratic function, and also the generation during each year of the critical period is uniform. In the first part of this chapter we do not take into account the load on the system; accordingly the generation during each year of the critical period must be greater than the load on the system, and in this case the excess in energy may be exported to a neighboring consumer.

In Section 5.4 we apply the discrete maximum principle to find the optimal operation scheduling for the same system. Finally, in Section 5.6, we maximize the generation of the system, but in this case the load on the system is taken into account. Some results are given for a real system in operation.

5.2. Problem Formulation (Refs. 5.5–5.9)

5.2.1. The System under Study

The system under consideration consists of m rivers, with one or several reservoirs and power plants in series on each, and interconnection lines to the neighboring system through which energy may be exchanged (Figure 4.1).

5.2.2. The Objective Function

The optimization objective function is to find the discharge $u_{ij,k}$, $i = 1, \ldots, n_j$; $j = 1, \ldots, m$; $k = 1, \ldots, K$, as a function of time over the optimization interval under the following conditions:

(1) The total generation from the system during each year of the critical period is a maximum, and uniform.

(2) The water conservation equation for each reservoir may be adequately described by the continuity-type equation

$$x_{ij,k} = x_{ij,k-1} + I_{ij,k} + z_{ij,k} - u_{ij,k} - s_{ij,k} \tag{5.1}$$

where $z_{ij,k}$ is the dependent water inflow connecting hydroelectric plants over the same hydraulic valley and is given by

$$z_{ij,k} = \sum_{\substack{l \in R_u \\ j \in R_r}} (u_{lj,k} + s_{lj,k}) \tag{5.2}$$

where R_u is the set of plants immediately upstream from plant i on river j and R_r is the set of upstream rivers. $s_{ij,k}$ is the spill from reservoir i on river j and is given by

$$s_{ij,k} = \begin{cases} (x_{ij,k-1} + I_{ij,k} + z_{ij,k} - x_{ij,k}) - u_{ij,k}^M \\ \quad \text{if } (x_{ij,k-1} + I_{ij,k} + z_{ij,k} - x_{ij,k}) > u_{ij,k}^M \quad \text{and} \quad x_{ij,k} = x_{ij}^M \\ 0 \quad \text{otherwise} \end{cases} \quad (5.3)$$

As we said earlier, in Chapter 4, if the rivers are independent of each other, then $z_{ij,k}$ is given by

$$z_{ij,k} = u_{(i-1)j,k} + s_{(i-1)j,k} \tag{5.4}$$

(3) The water conversion factor, MWh/Mm^3, as a function of the storage is given by

$$h_{ij,k} = a_{ij} + b_{ij}x_{ij,k} \text{ MWh/Mm}^3 \tag{5.5}$$

where a_{ij} and b_{ij} are constants. These were obtained by least-squares curve fitting to typical plant data available. In equation (5.5), we assumed that the efficiency and tailwater elevation are constants, independent of the discharge.

(4) The reservoirs should be drawn down from full at the beginning of the optimization interval to minimum storage, x_{ij}^m, at the end of the critical period. This can be expressed as

$$x_{ij,0} = x_{ij}^M \tag{5.6a}$$

$$x_{ij,KK} = x_{ij}^m \tag{5.6b}$$

where KK is the last period studied in the critical period.

(5) In order to be realizable and also to satisfy multipurpose stream use requirements such as flood control, irrigation, fishing, navigation, and other purposes if any, the following upper and lower limits on the variables must be satisfied:

$$x_{ij}^m \leq x_{ij,k} \leq x_{ij}^M \tag{5.7}$$

$$u_{ij,k}^m \leq u_{ij,k} \leq u_{ij,k}^M \tag{5.8}$$

(5) The MWh generated from reservoir i on river j during a period k is given by

$$G_{ij,k}(u_{ij,k}, \tfrac{1}{2}(x_{ij,k} + x_{ij,k-1})) = a_{ij}u_{ij,k} + \tfrac{1}{2}b_{ij}u_{ij,k}(x_{ij,k} + x_{ij,k-1}) \tag{5.9}$$

In the above equation, equation (5.9), an average of begin and end of time storage is used to avoid underestimation in production for falling water levels and overestimation for rising water levels.

In mathematical terms, the problem is to find the discharge $u_{ij,k}$, $i = 1, \ldots, n_j$, $j = 1, \ldots, m$, $k = 1, \ldots, K$ that maximizes

$$J = E\left[\sum_{j=1}^{m} \sum_{i=1}^{n_j} \sum_{k=1}^{K} \{a_{ij}u_{ij,k} + \tfrac{1}{2}b_{ij}u_{ij,k}(x_{ij,k} + x_{ij,k-1})\}\right] \text{MWh} \quad (5.10)$$

Subject to satisfying the equality constraints given by equations (5.1)-(5.4) and (5.6) and the inequality constraints given by equations (5.7) and (5.8).

5.2.3. A Minimum Norm Approach

We can form an augmented cost functional by adjoining the equality constraints via Lagrange multipliers and the inequality constraints via Kuhn-Tucker multipliers as

$$\tilde{J}[x_{ij,k-1}, u_{ij,k}] = E\left[\sum_{j=1}^{m} \sum_{i=1}^{n_j} \sum_{k=1}^{K} \{a_{ij}u_{ij,k} + b_{ij}u_{ij,k}x_{ij,k-1} + \tfrac{1}{2}b_{ij}u_{ij,k}I_{ij,k}\right.$$

$$+ \tfrac{1}{2}b_{ij}u_{ij,k}z_{ij,k} - \tfrac{1}{2}b_{ij}(u_{ij,k})^2 - \tfrac{1}{2}b_{ij}u_{ij,k}s_{ij,k}$$

$$+ \lambda_{ij,k}(-x_{ij,k} + x_{ij,k-1} + I_{ij,k} + z_{ij,k} - u_{ij,k} - s_{ij,k})$$

$$+ e_{ij,k}(x_{ij}^m - x_{ij,k}) + e_{ij,k}^1(x_{ij,k} - x_{ij}^M)$$

$$\left. + g_{ij,k}(u_{ij,k}^m - u_{ij,k}) + g_{ij,k}^1(u_{ij,k} - u_{ij,k}^M)\}\right] \quad (5.11)$$

We shall use the following identities:

$$\sum_{k=1}^{K} \lambda_{ij,k}x_{ij,k} = \lambda_{ij,K}x_{ij,K} - \lambda_{ij,0}x_{ij,0} + \sum_{k=1}^{K} \lambda_{ij,k-1}x_{ij,k-1} \quad (5.12)$$

$$\mu_{ij,k} = e_{ij,k}^1 - e_{ij,k} \quad (5.13a)$$

$$\psi_{ij,k} = g_{ij,k}^1 - g_{ij,k} \quad (5.13b)$$

Now, the cost functional in equation (5.11) can be written as

$$\tilde{J}[x_{ij,k-1}, u_{ij,k}] = E\left[\sum_{j=1}^{m} \sum_{i=1}^{n_j} (\lambda_{ij,K}x_{ij,K} - \lambda_{ij,0}x_{ij,0})\right.$$

$$+ \sum_{j=1}^{m} \sum_{i=1}^{n_j} \sum_{k=1}^{K} \{q_{ij,k}u_{ij,k} + b_{ij}u_{ij,k}x_{ij,k-1}$$

$$+ \tfrac{1}{2}b_{ij}u_{ij,k}z_{ij,k} - \tfrac{1}{2}b_{ij}(u_{ij,k})^2$$

$$+ (\lambda_{ij,k} - \lambda_{ij,k-1} + \mu_{ij,k})x_{ij,k-1}$$

$$\left. + (\mu_{ij,k} + \lambda_{ij,k})z_{ij,k} - (\lambda_{ij,k} + \mu_{ij,k} - \psi_{ij,k})u_{ij,k}\}\right] \quad (5.14)$$

where

$$q_{ij,k} = (a_{ij} + \tfrac{1}{2}b_{ij}I_{ij,k} - \tfrac{1}{2}b_{ij}s_{ij,k}) \tag{5.15}$$

Note that constant terms are dropped in equation (5.14).

If we define the following vectors as

$$x(k) = \text{col}(x_{ij,k}, i \in S, j \in R) \tag{5.16}$$

$$u(k) = \text{col}(u_{ij,k}, i \in S, j \in R) \tag{5.17}$$

$$z(k) = \text{col}(z_{ij,k}, i \in S, j \in R) \tag{5.18}$$

$$\lambda(k) = \text{col}(\lambda_{ij,k}, i \in S, j \in R) \tag{5.19}$$

$$\mu(k) = \text{col}(\mu_{ij,k}, i \in S, j \in R) \tag{5.20}$$

$$\psi(k) = \text{col}(\psi_{ij,k}, i \in S, j \in R) \tag{5.21}$$

$$q(k) = \text{col}(q_{ij,k}, i \in S, j \in R) \tag{5.22}$$

Furthermore, if one defines the $N \times N$ diagonal matrix, $N = \sum_{j=1}^{m} n_j$, as

$$b = \text{diag}(b_{ij}, i \in S, j \in R) \tag{5.23}$$

then the cost functional in equation (5.15) can be written as

$$\begin{aligned}
\tilde{J}[x(k-1), u(k)] = E\Big[&\lambda^T(K)x(K) - \lambda^T(0)x(0) + \sum_{k=1}^{K} \{q^T(k)u(k) \\
&+ u^T(k)bx(k-1) + \tfrac{1}{2}u^T(k)bz(k) - \tfrac{1}{2}u^T(k)bu(k) \\
&+ (\lambda(k) - \lambda(k-1) + \mu(k))^T x(k-1) \\
&+ (\mu(k) + \lambda(k))^T z(k) \\
&- (\lambda(k) + \mu(k) - \psi(k))^T u(k)\} \Big]
\end{aligned} \tag{5.24}$$

where

$$z(k) = Mu(k) + Ms(k) \tag{5.25}$$

and M is an $N \times N$ matrix containing the topological arrangement of the system. For independent rivers, M will be

$$M = \begin{bmatrix} 0 & 0 & & & 0 \\ 1 & 0 & & & \\ 0 & 1 & 0 & & \\ \vdots & & & & \\ 0 & & & & 0 \end{bmatrix} \tag{5.26}$$

If we substitute for $z(k)$ from equation (5.15) into equation (5.24), and we use the substitution

$$Q = M - I \qquad (5.27)$$

where I is an identity matrix, we obtain

$$
\begin{aligned}
\tilde{J}[x(k-1), u(k)] = E\Big[& \lambda^T(K)x(K) - \lambda^T(0)x(0) + \sum_{k=1}^{K} \{u^T(k)bx(k-1) \\
& + \tfrac{1}{2}u^T(k)bMu(k) - \tfrac{1}{2}u^T(k)bu(k) + (\lambda(k) - \lambda(k-1) \\
& + \mu(k))^T x(k-1) + (q(k) + Q^T\lambda(k) + Q^T\mu(k) \\
& + \psi(k) + \tfrac{1}{2}bMs(k))^T u(k)\Big]
\end{aligned}
\qquad (5.28)
$$

If we define the new vectors

$$x^T(k) = [x^T(k-1), u^T(k)] \qquad (5.29)$$

$$
\begin{aligned}
R^T(k) = [&(\lambda(k) - \lambda(k-1) + \mu(k))^T, (q(k) + Q^T\lambda(k) + Q^T\mu(k) \\
& + \psi(k) + \tfrac{1}{2}bMs(k))^T]
\end{aligned}
\qquad (5.30)
$$

and

$$
L(k) = \begin{bmatrix} 0 & \tfrac{1}{2}b \\ \tfrac{1}{2}b & (\tfrac{1}{2}bM - \tfrac{1}{2}b) \end{bmatrix}
\qquad (5.31)
$$

Then, we can write the cost functional in equation (5.28) in compact form as

$$
\begin{aligned}
\tilde{J}[x(K), X(k)] = & E[\lambda^T(K)x(K) - \lambda^T(0)x(0)] \\
& + E\Big[\sum_{k=1}^{K} \{X^T(k)L(k)X(k) + R^T(k)X(k)\}\Big]
\end{aligned}
\qquad (5.32)
$$

If one defines the vector $V(k)$ such that

$$V(k) = L^{-1}(k)R(k) \qquad (5.33)$$

then the cost functional in equation (5.32) can be written in the following form similar to completing the squares as

$$
\begin{aligned}
\tilde{J}[x(K), XC(k)] = & E[\lambda^T(K)x(K) - \lambda^T(0)x(0)] \\
& + E\Big[\sum_{k=1}^{K} (X(k) + \tfrac{1}{2}V(k))^T L(k)(X(k) + \tfrac{1}{2}V(k)) \\
& - \tfrac{1}{4}V^T(k)L(k)V(k)\Big]
\end{aligned}
\qquad (5.34)
$$

Since $x(0)$ is specified and $V(k)$ is constant, independent of $X(k)$, we can write equation (5.34) as

$$\tilde{J}[x(K), X(k)]$$

$$= E\left[\lambda^T(K)x(K) + \sum_{k=1}^{K} \{(X(k) + \tfrac{1}{2}V(k))^T L(k)(X(k) + \tfrac{1}{2}V(k))\}\right]$$

(5.35)

It can be noticed that $\tilde{J}$ in equation (5.35) is composed of two parts, a boundary part and a discrete integral part, which are independent of each other. To maximize $\tilde{J}$ in equation (5.35), one maximizes each term separately as

$$\max_{[x(K),X(k)]} \tilde{J}[x(K), X(k)] = \max_{x(K)} E[\lambda^T(K)x(K)]$$

$$+ \max_{X(k)} E\left[\sum_{k=1}^{K} (X(k) + \tfrac{1}{2}V(k))^T L(k)(X(k) + \tfrac{1}{2}V(k))\right] \quad (5.36)$$

5.2.4. The Optimal Solution

There is only one solution to the problem formulated in equation (5.36), namely, the optimal solution. The maximization of the boundary part is clearly achieved when

$$E[\lambda(K)] = [0] \tag{5.37}$$

or in a component form, equation (5.37) can be written as

$$E[\lambda_{ij,K}] = 0 \tag{5.38}$$

The discrete integral part in equation (5.36) defines as a norm in Hilbert space; hence we can write this term as

$$\max_{X(k)} J_2 = \max_{X(k)} E\|X(k) + \tfrac{1}{2}V(k)\|_{L(k)} \tag{5.39}$$

The maximization of the discrete integral part in equation (5.39) is clearly achieved when the norm of this equation is equal to zero:

$$E\|X(k) + \tfrac{1}{2}V(k)\| = [0] \tag{5.40}$$

provided the matrix $L(k)$ is invertible. Substituting from equation (5.33) for $V(k)$ into equation (5.40), one obtains the following conditions for optimality:

$$E[R(k) + 2L(k)X(k)] = [0] \tag{5.41}$$

Writing equation (5.41) explicitly, by substituting from equations (5.29)-(5.31) into equation (5.41) and adding the equality constraints given by equations (5.1), (5.2), and (5.4), we get the following set of optimal equations:

$$E[-x(k) + x(k-1) + I(k) + z(k) - u(k) - s(k)] = [0] \quad (5.42)$$

$$E[z(k) - Mu(k) - Ms(k)] = [0] \quad (5.43)$$

$$E[\lambda(k) - \lambda(k-1) + \mu(k) + bu(k)] = [0] \quad (5.44)$$

$$E[Q^T\lambda(k)Q^T\mu(k) + q(k) + \psi(k) + \tfrac{1}{2}bMs(k) + bx(k-1)$$
$$+ bMu(k) - bu(k)] = [0] \quad (5.45)$$

Equations (5.42)-(5.45) can be written in the component form for independent rivers as

$$E(x_{ij,k} + x_{ij,k-1} + I_{ij,k} + u_{(i-1)j,k} + s_{(i-1)j,k} - u_{ij,k} - s_{ij,k}) = 0 \quad (5.46)$$

$$E(\lambda_{ij,k-1}\lambda_{ij,k} + \mu_{ij,k} + b_{ij}u_{ij,k}) = 0 \quad (5.47)$$

$$E(\lambda_{(i-1)j,k} - \lambda_{ij,k} + \mu_{(i-1)j,k} - \mu_{ij,k} + q_{ij,k} + \psi_{ij,k}$$
$$+ \tfrac{1}{2}b_{ij}s_{(i-1)j,k} + b_{ij}x_{ij,k-1} + b_{ij}(u_{(i-1)j,k} - u_{ij,k})) = 0 \quad (5.48)$$

Besides the above equations, one has the following limits on the variables as

$$\begin{array}{ll} \text{If } x(k) < x^m, & \text{then we put } x(k) = x^m \\ \text{If } x(k) > x^M, & \text{then we put } x(k) = x^M \\ \text{If } u(k) < u^m(k), & \text{then we put } u(k) = u^m(k) \\ \text{If } u(k) > u^M(k), & \text{then we put } u(k) = u^M(k) \end{array} \quad (5.49)$$

Also, we have the following Kuhn-Tucker exclusion equations, which must be satisfied at the optimum:

$$\begin{array}{l} e_{ij,k}(x_{ij}^m - x_{ij,k}) = 0 \\ e_{ij,k}^1(x_{ij,k} - x_{ij}^M) = 0 \\ g_{ij,k}(u_{ij,k}^m - u_{ij,k}) = 0 \\ g_{ij,k}^1(u_{ij,k} - u_{ij,k}^M) = 0 \end{array} \quad (5.50)$$

5.2.5. Computer Logic

Equations (5.46)-(5.50) constitute a two-point boundary value problem (TPBVP), since the initial states $x(0)$, and the final costates $\lambda(K) = 0$ are given. The problem can be solved by using any algorithm explained in Chapter 4 with some slight modifications included:

1. The total generation during each year of the critical period should be uniform.

2. The reservoirs should be drawn down from full at the beginning of the critical period to the minimum storage at the end of the critical period.

5.2.6. Practical Example

The computer algorithm described above is applied to a large system in operation. The system consists of 88 projects. 38 of these projects are storage reservoirs, while the 50 projects are run-of-river projects. The mathematical model obtained for this system is based on the following types of relationships and data:

a. Basic relationships between the variables, such as the water conservation equations;
b. Water conversion relationship;
c. Relationships that specify the generated power for given flows and head at each plant in the system;
d. Initial water level in all reservoirs;
e. Side flows into the system for all time periods;
f. Constraints on generation, flows, contents, and draft.

For the proposed system, we have the following:

(1) The hydraulic system model is based primarily on empirical tables derived from field measurements and on water balance equations. The latter are simple relationships between contents, total discharge, and inflow.

(2) The empirical tables basically consist of the following for each project:

i. Forebay elevation as a function of the reservoir contents and flow through the reservoir;
ii. Tailwater elevation as a function of the discharge and elevation of the water in the immediate downstream reservoir;
iii. Maximum generation at each plant as a function of full-gate flow restriction;
iv. Water-to-energy conversion factor as a nonlinear function of reservoir contents and turbine discharge.

(3) The constraints reflect physical limits, bank erosion considerations, coordination agreements among various ownerships, system safety, and multipurpose requirements such as irrigation, navigation, fishing, and flood control.

(4) Storage for run-of-river plants is considered to be constant so that outflow is equal to inflow.

(5) The expected side inflow data are given and are preadjusted for consumptive use.

Table 5.1. The Installations Characteristics (Sample of the Data Given)

Reservoir	a (MW/MCF)	b [MW(MCF)^2]	Discharges (CFS)^a Minimum	Maximum	Storage (KSFD)^a Minimum	Maximum
R_1	3.88	6.624×10^{-5}	1000	13370	0.0	654.3
R_2	5.696	1.3045×10^{-5}	0	9350	0.0	225.4
$R_3{}^b$	32.50	0	40	535	0.0	13.1
R_4	9.365	3.7943×10^{-5}	1000	10500	0.0	1016.0
R_5	3.313	5.7318×10^{-5}	$V_5{}^c$	34500	0.0	494.2
R_6	4.982	8.5213×10^{-6}	V_6	280000	0.0	2614.3
R_7	3.231	8.3776×10^{-6}	3200	14346	0.0	614.7
R_8	5.475	3.4895×10^{-5}	400	8900	0.0	1593.6

^{a} 1 CFS = one cubic foot/sec, 1 KSFD = 2.64×10^6 m^3.

^{b} R_e is a run-of-river plant.

^{c} $V = \alpha q_1 + \beta q_2 + \gamma q_3$, where $\alpha + \beta + \gamma = 24$ period, the total number of periods/year, and q_1, q_2, and q_3 are the corresponding discharges during these periods.

(6) Discharge in excess of full-gate flow results in spill.

(7) The expected evaporation and percolation losses accounted for by deducting them from the forecasted side.

(8) There are some projects that have (a) seasonable storage capability, and (b) full gate flow restriction affecting the power conversion factor.

Additional tables and approximating function are used to account for these added complications.

The solution procedure will deal with the highly nonlinear representation of the system equations and constraints step by step; e.g., we will first consider a linear relationship between the plant generation and reservoir contents at a given discharge, then later we will consider the higher-order

Table 5.2. Optimal Storage during the First Year of the Critical Period in MCF

Month k	$x_{1,k}$	$x_{2,k}$	$x_{4,k}$	$x_{5,k}$	$x_{6,k}$	$x_{7,k}$	$x_{8,k}$
1	33590	19475	86699	33213	225875	5310	129984
2	53817	19012	86022	33201	225875	51804	128463
3	54253	18000	84755	42699	225875	50113	122665
4	56532	19475	82901	42699	225875	44091	112585
5	56532	19475	80649	42699	225875	28283	967126
6	55420	19475	80673	42699	225875	21177	787533
7	46000	19475	74000	42699	225875	22157	793049
8	40000	19475	69500	34180	225875	0.0	791279
9	51500	19475	87782	42699	196667	22114	107037
10	56532	19475	87782	42699	225875	53110	133242
11	56532	19475	87782	33823	225875	53110	137687
12	56532	19475	86943	26128	225875	48503	123612

Table 5.3. Optimal Storage during the Second Year of the Critical Period MCF

Month k	$x_{1,k}$	$x_{2,k}$	$x_{4,k}$	$x_{5,k}$	$x_{6,k}$	$x_{7,k}$	$x_{8,k}$
1	53073	19475	83430	23177	225875	43530	108614
2	49729	19475	76446	31651	225875	41978	95425
3	45550	11380	65572	42699	225875	37938	83441
4	48911	11378	61303	42699	225875	42599	65838
5	47000	11376	57710	42699	225875	23610	58134
6	56532	16346	56209	42699	225875	61545	58468
7	53690	17551	53646	31699	225875	59983	46254
8	55762	19475	77152	24232	225875	15635	67014
9	56532	19475	87782	31123	935564	25878	90255
10	56532	18982	86742	34192	222241	53110	10659
11	56532	16722	77670	22298	225875	53096	104720
12	55215	13978	66241	12389	225875	40397	96919

representations. Also, we will consider constant tail water elevation for each plant; next we will consider it as a function of the plant discharge; and later we will consider the relationship between tailwater elevation, the plant discharge, and the immediate downstream plant elevation and so forth.

The optimization is done on a half-monthly basis for a 42-month critical period. The system we are dealing with is a very large-scale power system, and we cannot give all the results obtained here; we supply here a sample of the results obtained using this proposed technique. Table 5.1 includes, for some of the reservoirs, the constants a and b obtained by least-squares curve fitting for the data available, the plant minimum, and maximum discharges and the reservoirs' minimum and maximum storages. Tables 5.2–5.5 include the optimal reservoir contents during the first year, the

Table 5.4. Optimal Storage during the Third Year of the Critical Period in MCF

Month k	$x_{1,k}$	$x_{2,k}$	$x_{4,k}$	$x_{5,k}$	$x_{6,k}$	$x_{7,k}$	$x_{8,k}$
1	51337	10950	54859	7945	225875	37401	83610
2	47600	8094	44743	16009	225875	37364	75199
3	45521	5699	35175	26325	225875	28204	75391
4	44116	2645	24336	35724	225875	13287	74021
5	51331	3877	16594	42699	225875	0.00	69485
6	52000	4990	9033	42699	225875	2939	53816
7	46000	10587	12211	42699	225875	8003	37411
8	52305	17598	30062	42699	225875	14769	36072
9	56531	19475	53290	42699	184250	53110	66392
10	56523	19311	51667	40457	225875	53110	79792
11	55889	18123	43408	26433	225875	42195	76614
12	53704	15039	33507	13881	223702	23022	70930

Table 5.5. Optimal Storage during the Rest of the Critical Period in MCF

Month k	$x_{1,k}$	$x_{2,k}$	$x_{4,k}$	$x_{5,k}$	$x_{6,k}$	$x_{7,k}$	$x_{8,k}$
1	49838	10742	23862	7606	179536	9746	64947
2	40961	7795	16224	13010	112193	1974	55741
3	29428	7649	10597	15909	52926	1974	41438
4	16851	7654	5571	13498	69840	2847	25464
5	74550	5513	4450	11105	27356	1958	89991
6	0.00	0.00	0.00	0.00	0.00	0.00	0.00

second, the third, and the rest of the critical period, respectively. Table 5.6 gives the GWh generated during each year of the critical period. The total GWh generated during the critical period is 378,510 GWh, with a yearly average of 108,146 GWh.

5.2.7. Concluding Remarks

A presentation has been made to illustrate the solution for the half-month operating policy of a multireservoir-tree connected hydroelectric power system with what is believed to be one of the largest hydroelectric nonlinear optimization problems attempted considering the number of variables and constraints.

Table 5.6. The Energy Generated during the Critical Period in GWh[a]

Period k	Year 1	Year 2	Year 3	Year 4
1	7552	7199	7902	9852
2	7797	7667	7828	9911
3	7664	8117	7423	10169
4	7887	8533	7314	10082
5	8081	8364	7845	9056
6	8144	9429	7953	7824
7	9934	10178	8961	
8	9378	11012	11170	
9	14579	13646	13519	
10	11785	7693	10199	
11	9259	11112	9250	
12	8036	7673	8813	
Total GWh	102816	110623	108177	56894

Total GWh = 378,510 GWh[a]
The average GWh/year = 108,146 GWh

[a] 1 GWh = 10^6 kWh = 10^9 W h.

The proposed solution has been done by a one-at-a-time method starting from the end reservoir or run-of-river plant of each branch and taking one branch at a time. In this method, the problem is formulated as a minimum norm problem and the equations obtained are nonlinear discrete equations. The time period used in the long-range modeling is half a month; therefore, short-range hydraulic and electrotechnique effects are not taken into consideration.

The tree system is a general case of the reservoir topology which adequately specifies a real system. It is an improvement over the previous methods, which deal with independent rivers with several reservoirs in series or in parallel, or those methods that employ nonlinear programming with the accompanying problems of using penalty functions and specifying good initial estimates.

The basic feature of this new procedure is its ability to automatically produce a maximum hydro generation while satisfying the system constraints. The technique has overcome the influence of starting points and combined the methodology and the experience to end with the system global maximum.

5.3. Nonlinear Storage Model

In Section 5.2 we discussed the maximization of the benefits from a multireservoir power system during the critical water conditions. The model used for this study was a linear model of the storage and water elevation. For the power systems in which the water head variations are small this model is adequate, but for power systems in which the water heads vary by a considerable amount, this model is not adequate. On the other hand, for these systems using a linear model of the storage and elevation may cause a great error in the storage, which for some reservoirs is greater than the natural inflow to this reservoir during a certain time of the critical period. In this section, we repeat the same study of Section 5.2, but the model used is a nonlinear model of the storage and elevation; we use a quadratic function for the storage elevation curve. The tail water elevation is assumed constant independent of the discharge.

5.3.1. Objective Function

We repeat here the objective function with a slight change for convenience. The object of the optimization is to find a long-term scheduling for the power system given in Figure 5.1, and at the same time satisfying some operational and hydraulic constraints. In other words, the optimization objective function is to find the discharge $u_{ij,k}$, $i = 1, \ldots, n_j, j = 1, \ldots, m,$

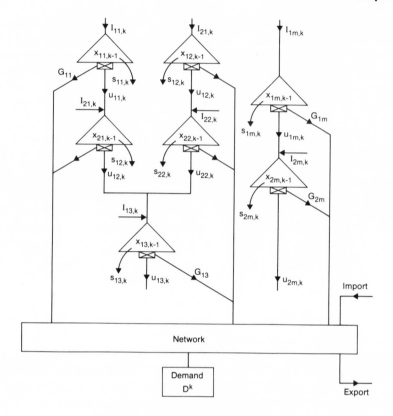

Figure 5.1. The hydroelectric system.

$k = 1, \ldots, K$ as a function of time over the optimization interval, subject to satisfying the following operation and hydraulic constraints:

(1) The total generation from the system, benefits obtained from the generation, during each year of the critical period is a maximum and equal.

(2) The water conservation equation for each reservoir may be adequately described by the continuity-type equation as

$$x_{ij,k} = x_{ij,k-1} + I_{ij,k} + z_{ij,k} - u_{ij,k} - s_{ij,k} \qquad (5.51)$$

where $z_{ij,k}$ is the dependent water inflow connecting hydroelectric plants over the same hydraulic valley and is given by

$$z_{ij,k} = \sum_{\substack{l \in R_u \\ j \in R_r}} (u_{lj,k} + s_{lj,k}) \qquad (5.52)$$

It can be noticed that for independent rivers, $z_{ij,k}$ is given by

$$z_{ij,k} = u_{(u-1)j,k} + s_{(i-1)j,k}, \qquad i \in R_s, \qquad j \in R_r \qquad (5.53)$$

where R_s and R_r are the sets of reservoirs and upstream rivers, respectively.

(3) The water conversion factor, MWh/Mm^3, as a function of the storage is given by

$$h_{ij,k} = a_{ij} + b_{ij}x_{ij,k} + c_{ij}(x_{ij,k})^2, \qquad i \in R_s, \qquad j \in R_r \qquad (5.54)$$

where a_{ij}, b_{ij}, and c_{ij} are constants that were obtained by least-squares curve fitting to typical plant data available. In this equation, we assumed that the efficiency and tailwater elevation are constant independent of the discharge.

(4) The hydro-operation constraints for the system are that the reservoirs should be drawn down from full at the beginning of the optimization interval to minimum storage, x_{ij}^m, at the end of the critical period. This can be mathematically expressed as

$$x_{ij,0} = x_{ij}^M, \qquad i \in R_s, \qquad j \in R_r \qquad (5.55)$$

$$x_{ij,KK} = x_{ij}^m, \qquad i \in R_s, \qquad j \in R_r \qquad (5.56)$$

where KK is the last period studied in the critical period.

(5) Another hydro-operation constraint for satisfying the multipurposes stream use requirements such as, flood control, irrigation, fishing, navigation, and other purposes if any, are the following upper and lower limits on the variables $x_{ij,k}$ and $u_{ij,k}$:

$$x_{ij}^m \leq x_{ij,k} \leq x_{ij}^M, \qquad i \in R_s, \qquad j \in R_r \qquad (5.57)$$

$$u_{ij,k}^m \leq u_{ij,k} \leq u_{ij,k}^M, \qquad i \in R_s, \qquad j \in R_r \qquad (5.58)$$

(6) The MWh generated from reservoir i on river j during a period k is given by

$$G_{ij,k}(u_{ij,k}, \tfrac{1}{2}(x_{ij,k} + x_{ij,k-1})) = a_{ij}u_{ij,k} + \tfrac{1}{2}b_{ij}u_{ij,k}(x_{ij,k} + x_{ij,k-1})$$
$$+ \tfrac{1}{4}c_{ij}u_{ij,k}(x_{ij,k} + x_{ij,k-1})^2 \qquad (5.59)$$

In the above equation, an average of begin and end of time storage is used to avoid underestimation in the production for falling water levels and overestimation for rising water levels.

If we substitute for $x_{ij,k}$ from equation (5.51) into equation (5.59), we obtain

$$G_{ij,k}(u_{ij,k}, \tfrac{1}{2}(x_{ij,k} + x_{ij,k-1})) = a_{ij}u_{ij,k} + \tfrac{1}{2}b_{ij}u_{ij,k}(2x_{ij,k-1}q_{ij,k} + z_{ij,k} - u_{ij,k})$$
$$+ \tfrac{1}{4}c_{ij}u_{ij,k}(2x_{ij,k-1} + q_{ij,k} + z_{ij,k} - u_{ij,k})^2 \qquad (5.60)$$

The above equation can be reduced to

$$G_{ij,k} = \alpha_{ij,k}u_{ij,k} + \beta_{ij,k}u_{ij,k}x_{ij,k-1} + \tfrac{1}{2}\beta_{ij,k}u_{ij,k}z_{ij,k}$$
$$+ \gamma_{ij,k}(u_{ij,k}^2 + c_{ij}u_{ij,k}y_{ij,k-1} + \tfrac{1}{4}c_{ij}z_{ij,k}r_{ij,k}$$
$$+ \tfrac{1}{4}c_{ij}u_{ij,k}w_{ij,k} + c_{ij}r_{ij,k}x_{ij,k-1} - c_{ij}x_{ij,k-1}w_{ij,k} \qquad (5.61)$$

where

$$q_{ij,k} = I_{ij,k} - s_{ij,k} \qquad (5.62a)$$

$$\alpha_{ij,k} = a_{ij} + \tfrac{1}{2}b_{ij}q_{ij,k} + \tfrac{1}{4}c_{ij}(q_{ij,k})^2 \qquad (5.62b)$$

$$\beta_{j,k} = b_{ij} + c_{ij}q_{ij,k} \qquad (5.62c)$$

$$\gamma_{ij,k} = \tfrac{1}{2}(b_{ij} + c_{ij}q_{ij,k} + \tfrac{1}{2}c_{ij}q_{ij,k}) \qquad (5.62d)$$

and the following are pseudostate variables (Refs. 5.5 and 5.6):

$$y_{ij,k} = (x_{ij,k})^2 \qquad (5.63)$$

$$r_{ij,k} = u_{ij,k}z_{ij,k} \qquad (5.64)$$

$$w_{ij,k} = (u_{ij,k})^2 \qquad (5.65)$$

In mathematical terms, the object of the optimization computation is to find the discharge $u_{ij,k}$, $i \in R_s$, $j \in R_r$, that maximizes the total benefits from the system, which is given by the cost functional

$$J = E\left[\sum_{j \in R_r} \sum_{i \in R_s} \sum_{k=1}^{KK} \{\alpha_{ij,k}u_{ij,k} + \beta_{ij,k}u_{ij,k}x_{ij,k-1} + \tfrac{1}{2}\beta_{ij,k}u_{ij,k}z_{ij,k} \right.$$
$$+ \gamma_{ij,k}(u_{ij,k})^2 + c_{ij}u_{ij,k}y_{ij,k-1} + \tfrac{1}{4}c_{ij}z_{ij,k}r_{ij,k}$$
$$\left. + \tfrac{1}{4}c_{ij}u_{ij,k}w_{ij,k} + c_{ij}r_{ij,k}x_{ij,k-1} - c_{ij}x_{ij,k-1}w_{ij,k}\} \right] \qquad (5.66)$$

Subject to satisfying the equality constraints given by equations (5.51), (5.52), (5.53), and (5.63)–(5.65) and the inequality constraints given by equations (5.57) and (5.58). The expectation E in equation (5.66) is taken over the random inflows $I_{ij,k}$. We assume that their probability properties are preestimated from past history.

5.3.2. A Minimum Norm Formulation

The cost functional in equation (5.66) and the equality constraints in equations (5.63)–(5.65) are nonlinear equations. We can form an augmented cost functional by adjoining the equality constraints via Lagrange multipliers and the inequality constraints via Kuhn–Tucker multipliers as

$$\tilde{J} = E\left[\sum_{j \in R_r} \sum_{i \in R_s} \sum_{k=1}^{KK} \{\alpha_{ij,k} u_{ij,k} + u_{ij,k}\beta_{ij,k}x_{ij,k-1} + \tfrac{1}{2}u_{ij,k}\beta_{ij,k}z_{ij,k} \right.$$

$$+ \gamma_{ij,k}(u_{ij,k})^2 u_{ij,k}c_{ij}y_{ij,k-1} + \tfrac{1}{4}c_{ij}z_{ij,k}r_{ij,k}$$

$$+ \tfrac{1}{4}c_{ij}u_{ij,k}w_{ij,k} + c_{ij}r_{ij,k}x_{ij,k-1} - c_{ij}x_{ij,k-1}w_{ij,k}$$

$$+ \lambda_{ij,k}(-x_{ij,k} + x_{ij,k-1} + q_{ij,k} + z_{ij,k} - u_{ij,k})$$

$$+ \mu_{ij,k}(-y_{ij,k} + (x_{ij,k})^2) + \psi_{ij,k}(-\gamma_{ij,k} + u_{ij,k}z_{ij,k})$$

$$+ \phi_{ij,k}(-w_{ij,k} + (u_{ij,k})^2) + e_{ij,k}(x_{ij}^m - x_{ij,k})$$

$$+ e_{ij,k}^1(x_{ij,k} - x_{ij}^M) + g_{ij,k}(u_{ij,k}^m - u_{ij,k})$$

$$\left. + g_{ij,k}^1(u_{ij,k} - u_{ij,k}^M)\} \right] \tag{5.67}$$

where $\lambda_{j,k}$, $\mu_{j,k}$, $\psi_{ij,k}$, and $\phi_{ij,k}$ are Lagrange multipliers. These are obtained in such a way that the corresponding equality constraints are satisfied, and $e_{ij,k}$, $e_{ij,k}^1$, $g_{ij,k}$, and $g_{ij,k}^1$ are Kuhn–Tucker multipliers. These are positive if the constraints are violated and they are zero if the constraints are not violated.

Define the $N \times 1$ column vectors, $N = \sum_{j \in R_r} n_j$

$$\alpha(k) = \mathrm{col}(\alpha_{ij,k}, j \in R_r, i \in R_s) \tag{5.68}$$

$$u(k) = \mathrm{col}(u_{ij,k}, j \in R_r, i \in R_s) \tag{5.69}$$

$$x(k) = \mathrm{col}(x_{ij,k}, j \in R_r, i \in R_s) \tag{5.70}$$

$$s(k) = \mathrm{col}(s_{ij,k}, j \in R_r, i \in R_s) \tag{5.71}$$

$$y(k) = \mathrm{col}(y_{ij,k}, j \in R_r, i \in R_s) \tag{5.72}$$

$$z(k) = \mathrm{col}(z_{ij,k}, j \in R_r, i \in R_s) \tag{5.73}$$

$$r(k) = \mathrm{col}(r_{ij,k}, j \in R_r, i \in R_s) \tag{5.74}$$

$$w(k) = \mathrm{col}(w_{ij,k}, j \in R_r, i \in R_s) \tag{5.75}$$

$$q(k) = \mathrm{col}(q_{ij,k}, j \in R_r, i \in R_s) \tag{5.76}$$

$$\lambda(k) = \mathrm{col}(\lambda_{ij,k}, j \in R_r, i \in R_s) \tag{5.77}$$

$$\mu(k) = \mathrm{col}(\mu_{ij,k}, j \in R_r, i \in R_s) \tag{5.78}$$

$$\psi(k) = \mathrm{col}(\psi_{ij,k}, j \in R_r, i \in R_s) \tag{5.79}$$

$$\phi(k) = \mathrm{col}(\phi_{ij,k}, j \in R_r, i \in R_s) \tag{5.80}$$

$$\nu_{ij,k} = e_{ij,k}^1 - e_{ij,k}, i \in R_s, j \in R_r \tag{5.81}$$

$$\nu(k) = \mathrm{col}(\nu_{ij,k}, j \in R_r, i \in R_s) \tag{5.82}$$

$$\sigma_{ij,k} = g_{ij,k}^1 - g_{ij,k} \tag{5.83}$$

$$\sigma(k) = \mathrm{col}(\sigma_{ij,k}, j \in R_r, i \in R_s) \tag{5.84}$$

Furthermore, define $N \times N$ diagonal vectors

$$\beta(k) = \text{diag}(\beta_{ij,k}, j \in R_r, i \in R_s) \tag{5.85}$$

$$\gamma(k) = \text{diag}(\gamma_{ij,k}, j \in R_r, i \in R_s) \tag{5.86}$$

$$C = \text{diag}(c_{ij}, j \in R_r, i \in R_s) \tag{5.87}$$

Then, the augmented cost functional in equation (5.67) becomes

$$
\begin{aligned}
\tilde{J} = E\Bigg[\sum_{k=1}^{KK} &\{\alpha^T(k)u(k) + u^T(k)\beta(k)x(k-1) + \tfrac{1}{2}u^T(k)\beta(k)Mu(k) \\
&+ \tfrac{1}{2}s^T(k)M^T\beta(k)u(k) + u^T(k)\gamma(k)u(k) + u^T(k)Cy(k-1) \\
&+ \tfrac{1}{4}r^T(k)CMu(k) + \tfrac{1}{4}s^T(k)M^TCr(k) + \tfrac{1}{4}u^T(k)Cw(k) \\
&+ r^T(k)Cx(k-1) - x^T(k-1)Cw(k) + \lambda^T(k)(-x(k) \\
&+ x(k-1) + q(k) + Mu(k) - u(k)) - \mu^T(k)y(k) \\
&+ x^T(k)\mu^T \mathbf{H}x(k) - \psi^T(k)r(k) + u^T(k)\psi^T(k)\mathbf{H}Mu(k) \\
&+ s^T(k)M^T\psi^T(k)\mathbf{H}u(k) - \phi^T(k)w(k) + u^T(k)\phi^T(k)\mathbf{H}u(k) \\
&+ v^T(k)x(k-1) + v^T(k)Mu(k) - v^T(k)u(k) + \sigma^T(k)u(k)\}\Bigg]
\end{aligned}
\tag{5.88}
$$

We shall use the following identities:

$$\sum_{k=1}^{K} \lambda^T(k)x(k) = \lambda^T(K)x(K) - \lambda^T(0)x(0) + \sum_{k=1}^{K} \lambda^T(k-1)x(k-1)$$

$$\sum_{k=1}^{K} \mu^T(k)y(k) = \mu^T(K)y(K) - \mu^T(0)y(0) + \sum_{k=1}^{K} \mu^T(k-1)y(k-1)$$

and

$$\sum_{k=1}^{K} x^T(k)\mu^T(k)\mathbf{H}x(k) = x^T(K)\mu^T(K)\mathbf{H}x(K) - x^T(0)\mu^T(0)\mathbf{H}x(0)$$

$$+ \sum_{k=1}^{K} x^T(k-1)\mu^T(k-1)\mathbf{H}x(k-1)$$

In the above equation M, an $N \times N$, vector contains the topological arrangement of the system and $\mathbf{H}$ is a vector matrix whose index varies from 1 to N while the matrix dimension of $\mathbf{H}$ is $N \times N$.

Note that in equation (5.88) we used the substitution for $z(k)$, which is given by

$$z(k) = Mu(k) + Ms(k) \tag{5.89}$$

and also, constant terms are dropped in equation (5.88).

Define the augmented vectors

$$X^T(k) = [x^T(k-1), u^T(k), y^T(k-1), r^T(k), w^T(k)]5Nx1 \tag{5.90}$$

$$L(k) = \begin{bmatrix} \mu^T(k-1)\mathbf{H} & \frac{1}{2}\beta(k) & 0 & \frac{1}{2}C & \frac{1}{2}C \\ \frac{1}{2}\beta(k) & \beta(k)M + \gamma(k) + \psi^T(k)\mathbf{H}M + \phi^T(k)\mathbf{H} & \frac{1}{2}C & \frac{1}{8}M^TC & \frac{1}{8}C \\ 0 & \frac{1}{2}C & 0 & 0 & 0 \\ \frac{1}{2}C & \frac{1}{8}CM & 0 & 0 & 0 \\ \frac{1}{2}C & \frac{1}{8}C & 0 & 0 & 0 \end{bmatrix};$$

$$5N \times 5N \tag{5.91}$$

and

$$\begin{aligned}
R^T(k) = &[(\lambda(k) - \lambda(k-1) + \nu(k))^T, (\alpha(k) + \tfrac{1}{2}\beta(k)Ms(k) + M^T\lambda(k) \\
&- \lambda(k) + \psi(k)\mathbf{H}Ms(k) + M^T\nu(k) - \nu(k) + \sigma(k))^T \\
&- \mu^T(k-1), -\psi^T(k), -\phi^T(k)]
\end{aligned} \tag{5.92}$$

Then, the cost functional in equation (5.88) can be written as

$$\begin{aligned}
\tilde{J} = E\Big[&-\lambda^T(K)x(K) + \lambda^T(0)x(0) - \mu^T(K)y(K) + \mu^T(0)y(0) \\
&+ x^T(K)\mu^T(K)\mathbf{H}x(K) - x^T(0)\mu^T(0)\mathbf{H}x(0) \\
&+ \sum_{k=1}^{k} \{X^T(k)L(k)X(k) + R^T(k)X(k)\}\Big]
\end{aligned} \tag{5.93}$$

Using the principle of completing the squares, and dropping constant terms, one obtains

$$\begin{aligned}
\tilde{J} = E\Big[&\{x^T(K)\mu^T(K)\mathbf{H}x(K) - \lambda^T(K)x(K) - \mu^T(K)y(K)\} \\
&+ \sum_{k=1}^{K} \{(X(k) + \tfrac{1}{2}V(k))^T L(k)(X(k) + \tfrac{1}{2}V(k))\}\Big]
\end{aligned} \tag{5.94}$$

where the augmented vector $V(k)$ is given by

$$V(k) = L^{-1}(k)R(k) \tag{5.95}$$

Note that in equation (5.94), we dropped terms containing $x(0)$ and $y(0)$, because these variables are constant independent of either $x(K)$ nor $X(k)$.

The cost functional in equation (5.94) consists of two parts. The first part is the boundary term, and the second part is the discrete integral part, which are independent of each other. To maximize $\tilde{J}$ in equation (5.94), one maximizes each term separately.

$$\max_{[x(K),X(k)]} \tilde{J} = \max_{x(K)} E[x^T(K)\mu^T(K)\mathbf{H}x(K) - \lambda^T(K)x(K) - \mu^T(K)y(K)]$$

$$+ \max_{X(k)} \left[\sum_{k=1}^{K} (X(k) + \tfrac{1}{2}V(k))^T L(k)(X(k) + \tfrac{1}{2}V(k)) \right] \quad (5.96)$$

5.3.3. The Optimal Solution

The optimal solution of the problem formulated in equation (5.96) is found by maximizing the boundary term, and the discrete integral part separately as

$$\max_{x(K),y(K)} J_1 = \max E[x^T(K)\mu^T(K)\mathbf{H}x(K) - \lambda^T(K)x(K) - \mu^T(K)y(K)]$$

$$(5.97)$$

The maximum of equation (5.97) is clearly achieved when

$$E[\lambda(K)] = [0] \quad (5.98)$$

and

$$E[\mu(K)] = [0] \quad (5.99)$$

The above two equations give the values of Lagrange multipliers at the end of the optimization interval. In component form, equations (5.98) and (5.99) can be written as

$$E[\lambda_{ij,K}] = 0, \quad i \in R_s, \quad j \in R_r \quad (5.100)$$

and

$$E[\mu_{ij,K}] = 0, \quad i \in R_s, \quad j \in R_r \quad (5.101)$$

The boundary part in equation (5.96) defines a norm in Hilbert space. This term can be written as

$$\max_{X(k)} J_2 = \max_{X(k)} \|X(k) + \tfrac{1}{2}V(k)\|_{L(k)} \quad (5.102)$$

The maximum of J_2 in the above equation is clearly achieved when the norm of this equation is equal to zero

$$E[X(k) + \tfrac{1}{2}V(k)] = [0] \quad (5.103)$$

provided the matrix $L(k)$ is invertible. If one substitutes for $V(k)$ from equation (5.95), then equation (5.103) can be written as

$$E[2L(k)X(k) + R(k)] = [0] \quad (5.104)$$

Substituting from equations (5.90), (5.91), and (5.92) into equation (5.104), and adding the equality constraints (5.51), (5.52), (5.63), (5.64), and (5.65), we get the following set of optimal equations:

$$E[-x(k) + x(k-1) + q(k) + z(k) - u(k)] = [0] \tag{5.105}$$

$$E[-z(k) + Mu(k) + Ms(k)] = [0] \tag{5.106}$$

$$E[-y(k) + x^T(k)\mathbf{H}x(k)] = [0] \tag{5.107}$$

$$E[-r(k) + u(k)\mathbf{H}z(k)] = [0] \tag{5.108}$$

$$E[-w(k) + u(k)\mathbf{H}u(k)] = [0] \tag{5.109}$$

$$E[2\mu^T(k-1)\mathbf{H}x(k-1) + \beta(k)u(k) + Cr(k) + Cw(k)$$
$$+ \lambda(k) - \lambda(k-1) + \nu(k)] = [0] \tag{5.110}$$

$$E[\alpha(k) + \tfrac{1}{2}\beta(k)Ms(k) + M^T\lambda(k) - \lambda(k) + \psi(k)\mathbf{H}Ms(k) + M^T\nu(k)$$
$$- \nu(k) + \sigma(k) + \beta(k)x(k-1) + 2(\beta(k)M + \gamma(k) + \psi^T(k)\mathbf{H}M$$
$$+ \phi^T(k)\mathbf{H})u(k) + Cy(k-1) + \tfrac{1}{4}M^TCr(k) + \tfrac{1}{4}Cw(k)] = [0] \tag{5.111}$$

$$E[-\mu(k-1) + Cu(k)] = [0] \tag{5.112}$$

$$E[-\psi(k) + Cx(k-1) + \tfrac{1}{4}CMu(k)] = [0] \tag{5.113}$$

$$E[-\phi(k) + Cx(k-1) + \tfrac{1}{4}Cu(k)] = [0] \tag{5.114}$$

Besides the above equations, we have the following limits on the variables, as stated earlier in the previous section:

$$\begin{aligned}
&\text{If } x(k) < x^m, &&\text{then we put } x(k) = x^m \\
&\text{If } x(k) > x^M, &&\text{then we put } x(k) = x^M
\end{aligned} \tag{5.115a}$$

$$\begin{aligned}
&\text{If } u(k) < u^m(k), &&\text{then we put } u(k) = u^m(k) \\
&\text{If } u(k) > u^M(k), &&\text{then we put } u(k) = u^m(k)
\end{aligned} \tag{5.115b}$$

Finally, we are using the Kuhn–Tucker theorem so that the following exclusion equations must be satisfied at the optimum:

$$e_{ij,k}(x_{ij}^m - x_{ij,k}) = 0 \tag{5.116}$$

$$e_{ij,k}^1(x_{ij,k} - x_{ij}^M) = 0 \tag{5.117}$$

$$g_{ij,k}(u_{ij,k}^m - u_{ij,k}) = 0 \tag{5.118}$$

$$g_{ij,k}^1(u_{ij,k} - u_{ij,k}^M) = 0 \tag{5.119}$$

Equations (5.105)–(5.119) completely specify the optimal solution for the long-term scheduling of a multichain power system under critical water conditions.

5.4. A Discrete Maximum Principle Approach (Linear Model)

In this section, we consider the application of the discrete maximum principle to the optimization of the production of the hydro energy from a multireservoir power system during the critical period. We repeat here, for convenience, the problem formulation.

5.4.1. Problem Formulation

The problem of the power system given in Figure 4.1 is to find the discharge $u_{ij,k}$, $i \in R_s$, $j \in R_r$ that maximizes the total hydro energy production given by

$$J = E \left[\sum_{j \in R_r} \sum_{i \in R_s} \sum_{k_0}^{k_f - 1} G_{ij,k}(u_{ij,k}, (x_{ij,k+1}, x_{ij,k})) \right] \tag{5.120}$$

subject to satisfying the following constraints:

(1) The reservoir's dynamic equation is given by the discrete difference equation as

$$x_{ij,k+1} = x_{ij,k} + I_{ij,k} + z_{ij,k} - u_{ij,k} - s_{ij,k} \tag{5.121}$$

where

$$z_{ij,k} = \sum_{\substack{\sigma \in R_u \\ j \in R_r}} (u_{\sigma j,k} + s_{\sigma j,k}) \tag{5.122}$$

with

$$x_{ij,k} = x_{ij,0}, \qquad i \in R_s, \qquad j \in R_r \tag{5.123}$$

and

$$x_{ij,k_f} = x_{ij}^m, \qquad i \in R_s, \qquad j \in R_r \tag{5.124}$$

Note that, in the above equations, we used a forward difference equation in describing the reservoir's dynamics.

(2) There are constraints on the states and control variables given by

$$x_{ij}^m \le x_{ij,k} \le x_{ij}^M, \qquad i \in R_s, \qquad j \in R_r \tag{5.125}$$

$$u_{ij,k}^m \le u_{ij,k} \le u_{ij,k}^M, \qquad i \in R_s, \qquad j \in R_r \tag{5.126}$$

The above inequality constraints can be included in the cost functional (5.120) by using weighting matrices as

$$\tilde{J} = J + \tfrac{1}{2}(x_{ij}^m - x_{ij,k})_Q^2 + \tfrac{1}{2}(x_{ij,k} - x_{ij}^M)_R^2 + \tfrac{1}{2}(u_{ij,k}^m - u_{ij,k})_S^2 + \tfrac{1}{2}(u_{ij,k} - u_{ij,k}^M)_T^2 \tag{5.127}$$

Since the final states are fixed, storages at the end of the critical period, then equation (5.124) can be included in the cost function as

$$\tilde{\tilde{J}} = \tilde{J} + \tfrac{1}{2}(x_{ij,k} - x_{ij}^m)_\psi^2 \tag{5.128}$$

In the above equations, Q, R, S, T, and ψ are weighting diagonal matrices. It is well known that higher values of the weighting matrices lead to shorter periods of exceeding of constraints (5.124)–(5.126).

(3) The MWh generated from a reservoir i on river j during a period k is given by

$$G_{ij,k}(u_{ij,k}, \tfrac{1}{2}(x_{ij,k} + x_{ij,k+1})) = a_{ij}u_{ij,k} + \tfrac{1}{2}b_{ij}u_{ij,k}(x_{ij,k} + x_{ij,k+1}) \quad (5.129)$$

where a_{ij}, b_{ij}, are constants. These are obtained by least-squares curve fitting to typical plant data available. In equation (5.129) an average of begin and end of time storage is used to avoid underestimation in production for rising water levels and overestimation for falling water levels. Also, we assume that tail water elevation is constant independent of the discharge. Furthermore, the efficiency of the turbines is constant.

If one substitutes for $x_{ij,k+1}$ from equation (5.121) into equation (5.129), we get

$$G_{ij,k}(u_{ij,k}, x_{ij,k}) = q_{ij,k}u_{ij,k} + b_{ij}u_{ij,k}x_{ij,k} + \tfrac{1}{2}b_{ij}u_{ij,k}\left(\sum_{\substack{\sigma \in R_u \\ j \in R_r}} u_{\sigma j,k} - u_{ij,k} \right) \quad (5.130)$$

where

$$q_{ij,k} = a_{ij} + \tfrac{1}{2}b_{ij}I_{ij,k} + \tfrac{1}{2}b_{ij}\left(\sum_{\substack{\sigma \in R_u \\ j \in R_r}} s_{\sigma j,k} - s_{ij,k} \right) \quad (5.131)$$

Now, the cost functional in equation (5.128) can be written as

$$\tilde{J} = E\left[\sum_{j \in R_r} \sum_{i \in R_s} \tfrac{1}{2}\psi_{ij}(x_{ij,k} - x_{ij}^m)^2 \right.$$

$$+ \sum_{j \in R_r} \sum_{i \in R_s} \sum_{k_0=0}^{k_f-1} \{q_{ij,k}u_{ij,k} + b_{ij}u_{ij,k}x_{ij,k}$$

$$+ \tfrac{1}{2}b_{ij}u_{ij,k}\left(\sum_{\substack{\sigma \in R_u \\ j \in R_r}} u_{\sigma j,k} - u_{ij,k} \right)$$

$$+ \tfrac{1}{2}b_{ij}u_{ij,k} \sum_{\substack{\sigma \in R_u \\ j \in R_r}} s_{\sigma j,k} + \tfrac{1}{2}(x_{ij}^m - x_{ij,k})^2 q_{ij}$$

$$+ \tfrac{1}{2}(x_{ij,k} - x_{ij}^M)^2 r_{ij} + \tfrac{1}{2}(u_{ij,k}^m - u_{ij,k})^2 s_{ij}$$

$$\left. + \tfrac{1}{2}(u_{ij,k} - u_{ij,k}^M)^2 t_{ij}\} \right] \quad (5.132)$$

Subject to satisfying

$$x_{ij,k+1} = x_{ij,k} + I_{ij,k} + \sum_{\substack{\sigma \in R_u \\ j \in R_r}} (u_{\sigma j,k} + s_{\sigma j,k}) - u_{ij,k} - s_{ij,k} \qquad (5.133)$$

The cost functional in equation (5.132) and the equality constraint in equation (5.133) can be written in vector form as

$$\tilde{J} = E\left[\tfrac{1}{2}(x(k_f) - x^m)^T \psi(x(k_f) - x^m) + \sum_{k_0=0}^{k_f-1} \{ q^T(k)u(k) + u^T(k)bx(k) \right.$$

$$+ \tfrac{1}{2}u^T(k)bMu(k) + \tfrac{1}{2}u^T(k)bMs(k) + x^T(k)Qx(k) - x^T(k)Qx^m$$

$$+ \tfrac{1}{2}x^T(k)Rx(k) - x^T(k)Rx^M + \tfrac{1}{2}u^T(k)Su(k) - u^T(k)Su^m(k)$$

$$\left. + \tfrac{1}{2}u^T(k)Tu(k) - u^T(k)Tu^M(k) \} \right] \qquad (5.134)$$

Subject to satisfying the equality constraint

$$x(k + 1) = x(k) + I(k) + Mu(k) + Ms(k) \qquad (5.135)$$

where the above vectors are defined as

$$x(k) = \text{col}(x_{ij,k}, i \in R_s, j \in R_r) \qquad (5.136)$$

$$u(k) = \text{col}(u_{ij,k}, i \in R_s, j \in R_r) \qquad (5.137)$$

$$q(k) = \text{col}(q_{ij,k}, i \in R_s, j \in R_r) \qquad (5.138)$$

$$x^m = \text{col}(x_{ij}^m, i \in R_s, j \in R_r) \qquad (5.139)$$

$$x^M = \text{col}(x_{ij}^M, i \in R_s, j \in R_r) \qquad (5.140)$$

$$x(k_f) = \text{col}(x_{ij,k_f}, i \in R_s, j \in R_r) \qquad (5.141)$$

$$I(k) = \text{col}(I_{ij,k}, i \in R_s, j \in R_r) \qquad (5.142)$$

$$s(k) = \text{col}(s_{ij,k}, i \in R_s, j \in R_r) \qquad (5.143)$$

Furthermore, the $N \times N$ diagonal matrices are defined as

$$N = \sum_{j \in R_r} n_j$$

$$Q = \text{diag}(q_{ij}, i \in R_s, j \in R_r) \qquad (5.144)$$

$$R = \text{diag}(r_{ij}, i \in R_s, j \in R_r) \qquad (5.145)$$

$$S = \text{diag}(s_{ij}, i \in R_s, j \in R_r) \qquad (5.146)$$

$$T = \text{diag}(t_{ij}, i \in R_s, j \in R_r) \qquad (5.147)$$

$$\psi = \text{diag}(\psi_{ij}, i \in R_s, j \in R_r) \qquad (5.148)$$

and M is an $N \times N$ matrix containing the topological arrangement of the system.

Define the Hamiltonian H as

$$
\begin{aligned}
H = E[&\tfrac{1}{2}x^T(k)(Q+R)x(k) + \tfrac{1}{2}u^T(k)(S+T+bM)u(k) \\
&+ \tfrac{1}{2}u^T(k)bx(k) + \tfrac{1}{2}x^T(k)bu(k) + u^T(k)(q(k) + \tfrac{1}{2}bMs(k)) \\
&- Su^m(k) - Tu^M(k) - x^T(k)(Rx^M + Qx^m) \\
&+ \lambda^T(k+1)(x(k) + I(k) + Mu(k) + Ms(k))]
\end{aligned}
\tag{5.149}
$$

5.4.2. Optimal Equations

By applying the discrete maximum principles as

$$
E[\lambda(k)] = E\left[\frac{\partial H}{\partial x(k)}\right]
\tag{5.150}
$$

we obtain the canonic equations, given by

$$
E[\lambda(k)] = E[\lambda(k+1) + (Q+R)x(k) + \tfrac{1}{2}bu(k) - (Rx^M + Qx^m)]
\tag{5.151}
$$

with the boundary condition given by

$$
E[\lambda(k_f)] = E[\psi(x(k_f) - x^m)]
\tag{5.152}
$$

Now, there are no explicit constraints on the control variables, $u(k)$, hence

$$
E\left[\frac{\partial H}{\partial u(k)}\right] = [0]
\tag{5.153}
$$

which yields to

$$
\begin{aligned}
E[&(S+T+bM)u(k) + \tfrac{1}{2}bx(k) + q(k) + \tfrac{1}{2}bMs(k) \\
&- Su^m(k) - Tu^M(k) + M^T\lambda(k+1)] = [0]
\end{aligned}
\tag{5.154}
$$

Finally, the state equations are given by

$$
E[x(k+1)] = E\left[\frac{\partial H}{\partial \lambda(k+1)}\right]
\tag{5.155}
$$

which gives

$$
E[x(k+1)] = E[x(k) + I(k) + Mu(k) + Ms(k)]
\tag{5.156}
$$

Equations (5.151), (5.152), (5.154), and (5.156) completely specify the optimal solution. These equations constitute a TPBVP, which can be solved using the algorithm explained in Section 4.4.2.

5.5. Optimization of Power System Operation with a Specified Monthly Generation

In the previous sections, we maximized the generation from a hydro-electric power system during the critical water conditions. In this maximization, the required load on the system was not accounted for; that load may be higher in winter than in summer. This section is devoted to the solution of the long-term optimal operating problem of the multireservoir power system for the critical period with a monthly variable load; this load is equal to a certain percentage of the total generation at the end of the year, and at the same time the total generation during each year of the critical period should be equal and maximum. To meet all these requirements, we maximize the generation from the system during each year of the critical period taking into account this specified load on the system. In other words, we minimize the difference between the monthly generation and the monthly load on the system, which is equal to the monthly percentage load required multiplied by the total generation at the end of the year.

5.5.1. Problem Formulation

5.5.1.1. The System under Study

The system under consideration consists of m independent rivers, with one or several reservoirs and power plants in series on each, and interconnection lines to the neighboring system through which energy may be exchanged, Figure 5.1.

5.5.1.2. The Objective Function

The long-term optimal operating problem, under critical water conditions, aims to find the discharge $u_{ij,k}$, $i \in R_h$, $j \in R_r$ as a function of time that minimizes

$$J(k) = E\left[\sum_{j=1}^{m} \sum_{i=1}^{n_j} G_{ij,k}(u_{ij,k}, \tfrac{1}{2}(x_{ij,k} + x_{ij,k-1})) \right.$$

$$\left. - a(k) \sum_{j \in R_r} \sum_{i \in R_h} \sum_{k=1}^{K} (G_{ij,k})(u_{ij,k}, \tfrac{1}{2}(x_{ij,k} + x_{ij,k-1})) \right] \quad (5.157)$$

In the above equation $a(k)$ is the given percentage load on the system during a period k subject to satisfying the hydro constraints given by the following:

(1) The water conservation equation, continuity equation, may be adequately described by the difference equation

$$x_{ij,k} = x_{ij,k-1} + I_{ij,k} + u_{(i-1)j,k} + s_{(i-1)j,k} - u_{ij,k} - s_{ij,k} \qquad (5.158)$$

where $s_{ij,k}$ is given by

$$s_{ij,k} = \begin{cases} (x_{ij,k-1} + I_{ij,k} + s_{(i-1)j,k} + u_{(i-1)j,k} - x_{ij,k}) - u_{ij,k}^M \\ \quad \text{if } (x_{ij,k-1} + I_{ij,k} + s_{(i-1)j,k} + u_{(i-1)j,k} - x_{ij,k}) > u_{ij,k}^M \\ \quad \text{and } x_{ij,k} > x_{ij}^M \\ 0 \quad \text{otherwise} \end{cases} \qquad (5.159)$$

Equation (5.159) states that water is spilt only when the reservoir is filled to capacity and the discharge exceeds $u_{ij,k}^M$. This spillage has nothing to do with the generation of that reservoir, and it is considered as a constant inflow to the downstream reservoir. The monthly inflows into the reservoirs form a random sequence that depends on natural phenomena such as rainfall and snowfall. The statistical parameters of the inflow can be determined from historical records. In reality, there is a statistical correlation between inflows of successive months. However, it is assumed here that the inflow in each month is statistically independent of previous inflows. In particular, it is assumed that each inflow $I_{ij,k}$ is characterized by a discrete distribution given by

$$\text{Prob}[I_{ij} = I_x] = p_x > 0$$

where

$$\sum_{x=1}^{N} p_x = 1$$

(2) To satisfy the multipurpose stream use requirements, such as flood control, irrigation, fishing, and other purposes if any, the following upper and lower limits on the variables should be satisfied at the optimum:

(a) Upper and lower bounds on the storages:

$$x_{ij}^m \le x_{ij,k} \le x_{ij}^M, \qquad i \in R_h, \qquad j \in R_r \qquad (5.160)$$

(b) Upper and lower bounds on the discharge:

$$u_{ij,k}^m \le u_{ij,k} \le u_{ij,k}^M \qquad (5.161)$$

In the cost functional given by equation (5.157) we simply minimize the difference between the monthly generation and the monthly load on the system, in such a way that the total generation at the end of the year is as large as possible since we can store water for the second year of the critical period. The expectation in equation (5.157) is taken with respect to the random inflow $I_{ij,k}$.

5.5.1.3. Modeling of the System

The generation of a hydroplant is a nonlinear function of the discharge $u_{ij,k}$ and the storage. To avoid overestimation of production for falling water levels and underestimation for rising water levels, an average of begin and end of time step storage is used. We may choose the generating function $G_{ij,k}$ as

$$G_{ij,k} = \alpha_{ij} u_{ij,k} + \tfrac{1}{2}\beta_{ij} u_{ij,k}(x_{ij,k} + x_{ij,k-1})$$
$$+ \tfrac{1}{4}\gamma_{ij} u_{ij,k}(x_{ij,k} + x_{ij,k-1})^2 \tag{5.162}$$

where α_{ij}, β_{ij}, and γ_{ij} are constants. These were obtained by least-squares curve fitting to typical plant data available. Substituting for $x_{ij,k}$ from equation (5.158) into equation (5.162) one obtains

$$G_{ij,k} = b_{ij,k} u_{ij,k} + u_{ij,k} d_{ij,k} x_{ij,k-1} + u_{ij,k} f_{ij,k}(u_{(i-1)j,k} - u_{ij,k})$$
$$+ \gamma_{ij} u_{ij,k} y_{ij,k} + \tfrac{1}{4} u_{ij,k} \gamma_{ij}(z_{(i-1)j,k} + z_{ij,k})$$
$$+ \gamma_{ij} r_{ij,k-1}(u_{(i-1)j,k} - u_{ij,k}) - \tfrac{1}{2}\gamma_{ij} z_{ij,k} u_{(i-1)j,k} \tag{5.163}$$

where

$$q_{ij,k} = I_{ij,k} + s_{(i-1)j,k} - s_{ij,k} \tag{5.164}$$

$$b_{ij,k} = \alpha_{ij} + \tfrac{1}{2}\beta_{ij} q_{ij,k} + \tfrac{1}{4}\gamma_{ij}(q_{ij,k})^2 \tag{5.165}$$

$$d_{ij,k} = \beta_{ij} + \gamma_{ij} q_{ij,k} \tag{5.166}$$

$$f_{ij,k} = \tfrac{1}{2} d_{ij,k} \tag{5.167}$$

and

$$y_{ij,k} = (x_{ij,k})^2 \tag{5.168}$$

$$z_{ij,k} = (u_{ij,k})^2 \tag{5.169}$$

$$r_{ij,k-1} = u_{ij,k} x_{ij,k-1} \tag{5.170}$$

5.5.2. The Optimal Solution

5.5.2.1. A Minimum Norm Formulation

The augmented cost functional is obtained by adjoining the equality constraints via Lagrange multipliers and the inequality constraints via

Kuhn–Tucker multipliers. One thus obtains

$$
\begin{aligned}
J(k) = E\Bigg[&\sum_{j \in R_r} \sum_{i \in R_h} \{ b_{ij,k} u_{ij,k} + u_{ij,k} d_{ij,k} x_{ij,k-1} \\
&+ u_{ij,k} f_{ij,k}(u_{(i-1)j,k} - u_{ij,k}) + \gamma_{ij} u_{ij,k} y_{ij,k-1} \\
&+ \tfrac{1}{4} u_{ij,k} \gamma_{ij}(z_{(i-1)j,k} + z_{ij,k}) + \gamma_{ij} r_{ij,k-1}(u_{(i-1)j,k} - u_{ij,k}) - \tfrac{1}{2}\gamma_{ij} z_{ij,k} u_{(i-1)j,k}\} \\
&- a(k) \sum_{j \in R_r} \sum_{i \in R_h} \sum_{k=1}^{K} \{ b_{ij,k} u_{ij,k} + u_{ij,k} d_{ij,k} x_{ij,k-1} \\
&+ u_{ij,k} f_{ij,k}(u_{(i-1)j,k} - u_{ij,k}) + \gamma_{ij} u_{ij,k} y_{ij,k-1} \\
&+ \tfrac{1}{4}\gamma_{ij} u_{ij,k}(z_{(i-1)j,k} + z_{ij,k}) + \gamma_{ij} r_{ij,k-1}(u_{(i-1)j,k} - u_{ij,k}) \\
&- \tfrac{1}{2}\gamma_{ij} z_{ij,k} u_{(i-1)j,k} + \mu_{ij,k}(-y_{ij,k} + (x_{ij,k})^2) \\
&+ \phi_{ij,k}(-z_{ij,k} + (u_{ij,k})^2) + \psi_{ij,k}(-r_{ij,k-1} + u_{ij,k} x_{ij,k-1}) \\
&+ \lambda_{ij,k}(-x_{ij,k} + x_{ij,k-1} + q_{ij,k} + u_{(i-1)j,k} - u_{ij,k}) + e_{ij,k}(x_{ij}^m - x_{ij,k}) \\
&+ e_{ij,k}^1(x_{ij,k} - x_{ij}^M) + g_{ij,k}(u_{ij,k}^m - u_{ij,k}) + g_{ij,k}^1(u_{ij,k} - u_{ij,k}^M)\} \Bigg]
\end{aligned} \qquad (5.171)
$$

where $\mu_{ij,k}$, $\phi_{ij,k}$, $\psi_{ij,k}$, and $\lambda_{ij,k}$ are Lagrange multipliers. They are determined such that the corresponding equality constraints are satisfied and $e_{ij,k}$, $e_{ij,k}^1$, $g_{ij,k}$, and $g_{ij,k}^1$ are Kuhn–Tucker multipliers. These are equal to zero if the constraints are not violated and greater than zero if the constraints are violated.

To formulate the problem as a norm, we define $Q \times 1$ column vectors as $Q = \sum_{j=1}^{m} n_j$:

$$B(k) = \mathrm{col}(b_{ij,k}; i \in R_h, j \in R_r) \qquad (5.172)$$

$$b(k) = \mathrm{col}(b_{ij,k}; i \in R_h, j \in R_r) \qquad (5.173)$$

$$u(k) = \mathrm{col}(u_{ij,k}; i \in R_h, j \in R_r) \qquad (5.174)$$

$$x(k) = \mathrm{col}(x_{ij,k}; i \in R_h, j \in R_r) \qquad (5.175)$$

$$y(k) = \mathrm{col}(y_{ij,k}; i \in R_h, j \in R_r) \qquad (5.176)$$

$$z(k) = \mathrm{col}(z_{ij,k}; i \in R_h, j \in R_r) \qquad (5.177)$$

$$r(k) = \mathrm{col}(r_{ij,k}; i \in R_h, j \in R_r) \qquad (5.178)$$

$$\mu(k) = \mathrm{col}(\mu_{ij,k}; i \in R_h, j \in R_r) \qquad (5.179)$$

$$\phi(k) = \mathrm{col}(\phi_{ij,k}; i \in R_h, j \in R_r) \qquad (5.180)$$

$$\lambda(k) = \mathrm{col}(\lambda_{ij,k}; i \in R_h, j \in R_r) \qquad (5.181)$$

$$\nu(k) = \mathrm{col}(\nu_{ij,k}; i \in R_h, j \in R_r) \qquad (5.182)$$

$$\sigma(k) = \mathrm{col}(\sigma_{ij,k}; i \in R_h, j \in R_r) \qquad (5.183)$$

where

$$\nu_{ij,k} = e_{ij,k} - e_{ij,k} \tag{5.184}$$

$$\sigma_{ij,k} = g_{ij,k} - g_{ij,k} \tag{5.185}$$

In the above equations R_h is the set of the hydro reservoirs connected in series on the same river and R_r is the set of the rivers.

Furthermore, define the $Q \times Q$ diagonal matrices as

$$d(k) = \text{diag}(d_{ij,k}; i \in R_h, j \in R_r) \tag{5.186}$$

$$D(k) = \text{diag}(d_{ij,k}/a(k); i \in R_h, j \in R_r) \tag{5.187}$$

$$f(k) = \text{diag}(f_{ij,k}; i \in R_h, j \in R_r) \tag{5.188}$$

$$F(k) = \text{diag}(f_{ij,k}/a(k); i \in R_h, j \in R_r) \tag{5.189}$$

$$C(k) = \text{diag}(\gamma_{ij}/a(k); i \in R_h, j \in R_r) \tag{5.190}$$

$$C = \text{diag}(\gamma_{ij}; i \in R_h, j \in R_r) \tag{5.191}$$

$$M = \text{diag}(M_j; j \in R_r) \tag{5.192}$$

where $M_1, \ldots, M_m$ are lower triangular matrices, whose elements are given by

1. $m_{ii} = -1, \qquad i \in R_n;$
2. $m_{(v+1)v} = 1, \qquad v = 1, \ldots, n_j - 1.$

$$N = \text{diag}(N_j, j \in R_r) \tag{5.193}$$

where $N_1, \ldots, N_m$ are lower triangular matrices whose elements are given by

1. $n_{ij} = 1, \qquad i \in R_r, \qquad j \in R_m;$
2. $n_{(v+1)v} = 1, \qquad v = 1, \ldots, n_j = 1.$

$$L = \text{diag}(L_j, j \in R_r) \tag{5.194}$$

where the elements of any matrix $L_j, j \in R_r$ are given by

1. $l_{(v+1)v} = 1, \qquad v = 1, \ldots, n_j - 1, \qquad j \in R_r;$
2. The rest of the elements are equal to zero.

Using the above definitions, the cost functional in equation (5.171) can be written as

$$J(k) = E\left[\{B^T(k)u(k) + u^T(k)D(k)x(k-1) + u^T(k)F(k)Mu(k) \right.$$

$$+ u^T(k)C(k)y(k-1) + \tfrac{1}{4}u^T(k)C(k)Nz(k)$$

$$\left. + r^T(k-1)C(k)Mu(k) - \tfrac{1}{2}z^T(k-1)C(k)Lu(k)\} \right.$$

$$- \sum_{k=1}^{K} \{b^T(k)u(k) + u^T(k)d(k)x(k-1)$$

$$+ u^T(k)f(k)Mu(k) + u^T(k)Cy(k-1) + \tfrac{1}{4}u^T(k)CNz(k)$$

$$+ r^T(k-1)CMu(k) - \tfrac{1}{2}z^T(k)CLu(k)$$

$$+ \mu^T(k)(-y(k) + x^T(k)\mathbf{H}x(k))$$

$$+ \phi^T(k)(-z(k) + u^T(k)\mathbf{H}u(k)) + \psi^T(k)(-r(k-1)$$

$$+ x^T(k-1)\mathbf{H}u(k)) + \lambda^T(k)(-x(k)$$

$$+ x(k-1) + q(k) + Mu(k)) + \nu^T(k)(x(k-1)$$

$$+ q(k) + Mu(k)) + \sigma^T(k)u(k)\}\Bigg] \tag{5.195}$$

In the above equation $\mathbf{H}$ is a vector matrix in which the vector index varies from 1 to Q while the matrix dimension of $\mathbf{H}$ is $Q \times Q$.

Employing the discrete version of integration by parts and dropping the constant terms, one obtains

$$J^k = E\Bigg[x^T(K)\mu^T(K)\mathbf{H}x(K) - \lambda^T(K)x(K) - \mu^T(K)y(K)$$

$$- x^T(0)\mu^T(0)\mathbf{H}x(0)$$

$$+ \lambda^T(0)x(0) + \mu^T(0)y(0) + \{B^T(k)u(k) + u^T(k)D(k)x(k-1)$$

$$+ u^T(k)F(k)Mu(k) + u^T(k)C(k)y(k-1) + \tfrac{1}{4}u^T(k)C(k)Nz(k)$$

$$+ r^T(k-1)C(k)Mu(k) - \tfrac{1}{2}z^T(k)C(k)Lu(k)\}$$

$$- \sum_{k=1}^{K} \{b^T(k)u(k) + u^T(k)d(k)x(k-1) + u^T(k)f(k)Mu(k)$$

$$+ u^T(k)Cy(k-1) + \tfrac{1}{4}u^T(k)CNz(k)$$

$$+ (\lambda(k) - \lambda(k-1))^T x(k-1) + \lambda^T(k)Mu(k) + r^T(k-1)CMu(k)$$

$$- \tfrac{1}{2}z^T(k)CLu(k) - \mu^T(k-1)y(k-1)$$

$$+ x^T(k-1)\mu^T(k-1)\mathbf{H}x(k-1) + \nu^T(k)X(k-1) - \phi^T(k)z(k)$$

$$+ u^T(k)\phi^T(k)\mathbf{H}u(k) - \psi^T(k)r(k-1)$$

$$+ x^T(k-1)\psi^T(k)\mathbf{H}u(k) + \sigma^T(k)u(k) + \nu^T(k)Mu(k)\}\Bigg] \tag{5.196}$$

We define the $5Q \times 1$ vectors as

$$X^T(k) = [x^T(k-1), y^T(k-1), u^T(k), z^T(k), r^T(k-1)] \tag{5.197}$$

$$R^T(k) = [(\lambda(k) - \lambda(k-1) + \nu(k))^T, -\mu^T(k-1), \lambda(b(k) + M^T\sigma(k) + (k)$$
$$+ M^T\nu(k))^T, -\phi^T(k), -\psi^T(k)] \tag{5.198}$$

and

$$Q^T(k) = [0, 0, B^T(k), 0, 0] \tag{5.199}$$

Furthermore, define the $5Q \times 5Q$ rectangular matrices as

$$L(k) = \begin{bmatrix} \mu^T(k-1)\mathbf{H} & 0 & \frac{1}{2}[d(k) + \psi^T(k)\mathbf{H}] & 0 & 0 \\ 0 & 0 & \frac{1}{2}C(k) & 0 & 0 \\ \frac{1}{2}[d(k) + \psi^T(k)\mathbf{H}] & \frac{1}{2}C & f(k)M + \phi^T(k)\mathbf{H} & \frac{1}{8}CN - L^TC & \frac{1}{2}M^TC \\ 0 & 0 & \frac{1}{8}N^TC - \frac{1}{4}CL & 0 & 0 \\ 0 & 0 & \frac{1}{2}CM & 0 & 0 \end{bmatrix} \tag{5.200}$$

and

$$W(k) = \begin{bmatrix} 0 & 0 & \frac{1}{2}D(k) & 0 & 0 \\ 0 & 0 & \frac{1}{2}C(k) & 0 & 0 \\ \frac{1}{2}D(k) & \frac{1}{2}C(k) & F(k)M & \frac{1}{8}C(k)N - \frac{1}{4}L^TC(k) & \frac{1}{2}M^TC(k) \\ 0 & 0 & \frac{1}{8}N^TC(k) - \frac{1}{4}C(k)L & 0 & 0 \\ 0 & 0 & \frac{1}{2}C(k)M & 0 & 0 \end{bmatrix} \tag{5.201}$$

Then, the cost functional in equation (5.196) becomes

$$J(k) = E[\{x^T(K)\mu^T(K)\mathbf{H}x(K) - \lambda^T(K)x(K) - \mu^T(K)y(K)$$

$$- x^T(0)\mu^T(0)\mathbf{H}x(0) + \lambda^T(0)x(0) + \mu^T(0)y(0)]$$

$$+ \left[X^T(k)W(k)X(k) + Q^T(k)X(k) \right.$$

$$\left. - \sum_{k=1}^{K} (X^T(k)L(k)X(k) + R^T(k)X(k))\} \right] \tag{5.202}$$

The cost functional in equation (5.202) can be written as

$$J(k) = J_1(K) + J_2(k) \tag{5.203}$$

where

$$J_1(K) = E[x^T(K)\mu^T(K)\mathbf{H}x(K) - \lambda^T(K)x(K) - \mu^T(K)y(k)$$

$$- x^T(0)\mu^T(0)\mathbf{H}x(0) + \lambda^T(0)x(0) + \mu^T(0)y(0)] \tag{5.204}$$

and

$$J_2(k) = E\left[X^T(k)W(k)X(k) + Q^T(k)X(k) \right.$$

$$\left. - \sum_{k=1}^{K} \{X^T(k)L(k)X(k) + R^T(k)X(k)\} \right] \tag{5.205}$$

5.5.2.2. Optimal Equations

To minimize $J(k)$ in equation (5.203), one minimizes $J_1(K)$ and $J_2(k)$ separately, because the variables $x(K)$ and $y(K)$ in $J_1(K)$ are independent of the variables $X(k)$ in $J_2(k)$. The minimum of $J_1(K)$ is clearly achieved when

$$E[\lambda(K)] = [0] \tag{5.206}$$

and

$$E[\mu(K)] = [0] \tag{5.207}$$

Since $x(K)$ and $y(K)$ are free, $\delta x(K) \neq 0$, $\delta y(K) \neq 0$, and $x(0)$, $y(0)$ are fixed (we assume that the storages of the reservoirs are known at the beginning of the optimization interval as in the real case). Equations (5.206) and (5.207) give the values of Lagrange multipliers at the last period studied.

If one defines

$$A(k) = W(k) - KL(k) \tag{5.208}$$

$$P(k) = Q(k) - KR(k) \tag{5.209}$$

then the discrete integral part in (5.205) can be written as

$$J_2 = \sum_{k=1}^{K} J_2(k) = E\left[\sum_{k=1}^{K} \{X^T(k)A(k)X(k) + P^T(k)X(k)\} \right] \tag{5.210}$$

In the above equation, if $J_2(k)$ is a minimum (equal to zero), which is our objective, then the sum of $J_2(k)$, J_2, over the optimization interval is also a minimum.

Now define

$$V(k) = A^{-1}(k)P(k) \tag{5.211}$$

Then equation (5.210) can be written in the following form by a process similar to completing the squares:

$$J_2 = E\left[\sum_{k=1}^{K} \{(X(k) + \tfrac{1}{2}V(k))^T A(k)(X(k) + \tfrac{1}{2}V(k)) - \tfrac{1}{4}V^T(k)A(k)V(k)\} \right] \tag{5.212}$$

The last term of the above equation does not depend explicitly on $X(k)$. Thus one needs only to consider minimizing

$$J_2 = E\left[\sum_{k=1}^{K} \{(X(k) + \tfrac{1}{2}V(k))^T A(k)(X(k) + \tfrac{1}{2}V(k))\} \right] \tag{5.213}$$

Equation (5.213) defines a norm in Hilbert space. Hence, one can write equation (5.213) as

$$\min_{X(k)} J_2 = \min E\|X(k) + \tfrac{1}{2}V(k)\|_{A(k)} \tag{5.214}$$

The minimum of J_2 is clearly achieved when the norm in equation (5.214) is equal to zero:

$$E[X(k) + \tfrac{1}{2}V(k)] = [0] \tag{5.215}$$

Substituting from equations (5.208), (5.209), and (5.211) into equation (5.215), one obtains the optimal solution as

$$E\left[2\left(L(k) - \frac{1}{K}W(k)\right)X(k) + \left(R(k) - \frac{1}{K}Q(k)\right)\right] = [0] \tag{5.216}$$

Writing equation (5.216) explicitly and adding equations (5.158) and (5.168)-(5.170), one obtains

$$E[-x(k) + x(k-1) + I(k) + Mu(k) + Ms(k)] = [0] \tag{5.217}$$

$$E[x^T(k)\mathbf{H}x(k) - y(k)] = [0] \tag{5.218}$$

$$E[u^T(k)\mathbf{H}u(k) - z(k)] = [0] \tag{5.219}$$

$$E[u^T(k)\mathbf{H}x(k-1) - r(k-1)] = [0] \tag{5.220}$$

$$E[\lambda(k) - \lambda(k-1) + 2\mu^T(k-1)\mathbf{H}x(k-1) + (\Delta(k)$$
$$+ \psi^T(k)\mathbf{H})u(k) + \nu(k)] = [0] \tag{5.221}$$

$$E[\Gamma(k)u(k) - \mu(k-1)] = [0] \tag{5.222}$$

$$E[\beta(k) + M^T\lambda(k) + M^T\nu(k) + \sigma(k) + (\Delta(k) + \psi^T(k)\mathbf{H})x(k-1)$$
$$+ \Gamma(k)y(k-1) + M^T\Gamma(k)r(k-1) + 2(\theta(k)M + \phi^T(k)\mathbf{H})u(k)$$
$$+ (\tfrac{1}{4}\Gamma(k)N - \tfrac{1}{2}L^T\Gamma(k))z(k)] = [0] \tag{5.223}$$

$$E[-\phi(k) + \tfrac{1}{4}N^T\Gamma(k)u(k) - \tfrac{1}{2}\Gamma(k)Lu(k)] = [0] \tag{5.224}$$

$$E[-\psi(k) + \Gamma(k)Mu(k)] = [0] \tag{5.225}$$

where

$$\beta(k) = b(k) - (1/K)B(k)$$
$$\Delta(k) = d(k) - (1/K)D(k)$$
$$\Gamma(k) = C - (1/K)C(k) \tag{5.226}$$
$$\theta(k) = f(k) - (1/K)F(k)$$

Besides the above equations, one has the following Kuhn-Tucker exclusion equations that must be satisfied at the optimum:

$$e_{ij,k}(x_{ij}^m - x_{ij,k}) = 0 \tag{5.227}$$

$$e_{ij,k}^1(x_{ij,k} - x_{ij}^M) = 0 \tag{5.228}$$

$$g_{ij,k}(u_{ij,k}^m - u_{ij,k}) = 0 \tag{5.229}$$

$$g_{ij,k}^1(u_{ij,k} - u_{ij,k}^M) = 0 \tag{5.230}$$

One also has the following limits on the variables:

$$\begin{aligned}
&\text{If } x(k) > x^M, &&\text{then we put } x(k) = x^M \\
&\text{If } x(k) < x^m, &&\text{then we put } x(k) = x^m \\
&\text{If } u(k) > u^M(k), &&\text{then we put } u(k) = u^M(k) \\
&\text{If } u(k) < u^m(k), &&\text{then we put } u(k) = u^m(k)
\end{aligned} \qquad (5.231)$$

Equations (5.217)-(5.231) with equations (5.206) and (5.207) completely specify the optimal long-term scheduling of the system during the critical period. The next section explains the algorithm of solution used to solve these equations.

5.5.3. Algorithm of Solution

The data required to solve the above equations are the numbers of rivers (m), the number of reservoirs on each river (n_j), the expected value for the natural inflows, $E[I(k)]$, the initial storage, $x(0)$, and the required percentage load on the system during each month $a(k)$.

Step 1. Assume initial feasible values for reservoir releases $u(k)$, such that

$$u^m(k) \le u^i(k) \le u^M(k)$$

where i is the iteration number; set $i = 0$ and go to step 2.

Step 2. Assume first that $s(k)$ is equal to zero. Solve equations (5.217)–(5.220) and equations (5.222), (5.224), and (5.225) forward in stages with $x(0)$ given.

Step 3. Check the limits on $x(k)$. If $x(k) < x^m$, put $x(k) = x^m$, and if $x(k) > x^M$, put $x(k) = x^M$, and go to step 4. Otherwise go to step 10.

Step 4. With the new values of $x(k)$ obtained in step 3, calculate the new discharges from the equation

$$E[u(k)] = E[[M]^{-1}(x(k) - x(k-1) - I(k))]$$

and go to step 5.

Step 5. Check the limits on $u(k)$. If $u(k) < u^m(k)$, then put $u(k) = u^m(k)$, and if $u(k) > u^M(k)$, put $u(k) = u^M(k)$. Now, if $x(k) = x^M$ and $u(k) = u^M(k)$ go to step 6; otherwise go to step 10.

Step 6. Calculate the spill at month k as

$$E[s(k)] = E[[M]^{-1}(x(k) - x(k-1) - I(k)) - u^M(k)]$$

Step 7. With $\nu(k) = 0$, solve equation (5.221) backward in stages with equation (5.206) as terminal condition.

Step 8. Calculate the Kuhn–Tucker multiplier for $u(k)$, $\sigma(k)$, from the following equation:

$$\begin{aligned}
E[\sigma(k)] = E[&2M^T\mu^T(k-1)Hx(k-1) + (M^T\Delta(k) + M^T\psi^T(k)H)u(k) \\
&- \beta(k) - M^T\lambda(k-1) - (\Delta(k) + \psi^T(k)H)x(k-1) \\
&- \Gamma(k)y(k-1) - M^T\Gamma(k)r(k-1) \\
&- 2(\theta(k)M + \phi^T(k)H)u(k) \\
&- (\tfrac{1}{4}\Gamma(k)N - \tfrac{1}{2}L^T\Gamma(k))z(k)]
\end{aligned}$$

Step 9. Determine a new control iterate from the following equation:

$$E[u^{i+1}(k)] = E[u^i(k) - \alpha D^i u(k)]$$

where α is a positive scalar that is chosen with consideration to such factors as convergence, and

$$\begin{aligned}
E[Du(k)] = E[&\beta(k) + M^T\lambda(k) + \sigma(k) + (\Delta(k) + \psi^T(k)H)x(k-1) \\
&+ M^T\Gamma(k)r(k-1) + \Gamma(k)y(k-1) \\
&+ 2(\theta(k)M + \phi^T(k)H)u(k) \\
&+ (\tfrac{1}{4}\Gamma(k)N - \tfrac{1}{2}L^T\Gamma(k))z(k)]
\end{aligned}$$

Step 10. Check the limits of $u^{i+1}(k)$. If $u^{i+1}(k)$ satisfies the inequality

$$u^m(k) < u^{i+1}(k) < u^M(k)$$

go to step 14; otherwise put $u^{i+1}(k)$ to its limits and go to step 6.

Step 11. Solve the following equation forward in stages:

$$\begin{aligned}
E[\lambda(k-1)] = E[&2\mu^T(k-1)Hx(k-1) + (\Delta(k) + \psi^T(k)H)u(k) \\
&- [M^T]^{-1}\beta(k) - ([M^T]^{-1}\Delta(k) + [M^T]^{-1}\psi^T(k)H)x(k-1) \\
&- \Gamma(k)r(k-1) - [M^T]^{-1}\Gamma(k)y(k-1) \\
&- 2([M^T]^{-1}\theta(k)M + [M^T]^{-1}\phi^T(k)H)u(k) \\
&- (\tfrac{1}{4}[M^T]^{-1}\Gamma(k)N - \tfrac{1}{2}[M^T]^{-1}L^T\Gamma(k))z(k)]
\end{aligned}$$

Step 12. Determine Kuhn–Tucker multipliers for $x(k)$, $\nu(k)$, from the following equation:

$$E[\nu(k)] = E[-[M^T]^{-1}\{\beta(k) + M^T\lambda(k) + (\Delta(k) + \psi^T(k)\mathbf{H})x(k-1)$$
$$+ \Gamma(k)y(k-1) + M^T(k)r(k-1)$$
$$+ 2(\theta(k)M + \psi^T(k)\mathbf{H})u(k)$$
$$+ (\tfrac{1}{4}\Gamma(k)N - \tfrac{1}{2}L^T\Gamma(k))z(k)\}]$$

Step 13. Determine a new state iterate from the following equation:
$$E[x^{i+1}(k)] = E[x^i(k) - \alpha Dx^i(k)]$$

where

$$E[Dx(k)] = E[\lambda(k) - \lambda(k-1) + 2\mu^T(k-1)\mathbf{H}x(k-1)$$
$$+ (\Delta(k) + \psi^T(k)\mathbf{H})u(k) + \nu(k)]$$

Step 14. Repeat the calculation from step 3. Continue until the stage $x(k)$ and the control $u(k)$ do not change significantly from iteration to iteration and $J(k)$ in equation (5.157) is a minimum.

5.5.4. Practical Example

The algorithm of the last section has been used to determine the optimal monthly operation of a real system in operation consisting of two rivers $(m = 2)$; each river has two series reservoirs $(n_j = 2; j = 1, 2)$. The characteristics of the installations are given in Table 5.7.

Table 5.7. Characteristics of the Installations

Site name	Capacity of the reservoirs (Mm^3)	Minimum storage (x) (Mm^3)	Maximum effective discharge (m^3/sec)	Minimum effective discharge (m^3/sec)
R_{11}	24763	9949	1119	85.0
R_{21}	5304	3734	1583	85.0
R_{12}	74255	33195	1877	283.2
R_{22}	0	0	1930.3	283.2

	Reservoirs constants		
Site name	α_{ij} (MWh/Mm^3)	β_{ij} $[MWh/(Mm^3)^2]$	γ_{ij} $[MWh/(Mm^3)^3]$
R_{11}	212.11	146.956×10^{-4}	$-20503142.65 \times 10^{-14}$
R_{21}	117.20	569.71×10^{-4}	$-368119890.482 \times 10^{-14}$
R_{12}	232.46	359.449×10^{-4}	$-1603544.3196 \times 10^{-14}$
R_{22}	100.74	0	0

Table 5.8. Expected Monthly Inflows to the Reservoirs in the Critical Period[a]

Month k	$I_{11,k}$ (Mm³)	$I_{21,k}$ (Mm³)	$I_{12,k}$ (Mm³)	$I_{22,k}$ (Mm³)
1	796	372	1805	23
2	369	184	910	7
3	288	140	645	8
4	207	74	781	8
5	190	81	452	6
6	313	70	485	6
7	947	521	866	7
8	1456	849	3898	53
9	2833	1307	9175	73
10	4611	1714	4877	61
11	3148	895	1798	23
12	1285	426	1585	22
13	811	387	1320	15
14	363	183	1299	15
15	219	97	872	8
16	188	67	880	8
17	84	102	493	6
18	213	76	473	6
19	411	288	1740	22
20	1798	1024	4103	53
21	3428	1666	7120	88
22	2950	1168	3989	53
23	2700	834	2184	23
24	1798	683	1549	22
25	918	493	2124	23
26	545	321	1431	22
27	293	225	1024	15
28	227	193	727	9
29	190	185	647	8
30	236	146	497	6
31	241	157	559	7
32	1623	956	4194	53
33	3346	1637	7560	73
34	3572	1229	3534	46
35	2526	789	1737	23
36	1270	433	1262	15
37	658	275	1380	15
38	340	162	837	7
39	224	161	689	8
40	149	105	465	6
41	149	84	351	5
42	234	80	411	5
43	465	291	837	7

[a] These values are obtained at any month k from the following equation:

$$I^k = \sum_{x=1}^{N} (i_x \cdot p_x)$$

where i_x is the value of the random variable, p_x is the probability that this random variable takes a value of i_x, and N is the number of measurements during a month k.

If we let $d(k)$ denote the number of days in month k, then

$u_{ij,k}^m = 0.0864d(k)$ (minimum effective discharge in m^3/sec) Mm^3

$u_{ij,k}^M = 0.0864d(k)$ (maximum effective discharge in m^3/sec) Mm^3

where the minimum and maximum discharges are given in Table 5.7.

The expected natural inflows to the sites during the critical period are given in Table 5.8.

In Tables 5.9–5.12 we give the optimal releases from the turbines, the profits realized, and the percentage required load. From these tables one can observe that the calculated percentage load is equal to the given percentage load. Also, one can observe that the benefits during each year of the critical period are also equal. In Tables 5.13–5.16 we give the optimal storage.

The dimensions of this example during the critical period are as follows:

$$\text{Number of states} = 860$$

$$\text{Number of dual variables} = 1032$$

The computing time to get the optimal solution during the 43-month critical period was 6.85 sec in CPU on the Amdahl 470V/6 computer, which is very small compared to what has been done so far using other approaches.

Table 5.9. Optimal Releases from the Turbines, the Profits Realized, and the Percentage Load Required during the First Year of the Critical Period

Month k	$u_{11,k}$ (Mm^3)	$u_{21,k}$ (Mm^3)	$u_{12,k}$ (Mm^3)	$u_{22,k}$ (Mm^3)	Profits (MWh)	Percent load Calculated	Percent load Given
1	1470	2119	2494	2517	2,619,563	8.340	8.34
2	2038	2048	2404	2411	2,786,490	8.871	8.87
3	1454	2552	2992	3001	2,949,777	9.391	9.39
4	1842	2424	3107	3114	3,068,649	9.769	9.77
5	1764	1837	2957	2963	2,748,784	8.751	8.75
6	1849	1906	3091	3097	2,839,532	9.040	9.04
7	1573	2064	2614	2622	2,510,435	7.992	7.99
8	1498	2079	2488	2541	2,430,745	7.738	7.74
9	1420	1966	2355	2428	2,352,717	7.490	7.49
10	1232	2455	2042	2103	2,333,564	7.429	7.43
11	1314	2209	2192	2215	2,399,492	7.639	7.64
12	1364	1906	2275	2297	2,371,382	7.549	7.55

Total benefits from the generation during the
first year of the critical period: 32,411,168

Table 5.10. Optimal Releases from the Turbines, Profits Realized, and the Percentage Load Required during the Second Year of the Critical Period

Month k	$u_{11,k}$ (Mm³)	$u_{21,k}$ (Mm³)	$u_{12,k}$ (Mm³)	$u_{22,k}$ (Mm³)	Profits (MWh)	Percent load	
						Calculated	Given
1	1491	2150	2529	2544	2,619,398	8.339	8.34
2	2077	2078	2438	2453	2,785,707	8.869	8.87
3	1475	2594	3041	3049	2.949,907	9.391	9.39
4	1910	2318	3221	3229	3,069,785	9.773	9.77
5	1793	1888	3007	3013	2,748,036	8.749	8.75
6	1872	1935	3179	3185	2,839,552	9.040	9.04
7	1653	1913	2751	2773	2,509,918	7.991	7.99
8	1529	2126	2542	2595	2,431,018	7.739	7.74
9	1435	2006	2383	2471	2,352,861	7.491	7.49
10	1267	2435	2101	2154	2,333,571	7.429	7.43
11	1367	2201	2268	2291	2,399,486	7.639	7.64
12	1370	2053	2288	2310	2,371,875	7.551	7.55

Total benefits from the generation during the
second year of the critical period: 31,411,056

Table 5.11. Optimal Releases from the Turbines, Profits Realized, and the Percentage Load Required during the Third Year of the Critical Period

Month k	$u_{11,k}$ (Mm³)	$u_{21,k}$ (Mm³)	$u_{12,k}$ (Mm³)	$u_{22,k}$ (Mm³)	Profits (MWh)	Percent load	
						Calculated	Given
1	1529	2184	2579	2602	2,619,493	8.339	8.34
2	2049	2251	2487	2509	2,786,245	8.870	8.87
3	1513	2630	3091	3106	2,949,134	9.389	9.39
4	1879	2698	3173	3182	3,068,526	9.769	9.77
5	1830	2023	3077	3085	2,748,968	8.752	8.75
6	1771	1909	3461	3467	2,840,465	9.143	9.04
7	255	390	4796	4803	2,509,759	7.990	7.99
8	1583	2203	2633	2686	2,432,202	7.743	7.74
9	1488	2066	2473	2546	2,353,636	7.493	7.49
10	1324	2407	2199	2245	2,332,929	7.427	7.43
11	1410	2199	2355	2378	2,399,289	7.638	7.64
12	1436	2012	2399	2414	2,370,568	7.547	7.55

Total benefits from the generation during the
third year of the critical period: 3,141,152

Table 5.12. Optimal Releases from the Turbines, Profits Realized, and the Percentage Load Required during the Rest of the Critical Period

Month k	$u_{11,k}$ (Mm³)	$u_{21,k}$ (Mm³)	$u_{12,k}$ (Mm³)	$u_{22,k}$ (Mm³)	Profits (MWh)	Percent load	
						Calculated	Given
1	779	1441	3519	3534	2,466,724	8.393	8.34
2	1180	2203	3099	3106	2,602,812	8.856	8.87
3	1444	1785	3579	3587	2,756,584	9.379	9.39
4	1207	1313	4373	4379	2,868,556	9.769	9.77
5	1624	1608	3314	3319	2,559,364	8.710	8.75
6	683	862	4826	4831	2,675,327	9.102	9.04
7	465	755	4410	4417	2,337,184	7.956	7.99

Total benefits from the generation during the
rest of the critical period: 18,266,544

Table 5.13. Optimal Reservoirs Storage during the First Year of the Critical Period

Month k	$x_{11,k}$ (Mm³)	$x_{21,k}$ (Mm³)	$x_{12,k}$ (Mm³)
1	24088	5027	73565
2	22419	5120	72071
3	21253	4242	69724
4	19618	3734	67398
5	18044	3742	64892
6	16507	3755	62285
7	15881	3784	60537
8	15838	4053	61947
9	17252	4813	68766
10	20630	5304	71600
11	22463	5304	71205
12	22384	5187	70515

Table 5.14. Optimal Reservoirs Storage during the Second Year of the Critical Period

Month k	$x_{11,k}$ (Mm^3)	$x_{21,k}$ (Mm^3)	$x_{12,k}$ (Mm^3)
1	21704	4915	69305
2	19989	5097	68165
3	18733	4075	65996
4	17011	3734	63654
5	15401	3742	61139
6	13742	3754	58433
7	12499	3782	57421
8	12766	4210	58982
9	14759	5304	63718
10	15442	5304	65606
11	17775	5304	65521
12	18203	5304	64810

Table 5.15. Optimal Reservoirs Storage during the Third Year of the Critical Period

Month k	$x_{11,k}$ (Mm^3)	$x_{21,k}$ (Mm^3)	$x_{12,k}$ (Mm^3)
1	17591	5142	64325
2	16085	5262	63269
3	14865	4369	610202
4	13213	3743	58755
5	11572	3734	56324
6	10037	3741	53360
7	10022	3764	49123
8	10061	4100	50683
9	11919	5158	55770
10	14167	5304	57105
11	15282	5304	65486
12	15115	5161	55348

Table 5.16. Optimal Reservoirs Storage during the Rest
of the Critical Period

Month k	$x_{11,k}$ (Mm³)	$x_{21,k}$ (Mm³)	$x_{12,k}$ (Mm³)
1	14994	4774	53210
2	14153	3914	50947
3	12932	3734	48057
4	11874	3734	44148
5	10399	3833	41184
6	9949	3734	36769
7	9949	3734	33196

5.5.5. Concluding Remarks

In the solution presented in this section we have presented an efficient approach for solving the long-term optimal operating problem of series-parallel reservoirs for the critical period with specified monthly load; this load is equal to a certain percentage of the total generation at each year of the critical period. The generating function used is a nonlinear function of the discharge and the average storage between two successive months k, $k - 1$. The resulting problem is a highly nonlinear problem; we defined a set of pseudostate variables to overcome these nonlinearities.

The proposed approach takes into account the stochasticity of the river flows; we assume that their probability properties were preestimated from past history. We use the expected values for the random inflow. The proposed approach also has the ability to deal with large-scale coupled power systems. Most of the current approaches either use composite reservoirs or do not consider the coupling of reservoirs.

The dimensions of the given example consisting of two rivers, each river having two series reservoirs and a 43-month critical period, are as follows:

Number of states = 860

Number of dual variables = 1032

The computing time to get the optimal operation for this example was 6.85 sec in CPU units on the Amdahl 470V/6 computer, which is very small compared to what has been done so far using other approaches.

References

5.1. ARVANITIDIES, N. V., and ROSING, J., "Optimal Operation of Multireservoir Systems Using a Composite Representation," *IEEE Transactions on Power Apparatus and Systems* **PAS-89**(2), 327–335 (1970).

5.2. TURGEON, A., "A Decomposition Method for the Long-Term Scheduling of Reservoirs in Series," *Water Resources Research* **17**(6), 1565–1570 (1981).

5.3. TURGEON, A., "Optimal Operation of Multireservoir Power Systems with Stochastic Inflows," *Water Resources Research* **16**(2), 275–283 (1980).

5.4. SHERKAT, V. T., CAMPO, R., MOSLEHI, K., and LO, E. O., "Stochastic Long-Term Hydrothermal Optimization for a Multireservoir System," *IEEE Transactions on Power Apparatus and Systems* **PAS-108**(8), 2040–2050 (1985).

5.5. SHAMALY, A., *et al.*, "A Transformation for Necessary Optimality Conditions for Systems with Polynomial Nonlinearities," *IEEE Transactions on Automatic Control* **AC-24**(6), 983–985 (1979).

5.6. SHAMALY, A., CHRISTENSEN, G. S., and EL-HAWARY, M. E., "Optimal Control of Large Turboalternator," *Journal of Optimization Theory and Applications* **34**(1), 83–97 (1981).

5.7. CHRISTENSEN, G. S., and SOLIMAN, S. A., "Long-Term Optimal Operation of a Parallel Multireservoir Power System Using Functional Analysis," *Journal of Optimization Theory and Applications* **50**(3), 000–000 (1986).

5.8. SOLIMAN, S. A., *et al.*, "Optimal Operation of Multireservoir Power System Using Functional Analysis," *Journal of Optimization Theory and Applications* **49**(3), 449–461 (1981).

5.9. SOLIMAN, S. A., and CHRISTENSEN, G. S., "Modelling and Optimization of Parallel Reservoirs Having Nonlinear Storage Curves under Critical Water Conditions for Long-Term Regulation Using Functional Analysis," *Journal of Optimization Theory and Applications* **55**(3) (1987).

5.10. SAGE, A. P., and WHITE, C. C., *Optimum System Control*, Prentice-Hall, Englewood Cliffs, New Jersey, 1977.

5.11. OLCER, S., HARSA, C., and ROCH, A., "Application of Linear and Dynamic Programming to the Optimization of the Production of Hydroelectric Power," *Optimal Control Application of Methods* **6**, 43–56 (1985).

5.12. MYRON, B. F., "Multivariable Techniques for Synthetic Hydrology," *Journal of Hydro Division, Proc. ASCE*, 43–50 (1964).

5.13. DAVIS, R. E., and PRONOVOST, R., "Two Stochastic Dynamic Programming Procedures for Long-Term Reservoir Management," *IEEE PAS Summer Meeting* **C72**, 493–495, July 9–14, 1972.

5.14. SAGE, A., and MELSA, J. L., *System Identification*, Academic Press, New York, 1971.

5.15. SJELVGREN, D., ANDERSON, S., and DILLON, T. S., "Optimal Operations Planning in a Large Hydrothermal Power System," *IEEE Transactions on Power Apparatus and Systems* **PAS-102**(11), 3644–3651 (1983).

6

Optimization of the Firm Hydro Energy Capability for Hydroelectric Systems

6.1. Introduction (Refs. 6.28–6.39)

In Chapter 5, we maximized the generation from hydroelectric power systems having a specified monthly load under critical water conditions. This load is equal to a certain percentage of the total generation at the end of each year of the critical period. The GWh for this load during each month varies according to the total generation at the end of the optimization period.

This chapter is devoted to the maximization of the firm hydro energy capability from multireservoir power systems under critical water conditions. In this chapter the load on the system is fixed, i.e., there is no relation between the load and the total generation at the end of optimization interval, and at the same time this firm energy should be uniform during each year of the critical period.

In the first section of this chapter, maximization of energy capability is formulated as a general nonlinear program. Certain soft constraints for which constraints parameter settings may eliminate the feasible region are accounted for. The objective function is modified by adding penalty functions; this results in a program with a nonlinear objective function, linear constraint equations, and simple bounds. In the second section functional analysis and minimum norm formulation have been used to maximize the surplus energy from a hydroelectric power system under critical water conditions. The optimal control problem is formulated by constructing a cost function in which we maximize the surplus energy and the difference between the total generation during a certain period and the load on the system during that period, and at the same time this surplus energy should be uniform over the planning period.

The shortest time period used in the modeling is half a month. Therefore, short-range hydraulic and electrotechnic effects are not taken into consideration.

6.2. Nonlinear Programming Model (Hicks *et al.* Approach) (Refs. 6.1–6.27)

Computing maximum energy capability of the hydro system is a discrete time problem, concerned with seasonal management of reservoir storage, provided that the streamflow period is specified, as are the hydro system loads, generator availability, irrigation, navigation, and flood control requirements. The problem is to determine, for all reservoirs, the discharge from the system that results in maximum energy capability for the system, and, at the same time, such that the system constraints are satisfied and with acceptable uniformity in the surplus of generation over load each month.

This section presents the formulation and solution of the maximization of energy capability in terms of nonlinear programming (NLP) methods.

6.2.1. System Modeling and Relationships

We assume that the optimization period consists of $k = 1, 2, \ldots, K$ discrete time intervals. Most of these intervals are one month, while the other intervals are half a month. The times of water travel from upstream reservoir to downstream reservoirs are assumed to be shorter than a month, ranging from 0 to about 24 hr; for this reason these times are neglected.

The system is assumed to have n hydro plants. m of these plants are run-of-river plants. Thus, plant interactions through the common river system are accounted for in the following water conservation equations.

For run-of-river plants:

$$u_{ij,k} = I_{ij,k} - s_{ij,k} + \sum_{e \in \alpha} u_{ej,k}, \qquad i = 1, \ldots, m, \qquad j = 1, \ldots, r,$$

$$k = 1, \ldots, K \tag{6.1}$$

where $u_{ij,k}$ is the average outflow from run-of-river plant i on river j in time interval k; $I_{ij,k}$ is the average natural inflow into plant i on river j in time interval k; α is the set of plants immediately upstream from plant i on river j; and $s_{ij,k}$ is the average spill from plant i on river j during a period k, and is given by

$$s_{ij} = \begin{cases} \left(I_{ik,k} + \sum_{e \in \alpha} u_{ej,k} \right) - u_{ij,k}^{M} & \text{if } \left(I_{ij,k} + \sum_{e \in \alpha} Q_{ej,k} \right) > u_{ij,k}^{M} \\ 0 & \text{otherwise} \end{cases} \tag{6.2}$$

For reservoir plants:

$$x_{ij,k} = x_{ij,k-1} + I_{ij,k} + \sum_{e \in \beta_i} u_{ej,k} - s_{ij,k} - u_{ij,k},$$

$$i = 1, \ldots, n_j - m_j, \qquad j = 1, \ldots, r, \qquad k = 1, \ldots, K \tag{6.3}$$

where $x_{ij,k}$ is the water content of reservoir plant i on river j at the end of time interval k; $I_{ij,k}$ is the average side-flow into plant i on river j during a time interval k from rainfall and snow melt; $u_{ij,k}$ is the average outflow from reservoir i on river j in time interval k; $x_{ij,0}$ is the beginning water content of reservoir plant i on river j; β_i is the set of plants immediately upstream from plant i; and $s_{ij,k}$ is the average spill from reservoir i on river j during a time interval k, and is given by

$$s_{ij,k} = \begin{cases} \left(x_{ij,k-1} + I_{ij,k} + \sum_{e \in \beta_i} u_{ej,k} - x_{ij,k} \right) - u^M_{ij,k} \\ \quad \text{if } \left(x_{ij,k-1} + I_{ij,k} + \sum_{e \in \beta_i} u_{ej,k} - x_{ij,k} \right) > u^M_{ij,k} \quad \text{and} \quad x_{ij,k} = x^M_{ij} \\ 0 \quad \text{otherwise} \end{cases}$$

The water conservation equations (6.1) and (6.3) can be written as a single matrix equation

$$AU + BX + C = 0 \tag{6.4}$$

where the matrix A is square and of order $[N \times K]$, $N = \sum_{j=1}^{r} n_j$, and B is rectangular with $[N \times K]$ rows and $[M \times K]$ columns, $M = \sum_{j=1}^{r} n_j - m_j$. Both matrices are sparse matrices with element values of 1, 0, −1; their element distribution is dependent upon the ordering of the $u_{ij,k}$, $i = 1, \ldots, n_j$, $j = 1, \ldots, r, k = 1, \ldots, K$ and C is the matrix of the inflows corresponding to the continuity equations.

The total power produced by the system is equal to the sum of the power produced by the reservoir plants and that produced by the run-of-river plants, and this can be expressed by

$$P_k = \sum_{j=1}^{r} \left[\sum_{i=1}^{n_j - m_j} P_{ij,k} + \sum_{i=1}^{m_j} P_{ij,k} \right] \tag{6.5}$$

where $\sum_{i=1}^{m_j} P_{ij,k}$ is the total power produced by run-of-river plants on river j during a time interval k; $\sum_{i=1}^{n_j - m_j} P_{ij,k}$ is the total power produced by reservoir plants on river j during a time interval k. The total energy from the system should supply the demands of the system. Average transmission losses are accounted for in the system loads. Variations in these losses are negligible compared to power generation variations resulting from storage management changes.

System average energy capability is defined by

$$P = \frac{1}{K} \sum_{k=1}^{K} P_k T_k \qquad (6.6)$$

where

$$K = \sum_{k=1}^{K} T_k$$

and T_k is the length of time interval k.

The power generated from reservoir plants is a nonlinear function of the system variables, the storage, the head, the tailwater, and the forebay elevations. The nonlinear relationships have been established by semiempirical means and are expressed in the form of tables and approximating functions.

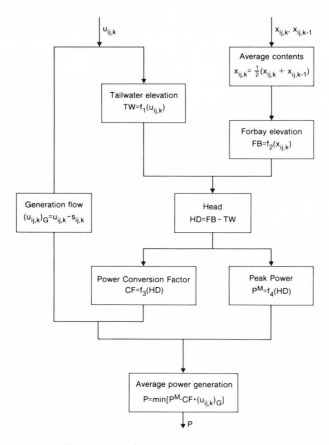

Figure 6.1. Typical reservoir plant generation model.

Figure 6.1 gives a typical model for reservoir plants. Head variations due to changes in the forebay and tailwater elevations are accounted for in this model.

If one or more of the following conditions exist, the model will be more complicated:

1. The forebay elevation is less than the elevation of the reservoir because of a flow restriction between the two;
2. The plant has a full-gate flow restriction affecting the power conversion factor (CF);
3. The tailwater elevation is dependent upon the downstream reservoir water elevation.

Additional table and approximating functions are used to account for these added complications.

In general reservoir plant generation can be expressed in the functional form as

$$P_{ij,k} = P_{ij}\left(u_{ij,k}, x_{ij,k}, x_{ij,k-1}, \sum_{e \in \beta_i} u_{ej,k} \right)$$

$$i = 1, \ldots, n_j - m_j, \qquad j = 1, \ldots, r, \qquad k = 1, \ldots, K \tag{6.7}$$

Power generation at a run-of-river plant is a function of the discharge of that plant only; as

$$P_{ij,k} = P_{ij}(u_{ij,k}), \qquad i = 1, \ldots, m_j, \qquad j = 1, \ldots, r \tag{6.8}$$

this function is obtained from a table—one for each plant. Note that for a constant head hydro plant, or a system with a small head variation, the power generated in either equation (6.7) (or (6.8) is equal to a constant times the discharge, and this constant is the water conversion factor.

6.2.2. Optimization Objective and Constraints

The optimization objective is to find the discharge $u_{ij,k}$, and consequently, a storage management schedule that maximizes the system energy capability as defined in (6.6), with acceptable uniformity in the surplus of power over load for each time interval. This objective can be expressed as

$$\min\left[F(X, U) = D + W \sum_{k=1}^{K} (D_k - D)^2 \right] \tag{6.9}$$

where $D_k = L_k - P_k$, L_k is the total system load for interval k,

$$D = \frac{1}{K} \sum_{k=1}^{K} D_k T_k$$

and where $W > 0$ is a suitable weight.

In order to be realizable and also to satisfy multipurpose stream use requirements, the storage management schedule must satisfy the following.

1. The water conservation equations

$$AU + BX + C = 0 \qquad (6.10)$$

It must also satisfy the following inequality constraints:

2. Upper and lower bounds on reservoir plant outflows

$$u_{ij,k}^{m} \leq u_{ij,k} \leq u_{ij,k}^{M}, \qquad i = 1, \dots, n_j - m_j, \qquad j = 1, \dots, r \qquad (6.11)$$

3. Lower bounds for run-of-river plant outflows

$$u_{ik,k}^{m} \leq u_{ij,k}, \qquad i = 1, \dots, m_j, \qquad j = 1, \dots, r \qquad (6.12)$$

4. Upper and lower bounds on reservoir contents

$$x_{ij}^{m} \leq x_{ij,k} \leq x_{ij}^{*} \leq x_{ij}^{M} \qquad (6.13)$$

where x_{ij}^{*} is the storage required for desired flood control levels.

5. Upper bounds on the decrease in reservoir water elevations

$$E_{ij,k-1} - E_{ij,k} \leq T_k \Delta E_{ij}^{M} \qquad (6.14)$$

where $E_{ij,k}$ is a nonlinear function of the contents $x_{ij,k}$ and is expressed by an empirical table and ΔE_{ij}^{M} is a given constant.

6. Upper bounds on plant generations

$$P_{ij,k} \leq 0.85 P_{ij,k}^{M}, \qquad i = 1, \dots, n_j, \qquad j = 1, \dots, r \qquad (6.15)$$

where $P_{ij,k}$ is a nonlinear function as given by equation (6.7).

Now, the optimization problem can be stated in a vector form as follows: Minimize

$$F(X, U) \qquad (6.16)$$

subject to

$$AU + BX + C = 0 \qquad (6.17)$$

and

$$u^{m} \leq U \leq u^{M} \qquad (6.18)$$

$$x^{m} \leq X \leq x^{*} \leq x^{M} \qquad (6.19)$$

$$h(X) \leq \Delta E^{M} \qquad (6.20)$$

$$P(X, U) \leq 0.85 P^{M}(X, U) \qquad (6.21)$$

where all the vectors in equations (6.17)–(6.21) are nonlinear functions of their arguments. The constraints on those equations are divided into "hard" and "soft" constraints, hard constraints being those that are due to physical limitations and that cannot be violated at any instance. For this problem, the hard constraints are the continuity equation and reservoir storage bounds. All of the remaining constraints are soft.

6.2.3. Method of Solution

The inequality constraints, soft constraints, are adjointed to the cost functional in equation (6.16) via penalty terms. The penalty term for any bounded variable, say $x \le x^M$, can be written as

$$V = \{\max[(x - x^M), 0]\} \tag{6.22}$$

Lower bounds are similarly defined. Thus, the modified objective function is

$$f(X, U) = F(X, U) + \sum_l W_l V_l \tag{6.23}$$

where l ranges over all constraints, and where $W_l > 0$ are suitable weights.

The modified objective function can be simplified by eliminating the discharge U from equation (6.17). Assume for the moment that the matrix A is nonsingular; then U from equation (6.17) can be written as

$$U = -A^{-1}(BX + C) \tag{6.24}$$

and the objective function in equation (6.23) becomes

$$\min\{f[X, U(x)]\} \tag{6.25}$$

subject to

$$x^m \le X \le x^M \tag{6.26}$$

where U and X are related to each other by (6.24).

The conjugate gradient method is used to solve the above objective function and is modified to satisfy (6.26), because of its low storage requirements and relatively good convergence properties. In this method, it is required to calculate the gradient of f with respect to X, which is given by

$$\nabla f_X = \frac{\nabla f}{\partial X} + \left(\frac{\partial U}{\partial X}\right)^T \frac{\partial f}{\partial U} \tag{6.27}$$

where $\partial U / \partial X$ is given by, from equation (6.24),

$$\left(\frac{\partial U}{\partial X}\right)^T = -B^T(A^T)^{-1} \tag{6.28}$$

Now the vector ∇f_x can be computed conveniently in two steps:

1. Solve the linear system equation

$$A^T \lambda = -\frac{\partial f}{\partial U} \tag{6.29}$$

for the Lagrange multipliers λ

2. and then, calculate

$$\nabla f_X = \frac{\partial f}{\partial X} + B^T \lambda \tag{6.30}$$

The following are the main steps of the solution:
(i) Start with an initial guess for X^0 satisfying

$$x^m \leq X^0 \leq x^M$$

and set $q = 0$, iteration number.
(ii) Solve

$$AU^q + BX^q + C = 0$$

for U^q.
(iii) Solve

$$A^T \lambda^q = -\frac{\partial f}{\partial U}$$

for λ^q, where $\partial f / \partial X$ is evaluated at U^q, X^q.
(iv) Compute

$$\nabla f_X^q = \frac{\partial f}{\partial X} + B^T \lambda^q$$

where $\partial f / \partial X$ is evaluated at U^q, X^q.
(v) Compute the conjugated search direction

$$\alpha^q = \begin{cases} -\nabla f_X^q + \dfrac{\|\nabla f_X^q\|^2}{\|\nabla f_X^{q-1}\|^2} \alpha^{q-1} & \text{for } q > 1 \\ -\nabla f_X^q & \text{for } q = 0 \end{cases}$$

where $\| \cdot \|$ is the norm.
(vi) Project the components of α^q onto the bounds for X

$$\alpha_r^q = \begin{cases} 0 & \text{if } (X_r^M - X_r^q \leq \varepsilon) \quad \text{and} \quad \alpha_r^q > 0 \\ 0 & \text{if } (X_r^q - X_r^m \leq \varepsilon) \quad \text{and} \quad \alpha_r^q < 0 \\ \alpha_r^q & \text{otherwise} \end{cases}$$

where r (the vector component) $= 1, 2, \ldots$, and where $\varepsilon > 0$.
(vii) Find the maximum step length θ^{max} that can be taken from X^q in direction α^q without violating the X bounds:

$$\theta^{max} = \min\left[\min_{\alpha_r^q > 0} \left(\frac{X_r^M - X_r^q}{\alpha_r^q} \right), \min_{\alpha_r^q < 0} \left(\frac{X_r^m - X_r^q}{\alpha_r^q} \right) \right]$$

(viii) Find $\theta^q \leq \theta^{max}$ such that

$$f[X^{q+1}, U(X^{q+1})] < f[X^q, U(X^q)]$$

where

$$X^{q+1} = X^q + \alpha^q \theta^q$$

and where $U(X^q) = U^q$ as obtained in step (ii).

(ix) Increase q by one and repeat the calculation starting with (ii).

Notes on Method of Solution

(1) The vectors of partial derivatives are obtained through the chain rule using slopes of the tabulated functions.

(2) The A and B matrices are introduced in the foregoing descriptions only for notational convenience; explicit computer storage and manipulation with these matrices is neither necessary nor desirable. Thus, equations (6.1) and (6.3) symbolized by the matrix equation in step (ii) can be solved by an appropriate ordering of the plants.

(3) The solution to the matrix equation in step (iii) can also be obtained in a simple manner. With the foregoing downstream plant ordering, the A matrix is unit lower-triangular, and the process of solving for the discharge is in effect a sequence of forward substitutions. Thus, the Lagrange multipliers can be calculated through a sequence of backward substitutions.

(4) For any row of A, the nonzero off-diagonal elements correspond to plants immediately upstream of the plant represented by the diagonal element. As a consequence, for any column of A (row of A^T), nonzero off-diagonal elements correspond to the plant immediately downstream of the diagonal element plant.

(5) Unique solutions to the foregoing equations exist by virtue of the fact that the A matrix is unit lower-triangular and therefore nonsingular. Furthermore, since the off-diagonal elements of A represent the sum of the outflows of upstream plant, severe cancelation cannot occur.

(6) The projection step (vi) uses $\varepsilon > 0$ to prevent "zigzagging" and also to prevent convergence to a nonoptimal point.

6.2.4. Practical Example

The algorithm described above was programmed for a CDC 6400 with extended core storage (ECS). The program requires approximately 55,000 octal words of central memory and 160,000 octal words of ECS.

The following is a typical size problem:

        13 reservoir plants
        10 control reservoirs
        26 run-of-river plants
        49 time periods

This problem has 2401 variables, 1911 equality constraints, and 4312 inequality constraints.

The problem required 2816 system evaluation and 41 minutes of CPU computer time unit. The program obtained the following results:

	Start	End
Average power deficit (MW)	-93.6	-109.7
Nonuniformity $\sum (D_j - D)^2$	6.10×10^7	4.0
Constraint violations $\sum W_l V_l$	1.03×10^9	5.61×10^5

Solutions are typically obtained in from 0.7 to 3 evaluations of the system performance (including constraint violations) per system available. The number of variables is defined here as all of the plant outflows for each time period and the control reservoir storage contents for each time period. A single evaluation of the system performance requires about 400 μsec per variable.

6.2.5. Conclusion

In this section, it has been shown how to formulate the problem of maximizing energy capability of multireservoir power systems as a nonlinear programming problem. In formulating the problem, the soft constraints are transferred to the objective function as penalty terms. The resulting problem was then solved by the conjugate gradient method extended to handle the remaining hard constraints. This method was adopted to take advantage of the special problem structure and was implemented on a digital computer. A very large practical example has been discussed to show the main features of the proposed algorithm.

6.3. A Minimum Norm Approach

In this section functional analysis and minimum norm formulation have been used to maximize the surplus energy from a hydroelectric power system under critical water conditions. The optimal control problem is formulated by constructing a cost function in which we maximize the surplus energy (the difference between the total generation during a certain month and the load on the system during that month), and at the same time this surplus energy should be uniform over the planning period. The cost functional is augmented by using Lagrange and Kuhn–Tucker multipliers to adjoin the equality and inequality constraints. The model used in this section for the water conversion factor is a linear model of the average storage, and the model used for each reservoir generation is a nonlinear model of the discharge through the turbines and the average storage between two successive periods to avoid underestimation for rising water levels and overestimation for falling water levels. Because of the nature of the planning objective, deterministic critical period type hydrological boundary data are used.

6.3.1. Optimization Objective and Constraints

Given the hydroelectric system in Figure 4.1, the problem of this system is to find the discharge $u_{ij,k}$, $i = 1, \ldots, n_j$, $j = 1, \ldots, m$, $k = 1, \ldots, K$ as a function of time over the optimization interval that satisfies the following conditions:

1. The system expected surplus energy over the load is maximum and uniform.

2. The elevation-storage curve may be described by

$$h_{ij,k} = \gamma_{ij} + \delta_{ij} x_{ij,k} \tag{6.31}$$

where γ_{ij} and δ_{ij} are constants; these were obtained by least-squares curve fitting to typical plant data available.

3. The water conversion factor, MWh/Mm^3, as a function of the storage is given by

$$C_{ij,k} = a_{ij} + b_{ij} x_{ij,k} \tag{6.32}$$

In equations (6.31) and (6.32), we assume that the tailwater elevation is constant independent of the discharge.

4. The water conservation equation for each reservoir may be adequately described by the continuity-type equation

$$x_{ij,k} = x_{ij,k-1} + I_{ij,k} - u_{ij,k} - s_{ij,k} + \sum_{\sigma \in R_i} u_{\sigma j,k} + \sum_{\sigma \in R_i} s_{\sigma j,k} \tag{6.33}$$

where R_i is the set of plants immediately upstream from plant i.

5. In order to be realizable, and also to satisfy multipurpose stream use requirements, such as flood control, navigation, fishing, and other purposes if any, the storage management schedule must satisfy the following inequality constraints:

(a) Upper and lower bounds on reservoir contents

$$x_{ij}^m \leq x_{ij,k} \leq x_{ij}^M \tag{6.34}$$

(b) Upper and lower bounds on reservoir plant outflow

$$u_{ij,k}^m \le u_{ij,k} \le u_{ij,k}^M \qquad (6.35)$$

6. The hydro power generating function can be expressed as

$$G_{ij}(u_{ij,k}, \tfrac{1}{2}(x_{ij,k} + x_{ij,k-1})) = \alpha_{ij}u_{ij,k} + \tfrac{1}{2}\beta_{ij}u_{ij,k}(x_{ij,k} + x_{ij,k-1}) \text{ MWh} \quad (6.36)$$

Substituting from equation (6.33) into equation (6.36) for $x_{ij,k}$, one obtains

$$G_{ij,k} = b_{ij,k}u_{ij,k} + \beta_{ij}r_{ij,k} + \tfrac{1}{2}\beta_{ij}w_{ij,k} \qquad (6.37)$$

where

$$\alpha_{ij} = a_{ij} + b_{ij}\gamma_{ij} \qquad (6.38a)$$

$$\beta_{ij} = b_{ij}\delta_{ij} \qquad (6.38b)$$

$$b_{ij,k} = \alpha_{ij} + \tfrac{1}{2}\beta_{ij}\left(I_{ij,k} - s_{ij,k} + \sum_{\sigma \in R_i} s_{\sigma j,k}\right) \qquad (6.38c)$$

and the following are pseudovariables:

$$r_{ij,k} = u_{ij,k}x_{ij,k-1} \qquad (6.39)$$

$$w_{ij,k} = u_{ij,k}\left(\sum_{\sigma \in R_i} u_{\sigma j,k} - u_{ij,k}\right) \qquad (6.40)$$

In mathematical terms, the optimization objective is to find the discharge $u_{ij,k}$ that maximizes

$$J = \sum_{k=1}^K \left(\sum_{i=1}^{n_j}\sum_{j=1}^m G_{ij,k} - D_k\right)^2 \qquad (6.41)$$

subject to satisfying the equality constraints given by equations (6.33), (6.39), and (6.40) and the inequality constraints given by equations (6.34) and (6.35). In equation (6.41) D_k is the firm monthly load on the system during period k.

If one substitutes from equation (6.37) into equation (6.41), one obtains the objective function as

$$J = \sum_{k=1}^K \left[\sum_{j=1}^m\sum_{i=1}^{n_j}(b_{ij,k}u_{ij,k} + \beta_{ij}r_{ij,k} + \tfrac{1}{2}\beta_{ij}w_{ij,k}) - D_k\right]^2 \qquad (6.42)$$

6.3.2. A Minimum Norm Formulation

In this section, we explain the minimum norm formulation in the framework of functional analysis. The augmented cost functional, $\tilde{J}$, is obtained by adjoining to the cost functional in (6.42) the equality constraints (6.33), (6.39), and (6.40) via Lagrange multipliers, and the inequality constraints (6.34) and (6.35) via Kuhn–Tucker multipliers. One thus obtains

$$
\begin{aligned}
\tilde{J} = \sum_{k=1}^{K} &\left\{ \left(\sum_{j=1}^{m} \sum_{i=1}^{n_j} b_{ij,k} u_{ij,k} \right)^2 + \left(\sum_{j=1}^{m} \sum_{i=1}^{n_j} \beta_{ij} r_{ij,k} \right)^2 \right. \\
&+ \tfrac{1}{4} \left(\sum_{j=1}^{m} \sum_{i=1}^{n_j} \beta_{ij} w_{ij,k} \right)^2 + (D_k)^2 \\
&+ 2 \left(\sum_{j=1}^{m} \sum_{i=1}^{n_j} b_{ij,k} u_{ij,k} \right) \left(\sum_{j=1}^{m} \sum_{i=1}^{n_j} \beta_{ij} r_{ij,k} \right) \\
&+ \left(\sum_{j=1}^{m} \sum_{i=1}^{n_j} b_{ij,k} u_{ij,k} \right) \left(\sum_{j=1}^{m} \sum_{i=1}^{n_j} \beta_{ij}, w_{ij,k} \right) \\
&- 2D_k \left(\sum_{j=1}^{m} \sum_{i=1}^{n_j} b_{ij,k} u_{ij,k} \right) + \left(\sum_{j=1}^{m} \sum_{i=1}^{n_j} \beta_{ij} r_{ij,k} \right) \left(\sum_{j=1}^{m} \sum_{i=1}^{n_j} \beta_{ij} w_{ij,k} \right) \\
&- 2D_k \left(\sum_{j=1}^{m} \sum_{i=1}^{n_j} \beta_{ij} r_{ij,k} \right) - D_k \left(\sum_{j=1}^{m} \sum_{i=1}^{n_j} \beta_{ij} w_{ij,k} \right) \\
&+ \sum_{j=1}^{m} \sum_{i=1}^{n_j} \left[\lambda_{ij,k} \left(-x_{ij,k} + x_{ij,k-1} + I_{ij,k} \right. \right. \\
&+ \sum_{\sigma \in R_i} u_{\sigma j,k} - u_{ij,k} - s_{ij,k} + \left. \sum_{\sigma \in R_i} s_{\sigma j,k} \right) + \mu_{ij,k}(-r_{ij,k} + u_{ij,k} x_{ij,k-1}) \\
&+ \phi_{ij,k} \left(-w_{ij,k} + u_{ij,k} \left(\sum_{\sigma \in R_i} u_{\sigma j,k} - u_{ij.k} \right) \right) + e^1_{ij,k}(u^m_{ij,k} - u_{ij,k}) \\
&\left. \left. + e_{ij,k}(u_{ij,k} - u^M_{ij,k}) + g^1_{ij,k}(x^m_{ij} - x_{ij,k}) + g_{ij,k}(x_{ij,k} - x^M_{ij,k}) \right] \right\} \quad (6.32)
\end{aligned}
$$

In the above equation $\lambda_{ij,k}$, $\mu_{ij,k}$, and $\phi_{ij,k}$ are Lagrange multipliers. These are to be determined such that the corresponding equality constraints must be satisfied, and $e^1_{ij,k}$, $e_{ij,k}$, $g^1_{ij,k}$, and $g_{ij,k}$ are Kuhn–Tucker multipliers. They are equal to zero if the constraints are not violated, and greater than zero if the constraints are violated.

Equation (6.43) can be written in vector form as

$$\tilde{J} = \sum_{k=1}^{K} [u^T(k)b(k)b^T(k)u(k) + r^T(k)\beta\beta^Tr(k) + \tfrac{1}{4}w^T(k)\beta\beta^Tw(k)$$

$$+ u^T(k)b(k)\beta^Tr(k) + r^T(k)\beta b^T(k)u(k) + \tfrac{1}{2}u^T(k)b(k)\beta^Tw(k)$$

$$+ \tfrac{1}{2}w^T(k)\beta b^T(k)u(k) - 2D_kb^T(k)u(k) - 2D_k\beta^Tr(k)$$

$$- D_k\beta^Tw(k) + \lambda^T(k)(-x(k) + x(k-1) + I(k) + Mu(k) + Ms(k))$$

$$+ \mu^T(k)(-r(k) + \tfrac{1}{2}u^T(k)Hx(k-1) + \tfrac{1}{2}x^T(k-1)Hu(k))$$

$$+ \phi^T(k)(-w(k) + u^T(k)HMu(k)) + \psi^T(k)u(k)$$

$$+ v^T(k)(x(k-1) + I(k) + Mu(k) + Ms(k))] \tag{6.44}$$

The vectors in equation (6.44) are $N \times 1$ column vector, $N = \sum_{j=1}^{m} n_j$, and H is an $N \times N$ vector matrix, in which the vector index varies from 1 to N, while the matrix dimension of H is $N \times N$. Furthermore, M is an $N \times N$ matrix whose elements depend on the topological arrangement of the reservoirs.

By employing the discrete version of integration by parts, dropping constant terms, and defining the following vectors:

$$X^T(k) = [x^T(k-1), u^T(k), r^T(k), w^T(k)] \tag{6.45}$$

$$R^T(k) = [(\lambda(k) - \lambda(k-1) + v(k))^T, (M^T\lambda(k) + \psi(k) - 2D_kb(k)$$

$$+ M^Tv(k))^T, (-2D_k\beta - \mu(k))^T, (-D_k\beta - \phi(k))^T] \tag{6.46}$$

and

$$L(k) = \begin{bmatrix} 0 & \tfrac{1}{2}\mu^T(k)H & 0 & 0 \\ \tfrac{1}{2}\mu^T(k)H & b(k)b^T(k) + \phi^T(k)HM & b(k)\beta^T & \tfrac{1}{2}b(k)\beta^T \\ 0 & \beta b^T(k) & \beta\beta^T & \tfrac{1}{2}\beta\beta^T \\ 0 & \tfrac{1}{2}\beta b^T(k) & \tfrac{1}{2}\beta\beta^T & \tfrac{1}{4}\beta\beta^T \end{bmatrix}$$

$$\tag{6.47}$$

one obtains the augmented cost functional as

$$\tilde{J} = -\lambda^T(K)x(K) + \lambda^T(0)x(0) + \sum_{k=1}^{K} [X^T(k)L(k)X(k) + R^T(k)X(k)] \tag{6.48}$$

Equation (6.48) is composed of a boundary part and a discrete integral part, which are independent of each other. If we define the vector $V(k)$ such that

$$V(k) = L^{-1}(k)R(k) \tag{6.49}$$

then, the cost functional in equation (6.48) can be written as

$$\tilde{J} = -\lambda^T(K)x(K) + \lambda^T(0)x(0)$$

$$+ \sum_{k=1}^{K} \{[X(k) + \tfrac{1}{2}V(k)]^T L(k)[X(k) + \tfrac{1}{2}V(k)] - \tfrac{1}{4}V^T(k)L(k)V(k)\} \tag{6.50}$$

Since $V(k)$ is independent of $X(k)$, equation (6.50) can be written as

$$\tilde{J} = [-\lambda^T(K)x(K) + \lambda^T(0)x(0)$$

$$+ \sum_{k=1}^{K} \{[X(k) + \tfrac{1}{2}V(k)]^T L(k)[X(k) + \tfrac{1}{2}V(k)]\} \tag{6.51}$$

$\tilde{J}$ in equation (6.51) can be written as

$$\tilde{J} = J_1(K) + \sum_{k=1}^{K} J_2(k) \tag{6.52}$$

where

$$J_1(K) = -\lambda^T(K)x(K) + \lambda^T(0)x(0) \tag{6.53}$$

and

$$J_2(k) = [X(k) + \tfrac{1}{2}V(k)]^T L(k)[X(k) + \tfrac{1}{2}V(k)] \tag{6.54}$$

Equation (6.54) defines a norm in Hilbert space; hence, we can write this equation as

$$J_2(k) = \|X(k) + \tfrac{1}{2}V(k)\|_{L(k)} \tag{6.55}$$

6.3.3. The Optimal Solution

To maximize $\tilde{J}$ in equation (6.52), one maximizes each term separately:

$$\max_{[x(K),X(k)]} \tilde{J} = \max_{x(K)} J_1(K) + \max_{X(k)} \sum_{k=1}^{K} J_2(k) \tag{6.56}$$

The maximum of $J_1(K)$ is clearly achieved when

$$\lambda(K) = 0 \tag{6.57}$$

because $x(0)$ is constant and $\delta x(K)$ is arbitrary. Equation (6.57) gives the value of Lagrange multiplier at the last period studied.

The maximization of equation (6.5) is mathematically equivalent to minimizing the norm of this equation. The minimum of equation (6.55) is clearly achieved when the norm of this equation is equal to zero

$$\|X(k) + \tfrac{1}{2}V(k)\|_{L(k)} = 0 \tag{6.58}$$

Substituting from equation (6.49) into equation (6.58), one obtains

$$[R(k) + 2L(k)X(k)] = 0 \qquad (6.59)$$

Equation (6.59) is the condition of optimality. Writing equation (6.59) explicitly and adding the equality constraints, one obtains

$$x(k) = x(k-1) + I(k) + Mu(k) + Ms(k) \qquad (6.60)$$

$$r(k) = x^T(k-1)\mathbf{H}u(k) \qquad (6.61)$$

$$w(k) = u^T(k)\mathbf{H}Mu(k) \qquad (6.62)$$

$$\lambda(k-1) = \lambda(k) + \nu(k) + \mu^T(k)\mathbf{H}u(k) \qquad (6.63)$$

$$M^T\lambda(k) + \psi(k) - 2D_kb(k) + M^T\nu(k) + \mu^T(k)\mathbf{H}x(k-1)$$

$$+ 2b(k)b^T(k)u(k)$$

$$+ 2\phi^T(k)\mathbf{H}Mu(k) + 2b(k)\beta^Tr(k) + b(k)\beta^Tw(k) = 0$$

$$\qquad (6.64)$$

$$\mu(k) = -2D_k\beta + 2\beta b^T(k)u(k) + 2\beta\beta^Tr(k) + \beta\beta^Tw(k) \qquad (6.65)$$

$$\phi(k) = -D_k\beta + \beta b^T(k)u(k) + \beta\beta^Tr(k) + \tfrac{1}{2}\beta\beta^Tw(k) \qquad (6.66)$$

According to Kuhn-Tucker theory, the following exclusion equations must be satisfied at the optimum:

$$e^1_{ij,k}(u^m_{ij,k} - u_{ij,k}) = 0 \qquad (6.67)$$

$$e_{ij,k}(u_{ij,k} - u^M_{ij,k}) = 0 \qquad (6.68)$$

$$g^1_{ij,k}(x^m_{ij} - x_{ij,k}) = 0 \qquad (6.69)$$

$$g_{ij,k}(x_{ij,k} - x^M_{ij}) = 0 \qquad (6.70)$$

One also has the following limits on the variables:

$$\begin{array}{lll} \text{If } x_{ij,k} < x^m_{ij}, & \text{then we put } x_{ij,k} = x^m_{ij} \\ \text{If } x_{ij,k} > x^M_{ij}, & \text{then we put } x_{ij,k} = x^M_{ij} \\ \text{If } u_{ij,k} < u^m_{ij,k}, & \text{then we put } u_{ij,k} = u^m_{ij,k} \\ \text{If } u_{ij,k} > u^M_{ij,k}, & \text{then we put } u_{ij,k} = u^M_{ij,k} \end{array} \qquad (6.71)$$

Equations (6.60)–(6.71) with equation (6.57) completely specify the optimal solution. The next section explains the algorithm of solution for these equations.

6.3.4. Algorithm of Solution

Assume we are given the natural inflows to the sites during the optimization interval, the initial storage for each reservoir $x(0)$, the required load on the system D_k, and the characteristics of each reservoir (the variation of the water conversion factor with the neat head, and the head-storage characteristic). The following steps are used to solve the system equations:

Step 1. Assume a feasible initial guess for $u(k)$, such that $u^m(k) \le u(k) \le u^M(k)$, and set the iteration counter i equal to zero.

Step 2. Assume first that $s(k)$ is equal to zero; solve equations (6.60)-(6.62) and equations (6.65) and (6.66) forward in stages with $x(0)$ given.

Step 3. Check the limits on $x(k)$. If $x(k) > x^M$, put $x(k) = x^M$ and if $x(k) < x^m$ put $x(k) = x^m$ and go to step 4. Otherwise go to step 8.

Step 4. Calculate the new discharge from the following equation:

$$u(k) = [M]^{-1}[x(k) - x(k-1) - I(k)]$$

Step 5. Solve again equations (6.61), (6.62), (6.65), and (6.66) forward in stages with the values of $u(k)$ given from step 4.

Step 6. Check the limits on $u(k)$. If $u(k) > u^M(k)$ put $u(k) = u^M(k)$ and go to step 7, and if $u(k) < u^m(k)$ put $u(k) = u^m(k)$ and go to step 2.

Step 7. Calculate the spill $s(k)$ from the following equation:

$$s(k) = [M^{-1}][x(k) - x(k-1) + I(k)] - u^M(k)$$

Step 8. Since $x(k)$ is within its limit, $x^m \le (k) \le x^M$, the Kuhn–Tucker multiplier for $x(k)$, $\nu(k)$, is equal to zero. Keeping this statement in mind, solve equation (6.62) backward in stages with equation (6.57) as a boundary condition.

Step 9. Calculate the Kuhn–Tucker multiplier for $u(k)$, $\psi(k)$, from the following equation:

$$\begin{aligned}
\psi(k) = -[&M^T\lambda(k-1) - M^T\mu^T(k)Hu(k) - 2D_k b(k) \\
&+ \mu^T(k)Hx(k-1) + 2b(k)b^T(k)u(k) \\
&+ 2\phi^T(k)HMu(k) + 2b(k)\beta^T r(k) + b(k)\beta^T w(k)]
\end{aligned}$$

Note that the above equation is obtained by multiplying equation (6.63) by M^T and subtracting the resultant equation from equation (6.64).

Step 10. Calculate a new control iterate as $u^{i+1}(k) = u^i(k) + \alpha Du^i(k)$, i is the iteration counter where $Du(k)$ is given by equation (6.64) with $\nu(k) = 0$ and α is a scalar, which is chosen with consideration to such factors as convergence.

Step 11. Check the limits on $u^{i+1}(k)$. If $u^{i+1}(k)$ is within its limits, i.e.,

$$u^m(k) < u^{i+1}(k) < u^M(k)$$

go to step 12; otherwise put $u(k)$ to its limit and go to step 2.

Step 12. Solve the following equation forward in stages:

$$\begin{aligned} \lambda(k-1) = -[M^T]^{-1}[&-M^T\mu^T(k)\mathbf{H}u(k) - 2D_kb(k) \\ &+ \mu^T(k)\mathbf{H}x(k-1) + 2b(k)b^T(k)u(k) \\ &+ 2\phi^T(k)\mathbf{H}Mu(k) + 2b(k)\beta^Tr(k) + b(k)\beta^Tw(k)] \end{aligned}$$

Step 13. Determine Kuhn–Tucker multiplier for $x(k)$, $\nu(k)$, from the following equation:

$$\begin{aligned} \nu(k) = -[M^T]^{-1}[&-M^T\lambda(k) + 2D_kb(k) - \mu^T(k)\mathbf{H}x(k-1) \\ &- 2b(k)b^T(k)u(k) - 2\phi^T(k)\mathbf{H}Mu(k) \\ &- 2b(k)\beta^Tr(k) - b(k)\beta^Tw(k)] \end{aligned}$$

Step 14. Determine a new state iterate from the following:

$$x^{i+1}(k) = x^i(k) + \alpha Dx^i(k)$$

and

$$[Dx(k)] = [\lambda(k) - \lambda(k-1) - \nu(k) + \mu^T(k)\mathbf{H}u(k)]$$

Step 15. Repeat the calculation starting from step 3. Continue until the state $x(k)$ and the control $u(k)$ do not change significantly from iteration to iteration and J in equation (6.41) is a maximum.

6.4. A Nonlinear Model (Minimum Norm Approach)

In Section 6.3, we discussed the applications of the minimum norm approach to the maximization of the firm hydro energy capability from a multireservoir power system; the model used for the water conversion factor was a linear model of the average storage. This section is devoted to the

applications of the minimum norm and functional analysis to the maximization of the firm hydro energy capability from the same system, but the
model used for the water conversion factor is a nonlinear model of the
average storage, and the model used for each reservoir generation is a highly
nonlinear model. In this section, we define a set of pseudo-state-variables
to cast the problem into a quadratic problem. We start this section, for the
reader's convenience, with a repetition of the optimization objective and
constraints; then we form an augmented cost functional by adjoining the
equality constraint to the cost function via Lagrange multipliers and the
inequality constraints via Kuhn–Tucker multipliers; and finally we form
the problem as a norm. Next, we obtain the optimal equations which
maximize the firm hydro energy capability from the system. Furthermore,
we explain the algorithm for solution of these equations.

6.4.1. Optimization Objective and Constraints

Given the hydroelectric system in Figure 4.1, the problem for this
system is to find the release, the discharge, $u_{ij,k} = i, \ldots, n_j; j = 1, \ldots, m,$
$k = 1, \ldots, K$ as a function of time over the critical period (the optimization
interval) under the following conditions:

(1) The total generation from the system over the optimization interval
is maximum.

(2) The system surplus energy over the load for each time interval
should be maximum and at the same time it should be uniform.

(3) The storage-elevation curve may be described by

$$h_{ij,k} = \phi_{ij} + \gamma_{ij}x_{ij,k} + \delta_{ij}(x_{ij,k})^2 \tag{6.72}$$

where ϕ_{ij}, γ_{ij} and δ_{ij} are constants. These were obtained by least-squares
curve fitting to typical plant data available.

(4) The water conversion factor, MWh/Mm^3, as a function of the
storage is given by

$$C_{ij,k} = a_{ij} + b_{ij}x_{ij,k} + c_{ij}(x_{ij,k})^2 \, \text{MWh/Mm}^3 \tag{6.73}$$

In equations (6.72) and (6.73), we assume that the tail-water elevation is
constant independent of the discharge.

(5) The water conservation equation for each reservoir may be
adequately described by the continuity-type discrete equation as

$$x_{ij,k} = x_{ij,k-1} + I_{ij,k} - u_{ij,k} - s_{ij,k} + \sum_{\sigma \in R_i} u_{\sigma j,k} + \sum_{\sigma \in R_i} s_{\sigma j,k} \tag{6.74}$$

where R_i is the set of plant immediately upstream from plant i.

(6) In order to be realizable, and also to satisfy multipurpose stream
use requirements, such as flood control, navigation, fishing, and other

purposes if any, the storage management schedule must satisfy the following inequality constraints:

a. Upper and lower bounds on reservoir contents:

$$x_{ij}^m \le x_{ij,k} \le x_{ij}^M \tag{6.75}$$

b. Upper and lower bounds on reservoir plant outflow:

$$u_{ij,k}^m \le u_{ij,k} \le u_{ij,k}^M \tag{6.76}$$

where $u_{ij,k}^m$ and $u_{ij,k}^M$ are the minimum and maximum discharges in Mm^3 ($1 \text{ Mm}^3 = 10^6 \text{ m}^3$), respectively; they are given by

$$u_{ij,k}^m = 0.0864 d^k \qquad \text{(minimum effective discharge in m}^3/\text{sec)} \tag{6.77}$$

$$u_{ij,k}^M = 0.0864 d^k \qquad \text{(maximum effective discharge in m}^3/\text{sec)} \tag{6.78}$$

In the above two equations d^k is the number of days in period k.

In mathematical terms, the object of the optimizing computation is to find the discharge $u_{ij,k}$ that maximizes

$$J = \sum_{k=1}^{K} \left[\sum_{j=1}^{m} \sum_{i=1}^{n_j} G_{ij}(u_{ij,k}, \tfrac{1}{2}(x_{ij,k} + x_{ij,k-1})) - D_k \right]^2 \text{ GWh} \tag{6.79}$$

Subject to satisfying the constraints (6.74), (6.75), and (6.76).

It is assumed that the initial storage $x_{ij,0}$, the value for the natural inflows to each stream during each period, and the value of the demand D_k are known.

6.4.2. Modeling of the System

The generation of a hydroelectric plant is a nonlinear function of the discharge $u_{ij,k}$ and the reservoir water head, which itself is a function of the storage. To avoid underestimation of production for rising water levels and overestimation for falling water levels, an average of begin and end of time step (month) storage is used; hence the generation of plant i on river j may be chosen as

$$G_{ij}(u_{ij,k}, \tfrac{1}{2}(x_{ij,k} + x_{ij,k-1})) = \alpha_{ij} u_{ij,k} + \tfrac{1}{2}\beta_{ij} u_{ij,k}(x_{ij,k} + x_{ij,k-1})$$

$$+ \tfrac{1}{4}\gamma_{ij} u_{ij,k}(x_{ij,k} + x_{ij,k-1})^2 \tag{6.80}$$

where α_{ij}, β_{ij}, and γ_{ij} are constants. These were obtained by least-squares curve fitting to typical plant data available.

Substituting for $x_{ij,k}$ from equation (6.74) into equation (6.80), we obtain

$$G_{ij}(u_{ij,k}, \tfrac{1}{2}(x_{ij,k} + x_{ij,k-1})) = b_{ij,k}u_{ij,k} + d_{ij,k}r_{ij,k} + f_{ij,k}w_{ij,k} + \gamma_{ij}g_{ij,k}$$
$$+ \tfrac{1}{4}\gamma_{ij}t_{ij,k} + \gamma_{ij}n_{ij,k} - \tfrac{1}{4}\gamma_{ij}m_{ij,k} \qquad (6.81)$$

where we defined the following pseudovariables:

$$q_{ij,k} = I_{ij,k} + \sum_{\sigma \in R_i} s_{\sigma j,k} - s_{ij,k} \qquad (6.82)$$

$$b_{ij,k} = \alpha_{ij} + \tfrac{1}{2}\beta_{ij}q_{ij,k} + \tfrac{1}{4}\gamma_{ij}(q_{ij,k})^2 \qquad (6.83)$$

$$d_{ij,k} = \beta_{ij} + \gamma_{ij}q_{ij,k} \qquad (6.84)$$

$$f_{ij,k} = \tfrac{1}{2}\beta_{ij} + \tfrac{1}{2}\gamma_{ij}q_{ij,k} \qquad (6.85)$$

$$y_{ij,k} = (x_{ij,k})^2 \qquad (6.86)$$

$$z_{ij,k} = (u_{ij,k})^2 \qquad (6.87)$$

$$r_{ij,k} = u_{ij,k}x_{ij,k-1} \qquad (6.88)$$

$$w_{ij,k} = u_{ij,k}\left(\sum_{\sigma \in R_i} u_{\sigma j,k} - u_{ij,k} \right) \qquad (6.89)$$

$$g_{ij,k} = u_{ij,k}y_{ij,k-1} \qquad (6.90)$$

$$t_{ij,k} = u_{ij,k}\left(\sum_{\sigma \in R_i} z_{\sigma j,k} + z_{ij,k} \right) \qquad (6.91)$$

$$n_{ij,k} = r_{ij,k}\left(\sum_{\sigma \in R_i} u_{\sigma j,k} - u_{ij,k} \right) \qquad (6.92)$$

$$m_{ij,k} = z_{ij,k}\left(\sum_{\sigma \in R_i} u_{\sigma j,k} \right) \qquad (6.93)$$

Now the cost functional in equation (6.79) can be written as

$$J = \sum_{k=1}^{K} \left[\sum_{j=1}^{m} \sum_{i=1}^{n_j} (b_{ij,k}u_{ij,k} + d_{ij,k}r_{ij,k} + f_{ij,k}w_{ij,k} + \gamma_{ij}g_{ij,k} \right.$$
$$\left. + \tfrac{1}{4}\gamma_{ij}t_{ij,k} + \gamma_{ij}n_{ij,k} - \tfrac{1}{4}\gamma_{ij}m_{ij,k} - D_k)^2 \right] \qquad (6.94)$$

Subject to satisfying the equality constraints given by equations (6.86)-(6.93) and the inequality constraints given by equations (6.75) and (6.76).

6.4.3. A Minimum Norm Formulation

The augmented cost functional J is obtained by adjoining to the cost functional the equality constraints (6.86)–(6.93) via Lagrange multipliers and the inequality constraints (6.75) and (6.76) via Kuhn–Tucker multipliers. One thus obtains

$$
\tilde{J} = \sum_{k=1}^{K} \left[\sum_{j=1}^{m} \sum_{i=1}^{n_j} (b_{ij,k} u_{ij,k} + d_{ij,k} r_{ij,k} + f_{ij,k} w_{ij,k} + \gamma_{ij} g_{ij,k} \right.
$$

$$
+ \tfrac{1}{4} \gamma_{ij} t_{ij,k} + {}_{ij} n_{ij,k} - \tfrac{1}{4} \gamma_{ij} m_{ij,k} - D_k)^2
$$

$$
+ \lambda_{ij,k} \left(-x_{ij,k} + x_{ij,k-1} + q_{ij,k} + \sum_{\sigma \in R_i} u_{\sigma j,k} - u_{ij,k} \right)
$$

$$
+ \mu_{ij,k}(-y_{ij,k} + (x_{ij,k})^2) + \phi_{ij,k}(-z_{ij,k} + (u_{ij,k})^2)
$$

$$
+ \psi_{ij,k}(-r_{ij,k} + u_{ij,k} x_{ij,k-1})
$$

$$
+ \theta_{ij,k} \left(-w_{ij,k} + u_{ij,k} \left(\sum_{\sigma \in R_i} u_{\sigma j,k} - u_{ij,k} \right) \right)
$$

$$
+ \delta_{ij,k}(-g_{ij,k} + u_{ij,k} y_{ij,k-1})
$$

$$
+ \Omega_{ij,k} \left(-t_{ij,k} + u_{ij,k} \left(\sum_{\sigma \in R_i} z_{\sigma j,k} + z_{ij,k} \right) \right)
$$

$$
+ \Gamma_{ij,k} \left(-n_{ij,k} + r_{ij,k} \left(\sum_{\sigma \in R_i} u_{\sigma j,k} - u_{ij,k} \right) \right)
$$

$$
+ \nu_{ij,k} \left(-m_{ij,k} + z_{ij,k} \sum_{\sigma \in R_i} u_{\sigma j,k} \right)
$$

$$
+ e_{ij,k}(x_{ij}^m - x_{ij,k}) + e_{ij,k}^1(x_{ij,k} - x_{ij}^M)
$$

$$
\left. + l_{ij,k}(u_{ij,k}^m - u_{ij,k}) + l_{ij,k}^1(u_{ij,k} - u_{ij,k}^M) \right] \tag{6.95}
$$

where $\lambda_{ij,k}$, $\mu_{ij,k}$, $\phi_{ij,k}$, $\psi_{ij,k}$, $\theta_{ij,k}$, $\delta_{ij,k}$, $\Omega_{ij,k}$, $\Gamma_{ij,k}$, and $\nu_{ij,k}$ are Lagrange multipliers. These are to be determined in such a way that the corresponding equality constraints are satisfied, and $e_{ij,k}$, $e_{ij,k}^1$, $l_{ij,k}$, and $l_{ij,k}^1$ are Kuhn–Tucker multipliers. They are equal to zero if the constraints are not violated, and greater than zero if the constraints are violated.

Equation (6.95) can be written in the following vector form:

$$
\tilde{J} = E \left[\sum_{k=1}^{K} \{ u^T(k) b(k) b^T(k) u(k) + r^T(k) d(k) d^T(k) r(k) \right.
$$

$$
+ \omega^T(k) f(k) f^T(k) \omega(k) + g^T(k) \gamma \gamma^T g(k) + \tfrac{1}{16} t^T(k) \gamma \gamma^T t(k)
$$

$$+ n^T(k)\gamma\gamma^T n(k) + \tfrac{1}{16}m^T(k)\gamma\gamma^T m(k) + 2u^T(k)b(k)d^T(k)r(k)$$

$$+ 2u^T(k)b(k)f^T(k)\omega(k) + 2u^T(k)b(k)\gamma^T g(k)$$

$$+ \tfrac{2}{4}u^T(k)b(k)\gamma^T t(k) + 2u^T(k)b(k)\gamma^T n(k)$$

$$- \tfrac{2}{4}u^T(k)b(k)\gamma^T m(k) + 2r^T(k)d(k)f^T(k)\omega(k)$$

$$+ 2r^T(k)d(k)\gamma^T g(k) + \tfrac{2}{4}r^T(k)d(k)\gamma^T(k)$$

$$+ 2r^T(k)d(k)\gamma^T n(k) - \tfrac{2}{4}r^T(k)d(k)\gamma^T m(k)$$

$$+ 2\omega^T(k)f(k)\gamma^T g(k) + \tfrac{2}{4}\omega^T(k)f(k)\gamma^T(k)$$

$$+ 2\omega^T(k)f(k)\gamma^T n(k) - \tfrac{2}{4}\omega^T(k)f(k)\gamma^T m(k)$$

$$+ \tfrac{2}{4}g^T(k)\gamma\gamma^T t(k) + 2g^T(k)\gamma\gamma^T n(k)$$

$$- \tfrac{2}{4}g^T(k)\gamma\gamma^T m(k) + \tfrac{2}{4}t^T(k)\gamma\gamma^T n(k)$$

$$- \tfrac{2}{16}t^T(k)\gamma\gamma^T m(k) - \tfrac{2}{4}n^T(k)\gamma\gamma^T m(k)$$

$$+ x^T(k)\mu^T(k)\mathbf{H}x(k) + u^T(k)\phi^T(k)\mathbf{H}u(k)$$

$$+ u^T(k)\psi^T(k)\mathbf{H}x(k-1) + u^T(k)\phi^T(k)\mathbf{A}u(k)$$

$$+ u^T(k)\delta^T(k)\mathbf{H}y(k-1) + u^T(k)\Omega^T(k)\mathbf{L}z(k)$$

$$+ r^T(k)\Gamma^T(k)\mathbf{F}u(k) + z^T(k)\gamma^T(k)\mathbf{Q}u(k)$$

$$+ (\xi^T(k) + \lambda^T(k)M - 2D_k b^T(k) + h^T(k)M)u(k) + (-\phi^T(k)z(k))$$

$$+ (-2D_k d^T(k) - \psi^T(k))r(k) + (-2D_k f^T(k) - \theta^T(k))\omega(k)$$

$$+ (-2D_k\gamma^T - \delta^T(k))g(k) + (-\tfrac{1}{2}D_k\gamma^T - \Omega^T(k))t(k)$$

$$+ (-2D_k\gamma^T - \Gamma^T(k))n(k) + (-\tfrac{1}{2}D_k\gamma^T - \nu^T(k))m(k)$$

$$+ (-\mu^T(k)y(k)) + \lambda^T(k)(-x(k) + x(k-1)) + h^T(k)x(k-1)\}\Bigr]$$

$$(6.96)$$

Note that constant terms are dropped in the above equation where we defined the vectors $h_{ij,k}$ and $\xi_{ij,k}$ as

$$h_{ij,k} = e_{ij,k}^1 - e_{ij,k} \tag{6.97}$$

$$\xi_{ij,k} = l_{ij,k}^1 - l_{ij,k} \tag{6.98}$$

The vectors in equation (6.96) are $N \times 1$ column vectors, $N = \sum_{j=1}^{m} n_j$, and $\mathbf{F}$, $\mathbf{H}$, $\mathbf{A}$, $\mathbf{L}$, and $\mathbf{Q}$ are $N \times N$ vector matrices in which the vector index varies from 1 to N, while the matrix dimension is $N \times N$. Furthermore, M is an $N \times N$ matrix whose elements depend on the topological arrangement of the reservoirs, then for an independent river system, M is given by

$$M = \text{diag}(M_1, M_2, \ldots, M_m) \tag{6.99}$$

where $M_1, \ldots, M_m$ are lower triangular matrices whose elements are given by

(i) $m_{ij} = -1, \qquad i = 1, \ldots, n_j, \qquad j = 1, \ldots, m$

(ii) $m_{(\sigma+1)\sigma} = 1, \qquad \sigma = 1, \ldots, n_j - 1, \qquad j = 1, \ldots, m$

$$(6.100)$$

Employing the discrete version of integration by parts, one obtains the following equation for the augmented cost functional:

$$
\mathbf{J} = E[x^T(K)\mu^T(K)\mathbf{H}x(K) - \lambda^T(K)x(K) + \mu^T(K)y(K)
$$

$$
- x^T(0)\mu^T(0)\mathbf{H}x(0) - \mu^T(0)y(0) + \lambda^T(0)x(0)
$$

$$
+ \sum_{k=1}^{K} \{X^T(k)L(k)X(k) + R^T(k)X(k)\}] \tag{6.101}
$$

where

$$
X^T(k) = [x^T(k-1), y^T(k-1), z^T(k), r^T(k), \omega^T(k), g^T(k),
$$

$$
t^T(k), n^T(k), m^T(k), u^T(k)] \tag{6.102}
$$

$$
R^T(k) = [(\lambda(k) - \lambda(k-1) + h(k))^T, -\mu^T(k-1),
$$

$$
-\phi^T(k), (-\psi(k) - 2D_k d(k))^T,
$$

$$
(-\theta(k) - 2D_k f(k))^{TT}, (-2D_k\gamma - \gamma(k))^T, (-\Omega(k) - \tfrac{1}{2}D_k\gamma)^T,
$$

$$
(-\Gamma(k) - 2D_k\gamma)^T,
$$

$$
(-\nu(k) - \tfrac{1}{2}D_k\gamma)^T, (\xi(k) + M^T\lambda(k) - 2D_k b(k) + M^T h(k))^T] \tag{6.103}
$$

and

$$
L(k) = \left[
\begin{array}{c|c}
L_{11}(k) & L_{21}(k) \\
\hline
L_{21}(k) & L_{22}(k)
\end{array}
\right] \tag{6.104}
$$

where

$$
L_{11}(k) = \begin{bmatrix}
\mu^T(k-1)\mathbf{H} & 0 & 0 & 0 & 0 \\
0 & 0 & 0 & 0 & 0 \\
0 & 0 & 0 & 0 & 0 \\
0 & 0 & 0 & d(k)d^T(k) & d(k)f^T(k) \\
0 & 0 & 0 & f(k)d^T(k) & f(k)f^T(k)
\end{bmatrix} \tag{6.105}
$$

$$
L_{21}(k) = \begin{bmatrix}
0 & 0 & 0 & 0 & \frac{1}{2}\psi^T(k)\mathbf{H} \\
0 & 0 & 0 & 0 & \frac{1}{2}\delta^T(k)\mathbf{H} \\
0 & 0 & 0 & 0 & \frac{1}{2}\nu^T(k)\mathbf{Q} + \frac{1}{2}\Omega^T(k)\mathbf{L} \\
d(k)\gamma^T & \frac{1}{4}d(k)\gamma^T & d(k)\gamma^T & -\frac{1}{4}d(k)\gamma^T & d(k)b^T(k) + \frac{1}{2}\Gamma^T(k)\mathbf{F} \\
f(k)\gamma^T & \frac{1}{4}f(k)\gamma^T & f(k)\gamma^T & -\frac{1}{4}f(k)\gamma^T & f(k)b^T(k)
\end{bmatrix}
$$

$$(6.106)$$

$$
L_{12}(k) = \begin{bmatrix}
0 & 0 & 0 & \gamma d^T(k) & \gamma f^T(k) \\
0 & 0 & 0 & \frac{1}{4}\gamma d^T(k) & \frac{1}{4}\gamma f^T(k) \\
0 & 0 & 0 & \gamma d^T(k) & \gamma f^T(k) \\
0 & 0 & 0 & -\frac{1}{4}\gamma d^T(k) & -\frac{1}{4}\gamma f^T(k) \\
\frac{1}{2}\psi^T(k)\mathbf{H} & \frac{1}{2}\delta^T(k)\mathbf{H} & \frac{1}{2}\nu^T(k)\mathbf{Q} + \frac{1}{2}\Omega^T(k)\mathbf{L} & b(k)d^T(k) + \frac{1}{2}\Gamma^T(k)\mathbf{F} & b(k)f^T(k)
\end{bmatrix}
$$

$$(6.107)$$

$$
L_{22}(k) = \begin{bmatrix}
\gamma\gamma^T & \frac{1}{4}\gamma\gamma^T & \gamma\gamma^T & -\frac{1}{4}\gamma\gamma^T & \gamma b^T(k) \\
\frac{1}{4}\gamma\gamma^T & \frac{1}{16}\gamma\gamma^T & \frac{1}{4}\gamma\gamma^T & -\frac{1}{16}\gamma\gamma^T & \frac{1}{4}\gamma b^T(k) \\
\gamma\gamma^T & \frac{1}{4}\gamma\gamma^T & \gamma\gamma^T & -\frac{1}{4}\gamma\gamma^T & \gamma b^T(k) \\
-\frac{1}{4}\gamma\gamma^T & -\frac{1}{16}\gamma\gamma^T & -\frac{1}{4}\gamma\gamma^T & \frac{1}{16}\gamma\gamma^T & -\frac{1}{4}\gamma b^T(k) \\
b(k)\gamma^T & \frac{1}{4}b(k)\gamma^T & b(k)\gamma^T & -\frac{1}{4}b(k)\gamma^T & \begin{array}{l} b(k)b^T(k) + \theta^T(k)\mathbf{A} \\ + \theta^T(k)\mathbf{H} \end{array}
\end{bmatrix}
$$

$$(6.108)$$

It will be noticed that, $\tilde{J}$, in equation (6.101) is composed of a boundary part and a discrete integral part which are independent of each other. To maximize $\tilde{J}$ in equation (6.101) one maximizes the boundary and the discrete integral parts separately:

$$
\max_{[x(K),y(K),X(k)]} \tilde{J} = \max_{[x(K),y(K)]} E[x^T(K)\mu^T(K)\mathbf{H}x(K) - \lambda^T(K)x(K)
$$
$$
+ \mu^T(K)y(K)
$$
$$
- x^T(0)\mu^T(0)\mathbf{H}(x(0) - \mu^T(0)y(0) + \lambda^T(0)x(0))]
$$
$$
- \max_{X(k)} \left[\sum_{k=1}^{K} (X^T(k)L(k)X(k) + R^T(k)X(k)) \right]
$$

$$(6.109)$$

If one defines the vector $V(k)$ such that

$$
V(k) = L^{-1}(k)R(k) \tag{6.110}
$$

then, the discrete integral part of equation (6.109) can be written as

$$
\max_{X(k)} J_2 = \max_{X(k)} \left\{ \sum_{k=1}^{K} \left[(X(k) + \tfrac{1}{2}V(k)^T L(k)(X(k) + \tfrac{1}{2}V(k)) \right. \right.
$$
$$
\left. \left. - \tfrac{1}{4}V^T(k)L(k)V(k) \right] \right\}
$$

$$(6.111)$$

Since it is desired to maximize J_2 with respect to $X(k)$, the problem is equivalent to

$$\max_{X(k)} J_2 = \max_{X(k)} \left\{ \sum_{k=1}^{K} [(X(k) + \tfrac{1}{2}V(k))^T L(k)(X(k) + \tfrac{1}{2}V(k))] \right\}$$

(6.112)

because $V(k)$ is independent of $X(k)$. Equation (6.112) defines a norm, one can write equation (6.112) as

$$\max_{X(k)} J_2 = \max_{X(k)} E[\|X(k) + \tfrac{1}{2}V(k)\|]_{L(k)}$$

(6.113)

6.4.4. The Optimal Solution

The condition of optimality for the problem formulated in equation (6.113) is that the norm of this equation should be equal to zero:

$$E[X(k) + \tfrac{1}{2}V(k)] = [0]$$

(6.114)

Substituting from equation (6.10) into equation (6.114), one obtains the following optimal equation:

$$E[R(k) + 2L(k)X(k)] = [0]$$

(6.115)

The boundary part in equation (6.109) is optimized when

$$E[\lambda(K)] = [0]$$

(6.116)

$$E[\mu(K)] = [0]$$

(6.117)

Equations (6.116) and (6.117) give the values of Lagrange multipliers at the last period studied.

Writing equation (6.115) explicitly and adding equations (6.86)–(6.93), one obtains the following optimal equations:

$$E[\lambda(k) - \lambda(k-1) + 2\mu^T(k-1)Hx(k-1) + \psi^T(k)Hu(k) + h(k)] = [0]$$

(6.118)

$$E[-\mu(k-1) + \delta^T(k)Hu(k)] = [0]$$

(6.119)

$$E[-\phi(k) + \nu^T(k)Qu(k) + \Omega^T(k)Lu(k)] = [0]$$

(6.120)

$$
\begin{aligned}
E[-\psi(k) - 2D_k d(k) &= 2d(k)d^T(k)r(k) \\
&+ 2d(k)f^T(k)\omega(k) + 2d(k)\gamma^T g(k) \\
&+ \tfrac{1}{2}d(k)\gamma^T t(k) + 2d(k)\gamma^T n(k) - \tfrac{1}{2}d(k)\gamma^T m(k) \\
&+ 2d(k)b^T(k)u(k) + \Gamma^T(k)Fu(k)] = [0]
\end{aligned}
$$

(6.121)

$$
\begin{aligned}
E[-\theta(k) - 2D_k f(k) &+ 2f(k)d^T(k)r(k) + 2f(k)f^T(k)\omega(k) \\
&+ 2f(k)\gamma^T g(k) + \tfrac{1}{2}f(k)\gamma^T t(k) + 2f(k)\gamma^T n(k) \\
&- \tfrac{1}{2}f(k)\gamma^T m(k) + 2f(k)b^T(k)u(k)] = [0]
\end{aligned}
$$

(6.122)

$$E[-\delta(k) - 2D_k\gamma + 2\gamma d^T(k)r(k) + 2\gamma f^T(k)\omega(k) + 2\gamma\gamma^T g(k) + \tfrac{1}{2}\gamma\gamma^T t(k)$$
$$+ 2\gamma\gamma^T n(k) - \tfrac{1}{2}\gamma\gamma^T m(k) + 2\gamma b^T(k)u(k)] = [0] \qquad (6.123)$$

$$E[-\Omega(k) - \tfrac{1}{2}D_k\gamma + \tfrac{1}{2}d^T(k)r(k) + \tfrac{1}{2}\gamma f^T(k)\mu(k) + \tfrac{1}{2}\gamma\gamma^T g(k) + \tfrac{1}{8}\gamma\gamma^T t(k)$$
$$+ \tfrac{1}{2}\gamma\gamma^T n(k) - \tfrac{1}{8}\gamma\gamma^T m(k) + \tfrac{1}{2}\gamma b^T(k)u(k)] = [0] \qquad (6.124)$$

$$E[-\Gamma(k) - 2D_k\gamma + 2\gamma d^T(k)r(k) + 2\gamma f^T(k)\omega(k) + 2\gamma\gamma^T g(k) + \tfrac{1}{2}\gamma\gamma^T t(k)$$
$$+ 2\gamma\gamma^T n(k) - \tfrac{1}{2}\gamma\gamma^T m(k) + 2\gamma b^T(k)u(k)] = [0] \qquad (6.125)$$

$$E[-\nu(k) - \tfrac{1}{2}D_k\gamma - \tfrac{1}{2}\gamma d^T(k)r(k) - \tfrac{1}{2}\gamma f^T(k)\omega(k) - \tfrac{1}{2}\gamma\gamma^T g(k) - \tfrac{1}{8}\gamma^\frown t(\kappa)$$
$$- \tfrac{1}{2}\gamma\gamma^T n(k) + \tfrac{1}{8}\gamma\gamma^T m(k) - \tfrac{1}{2}\gamma b^T(k)u(k)] = [0] \qquad (6.126)$$

$$E[\xi(k) + M^T\lambda(k) - 2D_k b(k) + M^T h(k) + \psi^T(k)Hx(k-1)$$
$$+ \delta^T(k)Hy(k-1) + \nu^T(k)Qz(k) + \Omega^T(k)Lz(k) + (2b(k)d^T(k)$$
$$+ \Gamma^T(k)F)r(k) + 2b(k)f^T(k)\omega(k) + 2b(k)\gamma^T g(k)$$
$$+ \tfrac{1}{2}b(k)\gamma^T t(k) + 2b(k)\gamma^T n(k) - \tfrac{1}{2}b(k)\gamma^T m(k)$$
$$+ 2b(k)b^T(k)u(k) + 2\theta^T(k)Au(k) + 2\phi^T(k)Hu(k)] = [0] \qquad (6.127)$$

$$E[-x(k) + x(k-1) + q(k) + Mu(k)] = [0] \qquad (6.128)$$

$$E[-y(k) + x^T(k)Hx(k)] = [0] \qquad (6.129)$$

$$E[-z(k) + u^T(k)Hu(k)] = [0] \qquad (6.130)$$

$$E[-r(k) + u^T(k)Hx(k-1)] = [0] \qquad (6.131)$$

$$E[-\omega(k) + u^T(k)Au(k)] = [0] \qquad (6.132)$$

$$E[-g(k) + u^T(k)Hy(k-1)] = [0] \qquad (6.133)$$

$$E[-t(k) + u^T(k)Lz(k)] = [0] \qquad (6.134)$$

$$E[-n(k) + r^T(k)Fu(k)] = [0] \qquad (6.135)$$

$$E[-m(k) + z^T(k)Qu(k)] = [0] \qquad (6.136)$$

According to Kuhn-Tucker theory, the following exclusion equations must be satisfied at the optimum:

$$e_{ij,k}(x_{ij}^m - x_{ij,k}) = 0 \qquad (6.137)$$

$$e_{ij,k}^1(x_{ij,k} - x_{ij}^M) = 0 \qquad (6.138)$$

$$l_{ij,k}(u_{ij,k}^m - u_{ij,k}) = 0 \qquad (6.139)$$

$$l_{ij,k}^1(u_{ij,k} - u_{ij,k}^M) = 0 \qquad (6.140)$$

One also has the following limits on the variables:

$$\begin{aligned}
&\text{If } x_{ij,k} < x_{ij}^m, &&\text{then we put } x_{ij,k} = x_{ij}^m \\
&\text{If } x_{ij,k} > x_{ij}^M, &&\text{then we put } x_{ij,k} = x_{ij}^m \\
&\text{If } u_{ij,k} < u_{ij,k}^m, &&\text{then we put } u_{ij,k} = u_{ij,k}^m \\
&\text{If } u_{ij,k} > u_{ij,k}^M, &&\text{then we put } u_{ij,k} = u_{ij,k}^M
\end{aligned} \qquad (6.141)$$

Equations (6.118)–(6.141) together with equations (6.116) and (6.117) completely specify the optimal solution for the system.

References

6.1. ALARCON, L., and MARKS, D., "A Stochastic Dynamic Programming Model for the Operation of the High Aswan Dam," Report No. JR 246, Ralph M. Parsons Laboratory for Water Resources and Hydrodynamics, Department of Civil Engineering, M.I.T., Cambridge, Massachusetts, 1979.

6.2. ARUNKUMAR, S., and YEH, W. W.-G, "Probabilistic Models in the Design and Operation of a Multi-purpose Reservoir System," Contribution No. 144, California, Davis, December 1973.

6.3. ASKEW, A., "Chance-Constrained Dynamic Programming and the Optimization of Water Resource Systems," *Water Resources Research* **10**(6), 1099–1106 (1974a).

6.4. ASKEW, A., "Optimum Reservoir Operating Policies and the Imposition of a Reliability Constraint," *Water Resources Research* **10**(1), 51–56 (1974b).

6.5. BERTSEKAS, D., *Dynamic Programming and Stochastic Control*, Academic, Orlando, Florida, 1976.

6.6. BERTSEKAS, D., *Constrained Optimization and Lagrange Multiplier Methods*, Academic, Orlando, Florida, 1982.

6.7. BRAS, R., BUCHANAN, R., and CURRY, K., "Realtime-Adaptive Closed-Loop Control of Reservoirs with the High Aswan Dam as a Case Study," *Water Resources Research* **19**(1), 33–52 (1983).

6.8. BUTCHER, W., "Stochastic Dynamic Programming for Optimum Reservoir Operation," *Water Resources Bulletin* **7**(1), 115–123 (1971).

6.9. CROLEY II, T. E., "Sequential Deterministic Optimization in Reservoir Operation," *Journal of Hydraulics Division of the American Society of Civil Engineers* **100**(HY3), 443–459, 1974.

6.10. EAGLESON. P. E., *Dynamic Hydrology*, McGraw-Hill, New York, 1970.

6.11. GEORGAKAKOS, A. P., and MARKS, D. H., "Real Time Control of Reservoir Systems," Technical Report 301, 313 pp. Ralph M. Parsons Laboratory for Hydrology and Water Resources, Department of Civil Engineering, M.I.T., Cambridge, Massachusetts, May 1985.

6.12. GEORGAKAKOS, K. P., and BRAS, R. L., "Real-Time, Statistically Linearized Adaptive Flood Routing," *Water Resources Research* **18**(3), 513–524 (1982).

6.13. GEORGAKAKOS, K. P., and MARKS, D. H., "A New Method for the Control of the River Nile," *International Journal of Water Resource Development*, in press, 1987.

6.14. HAIMES, Y., *Hierarchical Analysis of Water Resources Systems*, McGraw-Hill, New York, 1977.

6.15. HEIDARI, M., CHOW, V. T., KOKOTOVIC, P. V., and MEREDITH, D. D., "Discrete Differential Dynamic Programming Approach to Water Resources System Optimization," *Water Resources Research* **7**(2), 273–283 (1971).

6.16. HENDERSON, F., *Open Channel Flow*, MacMillan, New York, 1966.

6.17. JAMSHIDI, M., and HEIDARI, M., "Application of Dynamic Programming to Control Khuzestan Water Resources Systesm," *Automatica* **13**, 287–293 (1977).

6.18. JAZWINSKI, A., *Stochastic Processes and Filtering Theory*, Academic, Orlando, Florida, 1970.

6.19. KITANIDIS, P. K., and BRAS, R. L., "Real-Time Forecasting with a Conceptual Hydrologic Model. I., Analysis of Uncertainty," *Water Resources Research* **16**(6), 1025–1033 (1980).

6.20. KITANIDIS, P. K., and BRAS, R. L., "Real-Time Forecasting with a Conceptual Hydrologic Model. 2. Applications and Results," *Water Resources Research* **16**(6), 1034–1044 (1980b).

6.21. LARSON, R., and KECKLER, W., "Applications of Dynamic Programming to the Control of Water Resource Systems," *Automatica* **5**, 15–26 (1969).

6.22. LOAICIGA, H. A., and MARINO, M. A., "An Approach to Parameter Estimation and Stochastic Control in Water Resources with an Application to Reservoir Operation," *Water Resources Research* **21**(11), 1575–1584 (1985).

6.23. LOAICIGA, H. A., and MARINO, M. A., "Risk Analysis for Reservoir Operation," *Water Resources Research* **22**(4), 483–488 (1986).

6.24. MARINO, M. A., and LOAICIGA, H. A., "Dynamic Model for Multireservoir Operation," *Water Resources Research* **21**(5), 619–630 (1985).

6.25. MARINO, M. A., and LOAICIGA, H. A., "Quadratic Model for Multireservoir Management: Application to the Central Valley Project," *Water Resources Research* **21**(5), 631–641 (1985b).

6.26. MURRAY, D., and YAKOWITZ, S., "Constrained Differential Dynamic Programming and its Applications to Multireservoir Control," *Water Resources Research* **15**(5), 1017–1027 (1979).

6.27. OVEN-THOMPSON, K., ALARCON, L., and MARKS, D., "Agricultural versus Hydropower Trade-offs in the Operation of the High Aswan Dam," *Water Resources Research* **18**(6), 1605–1614 (1982).

6.28. PAPAGEORGIOU, M., "Optimal Multireservoir Network Control by the Discrete Maximum Principle," *Water Resources Research* **21**(12), 1824–1830 (1985).

6.29. ROSENTHAL, B., "The Status of Optimization Models for the Operation of Multireservoir Systems with Stochastic Inflows and Nonseparable Benefits," Report 75, Tennessee Water Resource Research Center, Knoxville, 1980.

6.30. SNIEDOVICH, M., "Reliability-Constrained Reservoir Control Problems 1. Methodological Issues," *Water Resources Research* **15**(6), 1574–1582 (1979).

6.31. SNIEDOVICH, M., "A Variance-Constrained Reservoir Control Problem," *Water Research* **16**(2), 271–274 (1980a).

6.32. SNIEDOVICH, M., "Analysis of a Chance-Constrained Reservoir Control Model," *Water Resources Research* **16**(5), 849–853 (1980b).

6.33. STEDINGER, J., SULE, B., and LOUCKS, D., "Stochastic Dynamic Programming Models for Reservoir Operation Optimization," *Water Resources Research* **20**(11), 1499–1505 (1984).

6.34. SU, S., and DEININGER, R., "Generalization of White's Method of Successive Approximations to Periodic Markovian Decision Processes," *Operations Research* **20**(2), 318–326 (1972).

6.35. WU, S., and DEININGER, R., "Modeling the Regulation of Lake Superior under Uncertainty of Future Water Supplies," *Water Resources Research* **10**(1), 11–25 (1974).

6.36. TURGEON, A., "Optimal Short-Term Hydro Scheduling from the Principle of Progressive Optimality," *Water Resources Research* **17**(3), 481–486 (1981).

6.37. WASIMI, S., and KITANIDIS, P. K., "Real-Time Forecasting and Daily Operation of a Multireservoir System during Floods by Linear Quadratic Gaussian Control," *Water Resources Research* **19**(6), 1511–1522 (1983).

6.38. YEH, W. W.-G., "State of the Art Review: Theories and Applications of Systems Analysis Techniques to the Optimal Management and Operation of a Reservoir System," Report No. UCLA-ENG-82-52, Engineering Foundation, University of California, Los Angeles, 1982.

6.39 YEH, W. W.-G., "Reservoir Management and Operations Models: A State-of-the-Art Review," *Water Resources Research* **21**, 1797–1818 (1985).

7

Long-Term Optimal Operation of Hydrothermal Power Systems

7.1. Introduction

The problem treated in the previous chapters was concerned with a pure hydro power system, where we maximized either the total benefits from the system (benefits from the hydro generation plus benefits from the amount of water left in storage at the end of the planning period) or the total firm hydro energy capability for the system. We discussed almost all techniques used to solve these problems. Most of the utility companies have a combination of hydro and thermal power plants to supply the required load on the system, where the hydro power plants are used to supply the base load, since they cost nothing to run (almost all the running costs are very small compared to those of thermal plants), while the thermal plants are used to supply the peak load, since these loads occur for a small period of time with peak demand.

The systematic coordination of the operation of a system of hydrothermal electric generation plants is usually more complex than the scheduling of an all-thermal generation system. The reason is that the hydroelectric plants may very well be coupled both electrically (i.e., they all serve the same load) and hydraulically (i.e., the water outflow from one plant may be a very significant portion of the inflow to one or more other, downstream plants) (Refs. 7.1–7.62).

No two hydroelectric systems in the world are alike. The reasons for the differences are the natural differences in the watersheds, the differences in the man-made storage and release elements used to control the water flows, and the very many different types of natural and man-made constraints imposed on the operation of hydroelectric systems. River systems may be simple, with relatively few tributaries with dams in series along the river, or they may extend over vast multinational areas, and include many tributaries and complex arrangements of storage reservoirs. Reservoirs may be developed with very large storage capacity with a few high head-plants along the river, or, alternatively, the river may have been developed with a larger number of dams and reservoirs, each with smaller storage capacity.

However, the one single aspect of hydroelectric plants that differentiates the coordination of their operation more than any other is the many and highly varied constraints. The hydro stream is a multistream use; in many hydro systems the generation of power is an adjunct to satisfying the multistream use requirements, such as flood control, irrigation, navigation, and fishing. Sudden changes with high-volume releases of water may be prohibited because the release may result in a large wave traveling downstream with potentially damaging effects. Water releases may be dictated by international treaty.

The coordination of the operation of hydroelectric plants involves the scheduling of water releases. The long-range hydro scheduling problem involves the long-range forecasting of water availability and the scheduling of reservoir water releases for an interval of time that depends on the reservoir capacities; typical long-range scheduling covers anywhere from one week to one year or several years.

The aim of this chaper is to study the optimal long-term operation of a hydrothermal power system, where the objective function is to minimize the thermal fuel cost and at the same time satisfy the hydro constraints on the system. Several mathematical techniques are at present available for solving the optimal scheduling problem, including Pontryagin's maximum principle, where it is applied for one thermal plant connected electrically with several reservoirs as we will see later on in this chapter. The dynamic programming approach restricts the size of the problem to be solved as it has the "curse of dimensionality." Nonlinear programming technique is used to solve the long-term optimal scheduling of hydrothermal power system together with the conjugate gradient method, where the equality constraints are adjoined using Lagrange multipliers and the inequality constraints using penalty functions. A direct method is applied to solve the problem; in this technique a direction of movement towards the optimum value is to be found. The active constraints are included in determining the direction of movement. Until now solutions to the hydrothermal scheduling problem have been carried out generally with the help of the methods that are normally used for unconstrained minimization.

The aggregation–decomposition approach has been applied to the optimal scheduling of large hydrothermal generation systems with multiple reservoirs. The problem (with N reservoirs) is decomposed into N subproblems with two state variables. Each subproblem finds the optimal operating policy for one of the reservoirs as a function of the energy content of the remaining reservoirs. The subproblems are solved by stochastic dynamic programming, taking into account the detailed models of the hydro chains as well as the stochasticity and correlation of the hydro inflows. Decomposition and coordination techniques are applied to minimize the operational cost of hydrothermal power systems. Through this formulation large systems are dealt with by a more general and precise model that includes thermal operation costs and hydroelectric power generation as nonlinear functions, water head variations, hydraulic networks with cascade plants, and spilling.

This chapter starts with the optimal long-term operation of all thermal power systems, where we discuss the optimal static dispatch with linear, quadratic, and nonlinear functions of the fuel costs. In these discussions, we neglect the transmission line losses. These losses may be considered as a constant during the optimization interval, which is at least a week or a month, and we consider it as a part of the load.

7.2. All-Thermal Power Systems (Refs. 7.1 and 7.2)

In this section we will discuss methods for solving the optimal static dispatch problem for thermal units. The transmission losses can be neglected. Consequently, these losses are not included in our problem formulation, reducing this problem considerably. In the unit commitment and optimal dynamic problems, the static dispatch problem is needed. We mean by the static dispatch problem how the power P (MW), demand, is distributed among the units in operation so that their fuel costs are minimal, while satisfying all the constraints.

The fuel cost function is derived from the fuel-consumption function, which can be measured. The unit is stabilized at a certain power point, after which the power and the fuel consumption can be measured during one or two hours. Based on these measurements a polynomial can be selected to fit these measurement points. Table 7.1 shows the relative error as a function

Table 7.1. Relative Error (%) as Function of the Order k and the Number of Measurements m

k	$m = 4$	$m = 7$
0	25.00	33.00
1	0.70	1.00
2	0.20	0.50
3	0.00	0.20

of the polynomial k and the number of measurements for one of the thermal units with a maximum power of 200 MW. Except for valve points, it is reasonable to assume that the fuel cost $F(P)$ will be a continuous function of P, so $F(P)$ is selected to be a polynomial of degree k. With the least-squares method the error between the measurements and this polynomial are minimized.

Suppose n units are selected, each having power P_i and fuel costs $F_i(P_i)$. Then the optimal dispatch problem can be formulated as

$$\min_{P_i} F(P) = \min_{P_i} \sum_{i=1}^{n} F_i(P_i) \tag{7.1a}$$

subject to satisfying the following constraints:

$$\sum_{i=1}^{n} P_{i,k} = D_k, \qquad k = 1, \ldots, K \tag{7.1b}$$

$$P_i^m \le P_{i,k} \le P_i^M, \qquad k = 1, \ldots, K \tag{7.1c}$$

Depending on the order of the polynomials $F_i(P_i)$, different methods are available for solving problem (7.1). In the following sections several different formulations of (7.1) will be discussed, namely:

- Piecewise linear F_i and lower and upper limits for the power of each unit;
- Convex, quadratic F_i and lower and upper limits for the power of each unit;
- General solution.

7.2.1. Linear Fuel-Cost Function

In this model the fuel-cost function $F_i(P_i)$ is approximated by piecewise linear functions, together with the active power balance equation and the inequality constraints on the limits of the power generated. In this case problem (7.1) can be written

$$\min_{P_i} i \sum_{k=1}^{K} \sum_{i=1}^{n} (\alpha_i P_{i,k} + \beta_i) \tag{7.2a}$$

Subject to

$$\sum_{i=1}^{n} P_{i,k} = D_k, \qquad k = 1, \ldots, K \tag{7.2b}$$

$$P_i^m \le P_{i,k} \le P_i^M, \qquad k = 1, \ldots, K \tag{7.2c}$$

The optimization is taken over the optimization interval, which may be a weekly or a monthly optimization.

The above equations are solved by using linear programming using what is called the simplex method. Reference 7.1 offers a faster solution method, which takes advantage of the special structure of this problem. In the following section we explain the algorithm of this method.

Algorithm of Solution (1a)

Step 1. $P_{i,k} = P_i^m$, $\qquad i = 1, \ldots, n$

$$\text{SUM} = \sum_{i=1}^{n} P_{i,k} = \sum_{i=1}^{n} P_i^m$$

Step 2. Determine i with $b_i \le b_j, j = 1, \ldots, n$

$$\text{SUM} = \text{SUM} - P_i^m$$

$$P_{i,k} = D_k - \text{SUM}$$

If $(P_{i,k} > P_i^M)$ THEN BEGIN

$$\text{SUM} = \text{SUM} + P_i^M$$
$$P_{i,k} = P_i^M$$
$$b_i = \text{BIG}$$

go to step 2

END

7.2.2. Quadratic Fuel-Cost Function

The model used for the fuel-cost function is a convex quadratic function of the power, together with the active power balance equation and the constraints on the power. The following problem arises:

$$\min_{P_i} \sum_{i=1}^{n} \sum_{k=1}^{K} (\alpha_i P_{i,k}^2 + \beta_i P_{i,k} + \gamma_i) \tag{7.3a}$$

$$\sum_{i=1}^{n} P_{i,k} = D_k, \qquad k = 1, \ldots, K \tag{7.3b}$$

$$P_i^m \le P_{i,k} \le P_i^M \tag{7.3c}$$

Assume, for instance, that the inequality constraints are neglected. Then, we can form an augmented cost functional by adjoining the equality constraints (7.3b) to (7.3a) via Lagrange multiplier λ_k as

$$L(P_{i,k}, \lambda_k) = \sum_{k=1}^{K} \left\{ \sum_{i=1}^{n} [(\alpha_i P_{i,k}^2 + \beta_i P_{i,k} + \gamma_i)] - \lambda_k \left(\sum_{i=1}^{n} P_{i,k} - D_k \right) \right\} \tag{7.4}$$

For $L(P_{i,k}, \lambda_k)$ to be minimum, $\partial L / \partial P_i = 0$, $i = 1, \ldots, n$, which can be written as

$$\lambda_k = \frac{\partial L}{\partial P_1} = \frac{\partial L}{\partial P_2} = \cdots = \frac{\partial P}{\partial P_i} \tag{7.5}$$

which yields to the following optimum equations:

$$\lambda_k = \left[D_k + \sum_{i=1}^{n} \left(\frac{\beta_i}{2\gamma_i} \right) \right] \bigg/ \left(\sum_{i=1}^{n} \frac{1}{2\gamma_i} \right) \tag{7.6}$$

$$P_{i,k} = (\lambda_k - \beta_i)/2\alpha_i \tag{7.7}$$

Equations (7.5)-(7.7) simply state that for L to be a minimum, each unit has the same incremental cost λ_k given by (7.6) or given by

$$\lambda_k = \beta_i + 2\alpha_i P_{i,k} \tag{7.8}$$

Two methods will be discussed to include the inequality constraints (7.3c). Both methods search for a solution $P_{i,k}$ from (7.7) that yields equal incremental costs for each unit, unless this is prevented by a lower or an upper limit.

i. The First Algorithm (2a)

In this algorithm, we search for a value of λ that satisfies

$$FL(\lambda_k) = \sum_{i=1}^{n} P'_{i,k}(\lambda) - D_k = 0$$

with

$$P'_{i,k}(\lambda) = \min\left[\max\left(\frac{\lambda_k - \beta_i}{2\alpha_i}, P_i^m \right), P_i^M \right]$$

The value of λ is limited between a lower value, λ_k^m, and an upper value, λ_k^M, with

$$\lambda_k^m = \beta_i + 2\alpha_i P_{i,k}^m, \qquad k = 1, \ldots, K \tag{7.9a}$$

$$\lambda_k^M = \beta_i + 2\alpha_i P_{i,k}^M, \qquad k = 1, \ldots, K \tag{7.9b}$$

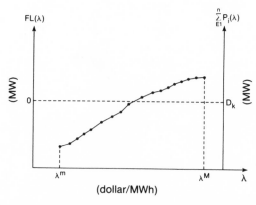

Figure 7.1. Cohesion among the incremental costs λ, the demand D_k, and the function $FL(\lambda)$.

Function $FL(\lambda)$ is a piecewise-linear, nondecreasing function of λ between the values of λ_k^m and λ_k^M, with $FL(\lambda^m) < 0$ and $FL(\lambda^M) \geq 0$ as illustrated in Figure 7.1. The solution of $FL(\lambda) = 0$ can be determined, for example, with the bisection method, by halving each time the interval in which the solution is present. This can be tested because $FL(\lambda)$ is positive if λ is too large and negative if λ is too small. The starting interval can be (λ^m, λ^M). This iterative process is stopped if $|FL(\lambda)| \leq \varepsilon$, with $\varepsilon > 0$ as a predefined accuracy measure.

ii. The Second Algorithm (2b)

The solution of (7.3) with equations (7.6) and (7.7) can be done as follows:

Step 1. Calculate λ_k from (7.6), $P_{i,k}$ from (7.7), and

$$PX_k = \sum_{i=1}^{n} \max[0, (P_{i,k} - P_i^M)] \qquad (7.10a)$$

$$PN_k = \sum_{i=1}^{n} \max[0, P_i^m - P_{i,k}] \qquad (7.10b)$$

Step 2. If $(PN_k = PX_k = 0)$
Then $P_{i,k} = , i = 1, \ldots, n$ is the solution of (7.3)
ELSE BEGIN
IF $(PX_k \geq PN_k)$ THEN
BEGIN IF $(P_{i,k} < P_i^M)$ THEN BEGIN
$\qquad\qquad P_{i,k} = P_i^M$
$\qquad\qquad D_k = D_k - P_i^M$
$\qquad\qquad$ remove Unit i
$\qquad\qquad n = n - 1$
$\qquad\qquad$ END

Step 3. Go to Start
END

Step 4. $\qquad\qquad$ ELSE
BEGIN IF $(P_{i,k} < P_i^m)$ THEN BEGIN
$\qquad\qquad P_{i,k} = P_i^m$
$\qquad\qquad D_k = D_k - P_i^m$
$\qquad\qquad$ remove Unit i
$\qquad\qquad$ END

Step 5. Go to Start
END

Some units can be removed from the problem formulation, because their power is fixed at their lower or upper limit. Then the algorithm proceeds with fewer units, and correspondingly, with a lower value of the demand, until $PN_k = PX_k = 0$.

To explain this algorithm in detail, we offer a simple example of two units; the calculations for this example can be done by hand.

Example 7.1. The fuel input costs for two units supplying a common load without transmission losses are given by

$$F_1(P_1) = 8.644P_1 + 0.010707P_1^2 \text{ (MBtu/hr)}$$

$$F_2(P_2) = 7.552P_2 + 0.014161P_2^2 \text{ (MBtu/hr)}$$

with the following limits:

$$0 \le P_1 \le 50 \text{ MW}$$

$$0 \le P_2 \le 200 \text{ MW}$$

Determine the output of each unit, if the common system load has 150 MW.

Step 1. Calculate λ and P_i, $i = 1, 2, \ldots$, neglecting the inequality constraints, from equations (7.6) and (7.8) as

$$\lambda = 11.4316$$

$$P_1 = 13 \text{ MW}$$

$$P_2 = 137 \text{ MW}$$

Step 2. Calculate PN and PX from (7.10a) and (7.10b) as

$$PX = 0, \qquad PN = 0$$

Step 3. Since $PX = PN = 0$, the values of P_1 and P_2 are the optimum. Suppose that the load increased to 250 MW; then

$$\lambda = 13.9295$$

and

$$P_1 = 24.68 \text{ MW}$$

$$P_2 = 225.3175 \text{ MW}$$

Step 4. Calculate PX and PN:

$$PX = \sum_{i=1}^{n} \max[0, (P_i - P_i^M)]$$

$$= 0 + 25.3175 = 25.3175$$

$$PN = \sum_{i=1}^{n} \max[0, (P_i^m - P_i)]$$

$$= 0$$

Since $PX > PN$, this means that one of the units violates limits; from step 1, P_2 violates the upper limit; then we put $P_2 = P_2^M$.

Step 5. Put $P_2 = P_2^M = 200$ MW, and hence

$$D = 250 - 200 = 50 \text{ MW}$$

this load should be supplied by the first unit:

$$P_1 = 50 \, M$$

Step 6. Calculate again PX and PN as

$$PX = 0$$
$$PN = 0$$

i.e., the values for P_1 and P_2 are the optimum.

The second algorithm (2b) finds the accurate solution with at most n iterations, each new iteration with fewer units. This algorithm is more accurate than algorithm (2a) and it can be faster.

In Table 7.2, the average calculation times are given for a number of algorithms for solving the dispatch problems with either a linear, a piecewise linear, or a quadratic fuel-cost function. One unit of calculation time is 1 msec for an Amdahl V7B mainframe computer. A PDP 11/60 minicomputer requires about seven times more calculation time. This table is illustrated in Figure 7.2. The dotted line in Figure 7.2 illustrates the calculation time requirements for solving (7.3) with the gradient projection method.

Table 7.2. Average Calculation Times (msec) for Solving the Static Dispatch Problem

Number of units n	Linear (2)		Piecewise linear		Quadratic (3)	
	1a	1b	1b	2a	2b	Gradient projection
5	0.17	0.12	0.4	0.7	0.15	3
10	0.4	0.3	1.8	1.4	0.3	13
15	0.7	0.4	4	2.0	0.4	28
25	2.4	1.2	9	3.6	0.8	210
50	9	4.8	39	7	1.6	2000[a]
100	34	17	152	13	3.5	
150	75	39	400[a]	21	6	
200	132	69	650[a]	30	8	
250	206	106	1000[a]	37	11	
300	296	152	1600[a]	43	13	

[a] Estimated.

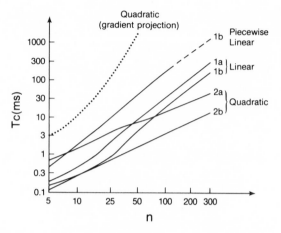

Figure 7.2. Average calculation time T_c for solving the dispatch problem.

7.2.3. General Solution Method for Nonlinear Functions

In this section, we will discuss a general solution method, which deals with a utility company with n units. n_1 of these units uses fuel one, while the rest $(n - n_1)$ use fuel two. Suppose there is a limitation in the use of fuel one, namely, Q_k (GJ/hr). Then the problem can be formulated as follows. Find $P_{i,k}$ over the optimization interval that

$$\min_{P_i} \sum_{k=1}^{K} \left\{ \sum_{i=1}^{n} [(\alpha_i P_{i,k}^2 + \beta_i P_{i,k} + \gamma_i) P_{i,k}] \right\} \qquad (7.11a)$$

Subject to satisfying the following constraints:

$$\sum_{i=1}^{n} P_{i,k} = D_k, \qquad k = 1, \ldots, K \qquad (7.11b)$$

$$P_i^m \leq P_{i,k} \leq P_i^M, \qquad k = 1, \ldots, K \qquad (7.11c)$$

$$\sum_{i=1}^{n_1} (\alpha_i P_{i,k}^2 + \beta_i P_{i,k} + \gamma_i) \leq Q_k, \qquad k = 1, \ldots, K \qquad (7.11d)$$

Problem (7.11) can be reformulated by using a penalty function $B(P)_k$ ($/hr) to include the nonlinear inequality constraint in the cost function. The influence of the penalty function is illustrated in Figure 7.3. Problem (7.11) can be written as a nonlinear programming problem with only linear constraints. Find $P_{i,k}$, as a function of time, which minimizes

$$\min_{P_i} \sum_{k=1}^{K} \left\{ \sum_{i=1}^{n} [(\alpha_i P_{i,k}^2 + \beta_i P_{i,k} + \gamma_i) P_{i,k} + B(P)_k] \right\} \qquad (7.12a)$$

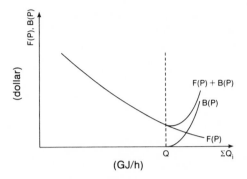

Figure 7.3. Influence of the penalty function $B(P)$ on the criterion $F(P) + B(P)$.

Subject to satisfying the following constraints:

$$\sum_{i=1}^{n} P_{i,k} = D_k, \qquad k = 1, \ldots, K \tag{7.12b}$$

$$P_i^m \le P_{i,k} \le P_i^M, \qquad i = 1, \ldots, n, \qquad k = 1, \ldots, K \tag{7.12c}$$

$$B(P)_k = \alpha\{\min[C, (QT_k - Q_k)]\}^2 \tag{7.12d}$$

with

$$QT_k = \sum_{i=1}^{n_1} (\alpha_i P_{i,k}^2 + \beta_i P_{i,k} + \gamma_i) \tag{7.12e}$$

The size of the penalty factor α [$\$hr/GJ^2$) determines the accuracy with which the constraint will be satisfied. α has been chosen to be 1, to avoid the iterative solution of (7.12) with increasing values of α. The majority of optimal dispatch problems deals with linear constraints. If nonlinear constraints are in the problem formulation, they are incorporated into the criterion by means of a penalty function (Ref. 7.1). Linear constraints can be dealt with advantageously in the projection and reduction methods. Bosch, in his paper, Ref. 7.1, was able to avoid the calculation of the project matrix, which favors the projection methods. Therefore he selected a modified conjugate gradient-projection method for the solution of the nonlinear programming problems with linear constraints. Such a nonlinear optimization problem has the appearance

$$\min_{P_i} \sum_{k=1}^{K} \left[\sum_{i=1}^{n} F_i(P_{i,k}) \right] \tag{7.13a}$$

$$\sum_{i=1}^{n} P_i = D_k, \qquad k = 1, \ldots, K \tag{7.13b}$$

$$P_i^m \le P_{i,k} \le P_i^M, \qquad i = 1, \ldots, n, \qquad k = 1, \ldots, K \tag{7.13c}$$

A conjugate search direction u can be derived that satisfies the equality constraint and all active inequality constraints.

If there are m active inequality constraints in (7.13), suppose for the first m units, then the projected negative gradient gP is

$$gP_i = 0, \qquad i = 1, \ldots, m \tag{7.14a}$$

$$gP_i = -g_i + \frac{1}{n-m} \sum_{j=m+1}^{n} g_j, \qquad i = m+1, \ldots, n \tag{7.14b}$$

Equations (7.14) are used to avoid the time-consuming calculation of the project matrix. The vectors u, with l the iteration number, is given by

$$u^l = -gp^l + b^l u^{l-1} \tag{7.15a}$$

with

$$\beta^l = \frac{(gp^l)^T g^l}{(gp^{l-1})^T (gp^{l-1})} \tag{7.15b}$$

$$\beta^1 = 0 \tag{7.15c}$$

are mutually conjugate in the subspace R^{n-m-1} spanned by the vectors of these equality constraints and the m active inequality constraints. If the set of active inequality constraints changes, β is reset to the value zero. The developed conjugate gradient projection allows a fast and accurate solution of the problem. Bosch in Ref. 7.1 has compared the computing times of the conjugate gradient-projection method and the gradient-projection method of Rosen for solving problem (7.13). Table 7.3 indicates these results. It is

Table 7.3. Comparison between the Conjugate Gradient Projection Method and the Gradient Projection Method for Solving (7.13) with the Nonlinear Constraint Violated and Not Violated[a]

| | Conjugate gradient projection method | | | | | | Gradient projection method violated | | |
| | Not violated | | | Violated | | | | | |
n	Tc	Ng	Nm	Tc	Ng	Nm	Tc	Ng	Nm
5	1	2	2.4	8	11	0.6	17	14	2
10	2	3	1.4	20	19	3	220	68	5
20	6	4	2.6	65	32	7	605	48	10
30	15	7	6.0	155	51	13	2005	71	16
40	30	10	8.5	240	60	16	6710	139	20
50	50	12	10.5	355	70	18	b		
60	70	14	12.0	650	108	24	b		

[a] Tc is the average calculation time in milliseconds; Ng is the average number of gradient calculations; and Nm is the average number of modifications of the projection matrix.
[b] Not calculated because of too many constraints (>100).

assumed that the amount of fuel for one third of the units is limited by an upper bound. With n units, there are $2n + 1$ linear constraints, and one nonlinear inequality constraint. This nonlinear constraint is incorporated into a penalty function. From Table 7.3 one can conclude that the gradient-projection method requires about 2, for $n = 5$, to about 30 times, for $n = 40$, more calculation time than the proposed projection method [(7.14) and (7.15)].

7.2.4. Dynamic Programming Approach for Nonmonotonic Heat Rate Curves (Ref. 7.2)

Using a third or higher-order polynomial curve to model a unit's input/output characteristics increases the accuracy of the model, but the resulting incremental heat rate may not be monotonically increasing. Figure 7.4 illustrates the results of a unit with multivalued incremental heat rate points from lambda of 8.2 to 8.7 MBtu/MW hr. The problem arises when many such units must be dispatched.

Classical methods of economic dispatch using the Lagrangian multiplier are valid only for monotonically increasing values of incremental heat rate. Previous methods of accounting for these aberrations included ignoring the lower regions of the curve since most units would normally be dispatched at their higher load levels. However, with the increasing number of less expensive base generation units on line, the other more expensive units are being operated in their lower load range more often. Past methods of either flattening out the lower portion of the IHC or ignoring it would no longer guarantee an accurate economic dispatch.

Thus, a method of calculating new pseudoincremental heat rate curves was developed using dynamic programming that would overcome this problem.

The first step to this process uses DP to calculate each unit's output over the range of the sum of minimums to the sum of the maximums on all units to be dispatched.

The results of each unit's output versus the range of system load were smoothed by a curve-fitting routine. Using another DP-based method, the curve fit was then divided into connected straight line segments that were monotonically increasing and immediately applicable to conventional dispatching algorithms. Loss coefficients can be applied in the same manner as in conventional coordination equations.

Economic dispatch using DP can be considered an allocation problem. The DP solution requires the following two assumptions:

1. Losses in the power system are ignored for the DP calculation only.
2. It is necessary to find the economic dispatch at discrete load steps rather than for continuous load levels.

The dynamic programming problem is characterized by stages, states, and decision variables at every stage, and a recursive relationship. The recursive relationship identifies the optimal policy for each state with n stages remaining, given the optimal policy for each state and $(n - 1)$ stages remaining.

Every generating unit is associated with a stage. Thus, the number of stages in the EDC problem is equal to the number of available units for dispatch. The states in the EDC problem are the total megawatts of generation at a particular stage. The range of values is from the sum of the minimum outputs of all units considered in that stage to the sum of the maximum of the units in that stage.

The decision made at a particular stage is the megawatts of generation that can be allocated to the generating unit associated with that stage. The megawatts allocated are assumed to be discrete.

The recursive relationship used is

$$f_n^*(X_n) = \min_{Y_n} [C_n(Y_n) + f_{n-1}^*(X_n - Y_n)], \qquad n = 2, N \qquad (7.16)$$

for

$$a_n \leq Y_n \leq b_n \qquad (7.17)$$

and

$$\sum_{i=1}^{n} a_i \leq X_n \leq \sum_{i=1}^{n} b_i \qquad (7.18)$$

where n is the stage number (i.e. units considered), X_n is the state of the system (i.e., MW of generation), Y_n is the decision being evaluated at stage n, $C_n(Y_n)$ is the operating cost of unit n at Y_n output, a_n is the minimum output of the nth unit, and b_n is the maximum output of the nth unit.

The DP solution to the economic dispatch problem requires setting up a stage corresponding to each generating unit, which involves finding the optimal f_n and Y_n corresponding to each state of a stage using the recursive relationship. A backward pass through all the stages determines the output of each unit for a particular system load.

Note that the DP solution determines each unit's optimal output without regard to its ability to reach those load levels within the time constraints of the system requirement.

Thus, reviewing the results of the DP based dispatch of the units versus the system generation indicates that the curves must be smoothed to make the dispatch practical.

The curve-fitting program uses a standard least-squares curve-fitting routine and calculates for each unit the best order of curve fit from a second order to a sixth order. The data used for the curve fit is the range of system load over which the unit responds from its minimum to maximum generation.

The statistical value of variance and R-squared were used to determine which order of curve fit best modeled the results of the DP dispatch:

$$\text{var} = \sum_{i=1}^{N} (\hat{Y}_i - Y_i)^2 / (N - \text{coef}) \tag{7.19}$$

and

$$R^2 = \sum_{i=1}^{N} (\hat{Y}_i - Y)^2 \Big/ \left[\sum_{i=1}^{N} (\hat{Y}_i - Y)^2 + \sum_{i=1}^{N} (\hat{Y}_i - Y_i)^2 \right] \tag{7.20}$$

where N is the number of samples, $\hat{Y}_i$ is the value of the smoothed point, Y_i is the value of the raw data point, Y is the average value of all raw data, and coef is the order of curve fit $+ 1$.

In real operation, some of the thermal units have monotonically increasing incremental heat rate curves (IHC), while other units do not have that. Reference 7.2 offered incremental heat rate curves for three thermal units; these are given in Figures 7.4, 7.5, and 7.6; also the dynamic programming dispatch (DP) of the units is shown in Figures 7.7, 7.8, and 7.9. These figures show unit MW output versus system generation. These results illustrate that DP is not dependent on the shape of the incremental heat rate curve. It has been shown in Ref. 7.2 that for these three units a fifth-order polynomial best represented the DP dispatch.

Reference 7.2 develops a program for DP determination of straight line segments; the program requires the coefficient for the selected order of curve fit for each unit as well as the unit's maximum and minimum output; also the program requires as input the number of discrete straight line segments desired to model the resultant curve developed. The program

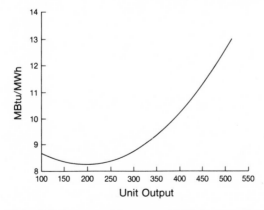

Figure 7.4. Incremental heat curve (IHC) for thermal unit 1.

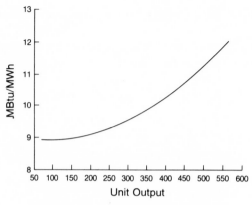

Figure 7.5. IHC curve for thermal unit 2.

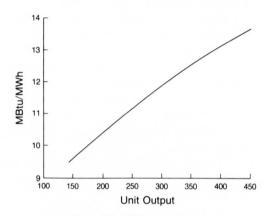

Figure 7.6. IHC curve for unit 3.

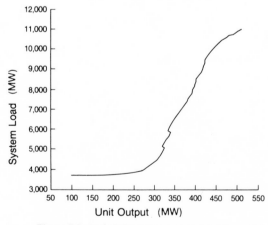

Figure 7.7. Unit MW versus system load for unit 1.

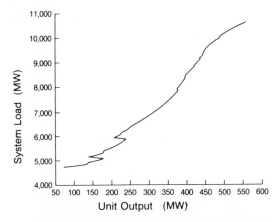

Figure 7.8. Unit MW versus system load for unit 2.

performs a DP evaluation to minimize the error of determining where along the curve the required break points should lie.

The piecewise linear representation determined for each unit is based on DP results, which consider all units available at the same time (i.e., all units committed).

7.2.4.1. Comparison of the Resultant Conventional Dispatch versus DP Economic Dispatch

The resultant piecewise linear curves for the units were installed in a conventional dispatch algorithm. A dispatch was performed over the entire

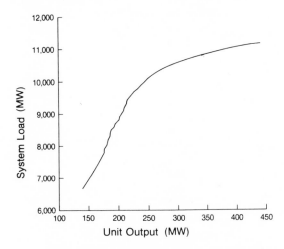

Figure 7.9. Unit MW versus system load for unit 3.

Table 7.4. Comparison of DP Dispatch versus Conventional Dispatch with Results of DP

Load (MW)	Dp dispatch–conventional dispatch (MBtu)	Percent error
2900	2	0.006
3100	16	0.041
3300	25	0.062
3500	133	0.317
3700	132	0.301
3900	84	0.183
4100	55	0.118
4300	2	0.003
4500	1	0.002
4700	13	0.025
4900	113	0.208
5100	65	0.115
5300	3	0.004
5500	1	0.002
5700	1	0.002
5900	1	0.002
6100	3	0.004
6300	2	0.004
6500	3	0.004
6700	4	0.005
6900	4	0.006
7100	6	0.007
7300	6	0.009
7500	6	0.008
7700	45	0.057
7900	16	0.019
8100	10	0.013
8300	9	0.011
8500	8	0.010
8700	8	0.009
8900	8	0.008
9100	7	0.008
9300	7	0.007
9500	5	0.005
9700	3	0.003
9900	3	0.002
10100	2	0.002
10300	2	0.002
10500	1	0.001
10700	1	0.001
10900	1	0.001
11100	0	0.000
11300	6	0.005
		Average error = 0.04%

range of available dispatch and the cost to dispatch these discrete system loads was compared to the cost to dispatch by an optimal DP solution. Table 7.4 shows the results of a DP dispatch compared to the results of a conventional dispatch using the derived pseudo-IHR curves in piecewise linear form. The percent error is shown to be very small.

As Table 7.4 illustrates, the errors for various levels of dispatch are extremely small, the largest error being only 0.317% and the average error over the range of the dispatch being only 0.04%.

A separate study was performed to evaluate the performance of the new technique against a more conventional method of obtaining incremental heat rate curves. The original input–output curves were modeled as a second-order rather than a third-order curve. Since the incremental of a second-order input–output curve is a straight line and in our case was constrained to be a monotonically increasing straight line, we were able to dispatch using these curves without modifications.

The cost of the resultant dispatch was again compared to the pure DP dispatch. We used the original third-order equations to price the unit's output at the new dispatch. Predictably the average error increased to 0.423%, approximately ten times the error of the DP process, and the maximum error was 0.755% in the lower range of the dispatch. Table 7.5 shows the results of this dispatch.

It is interesting to note that although the error is still small, a better dispatch was obtained using the third-order equations. The second-order equation model showed that for long-term future studies this model is satisfactory, but for real-time dispatch the new method provided is more accurate.

7.2.5. Conclusion

In this section we discussed the long-term optimal operation of all-thermal power systems. Most of the studies that have been done consider the transmission line losses as a constant for long-term study and a part of the electrical demand on the system. In the first part of this section we have reviewed some algorithms used for the optimal dispatch problem. It has been shown that these algorithms are faster and are more accurate as well, and require less memory storage than general-purpose solution methods, such as the gradient-projection method. It is advisable, from the point of view of calculation-time requirements, to use convex, quadratic instead of linear fuel-consumption functions.

In the second part of this section, we discussed how to handle the problem inherent in nonmonotonically increasing incremental heat rate curves by using dynamic programming. The results of the dispatch can be

Table 7.5. Comparison of DP Dispatch versus Second-Order Dispatch
with results of DP

Load (MW)	DP dispatch–second-order dispatch (MBtu)	Percent error
2900	3	0.008
3100	44	0.111
3300	158	0.390
2500	195	0.463
3700	185	0.421
3900	215	0.472
4100	234	0.497
4300	288	0.586
4500	330	0.650
4700	354	0.676
4900	373	0.686
5100	384	0.683
5300	415	0.716
5500	443	0.744
5700	463	0.755
5900	476	0.754
6100	484	0.744
6300	482	0.721
6500	475	0.691
6700	462	0.655
6900	440	0.608
7100	412	0.553
7300	394	0.518
7500	366	0.469
7700	399	0.500
7900	338	0.413
8100	344	0.411
8300	432	0.505
8500	353	0.403
8700	325	0.363
8900	293	0.320
9100	256	0.273
9300	225	0.235
9500	189	0.193
9700	153	0.154
9900	129	0.126
10100	123	0.118
10300	115	0.108
10500	80	0.074
10700	46	0.042
10900	23	0.020
11100	61	0.053
11300	68	0.058
		Average error = 0.423%

used effectively to create a set of curves that behave like conventional incremental heat rate curves.

7.3. Optimal Scheduling of Hydrothermal Power Systems

The optimal scheduling of hydrothermal power systems for long-term operation is basically a nonlinear programming problem with a nonlinear objective function and a mixture of linear and nonlinear constraints. The aim of this section is to discuss different techniques used to solve the problem. In the first part of this section, the problem is treated as a deterministic problem, and we discuss different techniques used to solve this deterministic problem including the maximum principle, a feasible direction approach, the conjugate gradient approach, and decomposition and coordination techniques, while in the second part of this section, we discuss different techniques used to solve the stochastic long-term optimal scheduling of hydrothermal power systems, stochastic load, and natural inflows.

The negligible marginal cost of hydroelectric generation makes the hydrothermal power system optimization a problem of how to use, in a given period of time, the water availability for hydroelectric generation to reduce thermal production. Until now solutions to the hydrothermal scheduling problem have been carried out generally with the help of methods that are normally used for unconstrained minimization.

7.3.1. A Direct Method Approach (Ref. 7.17)

This section presents the formulation and the optimal solution of the problem in terms of what may be called a direct method. The fundamental principle of this technique is to find a direction of movement towards the optimum value. The active constraints are included in determining the direction of movement such that it always remains in the feasible domain.

For the long-term scheduling problem the objective function is to schedule the water discharge from each hydroplant during each optimization interval in such a way that the total cost of thermal generation during the entire period is minimum.

In mathematical terms, the problem can be formulated as

$$\min F = \sum_{i \in R_s} \sum_{k=1}^{K} F_i(P_{s_{i,k}}) \tag{7.21}$$

where $P_{s_{i,k}}$ is the average thermal power generation from plant i during period k and R_s is a set of thermal plants.

This optimization should satisfy a number of equality and inequality constraints in both the thermal plants and the hydroplants. In other words, the following constraints should be satisfied.

(1) The active power balance equation between the generation and the demand should be satisfied at each optimization interval. This equation can be expressed as

$$\sum_{i \in R_h} P_{h_{i,k}} + \sum_{i \in R_s} P_{s_{i,k}} = D_k + P_{L,k}, \qquad k = 1, \ldots, K \qquad (7.22)$$

where $P_{h_{i,k}}$ is the average hydroelectric power generated by the ith plant, during period k; D_k is the system demand during period k; $P_{L,k}$ is the system transmission losses; and R_n is a set of hydroplants. As we said earlier, in the beginning of the chapter, we can consider the transmission lines losses as a constant, and add them to the system demand during the optimization interval.

(2) The reservoir dynamics may be adequately described by a continuous type difference equation as

$$x_{i,k} = x_{i,k-1} + I_{i,k} - u_{i,k} + u_{(i-1),k} \qquad (7.23)$$

The above equation is valid with the assumption that no spillage occurs during the optimization interval. From equation (7.23), we can find the final storage at the reservoir supplying the ith hydroplant as

$$x_{i,K} = x_{i,0} + \sum_{k=1}^{K} I_{i,k} - \sum_{k=1}^{K} u_{i,k} + \sum_{k=1}^{K} u_{(i-1),k} \qquad (7.24)$$

(3) The output of the ith hydroplant during a period k may be approximated by

$$P_{h_{i,k}} = H_i(1 + c_i x_{i,k})(u_{i,k} - s_{i,k}) \qquad (7.25)$$

where H_i is the basic water head, c_i is the correction factor of the water head for the change of water storage, $u_{i,k}$ is the discharge from plant i, and $s_{i,k}$ is the spill from reservoir i during a period k.

(4) For multipurpose stream use requirements such as flood control, navigation, irrigation, and other purposes, if any, the following limits on the variables should be satisfied.

i. Lower and upper bounds on the reservoir storage

$$x_i^m \leq x_{i,k} \leq x_i^M, \qquad k = 1, \ldots, n, \qquad k = 1, \ldots, K \qquad (7.26)$$

ii. Lower and upper bounds on the outflow

$$u_{i,k}^m \leq u_{i,k} \leq u_{i,k}^M, \qquad i = 1, \ldots, m, \qquad k = 1, \ldots, K \qquad (7.27)$$

(5) To control the temperature rise for the generating units, and at the same time control their efficiencies, there are upper and lower limits on

generator capacities. These are given by

$$P_{s_i}^m \le P_{s_{i,k}} \le P_{s_i}^M, \qquad i \in R_s \tag{7.28}$$

$$P_{h_i}^m \le P_{h_{i,k}} \le P_{h_i}^M, \qquad i \in R_h \tag{7.29}$$

Among the above constraints, those given by equations (7.22), (7.23), and (7.24) are hard constraints. The rest of the constraints are soft constraints. To find a feasible direction these hard constraints are taken into account, while the inequality constraints are checked during the solution process so that none of them are violated.

7.3.1.1. Method of Feasible Direction

This method is a result of the gradient technique, in which we seek a new set of variables given by

$$X^{i+1} = X^i + \alpha^* \cdot S^i \qquad i = \text{iteration counter} \tag{7.30}$$

such that

$$J(X^{i+1}) < J(X^i) \tag{7.31}$$

where α^* is the optimum step size length. It is determined from one-dimension minimization.

The efficiency of this method lies in finding the feasible direction S^i, which satisfies the following criteria.

(i) If we take a small step along the direction S, without leaving the feasible domain, S is called a feasible direction. In mathematical terms this can be expressed as

$$S^T \nabla g^i \le 0 \tag{7.32}$$

In other words, there is some $\alpha > 0$ for which x^{i+1} is the feasible domain as shown in Figure 7.10.

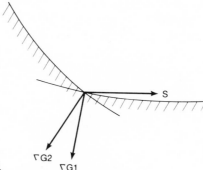

Figure 7.10. The feasible direction S.

(ii) S is a usable feasible direction if

$$S^T \nabla F \le 0 \qquad (7.33)$$

Any vector satisfying this condition is guaranteed to produce a new feasible point that reduces the value of the function, which can be observed from Figure 7.11.

A direction vector S satisfying the above two conditions can be obtained through the projection approach, which subtracts all components parallel to the normals of the active constraints from the negative of the gradient of the objective function:

$$S = -\nabla F - \sum_{j \in J} l_j \cdot \nabla g_j \qquad (7.34)$$

where l_j are scalar multipliers.

Equation (7.34) can be written as

$$S = -P\nabla F \qquad (7.35)$$

where P is the projection matrix as described in Appendix B.

7.3.1.2. Computation Logic

A simple flow chart for the method of feasible direction is given in Figure 7.12. A few important points are as follows:

The value of the initial step size in stage 2 can be calculated as equal to

$$\frac{-2(f^{est} - f^0)}{\nabla F^T \cdot \nabla F} \qquad (7.36)$$

where f^{est} is the estimated value of the function.

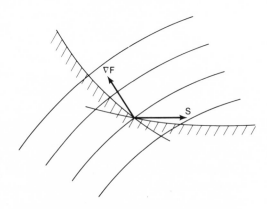

Figure 7.11. The feasible directions S and ∇F.

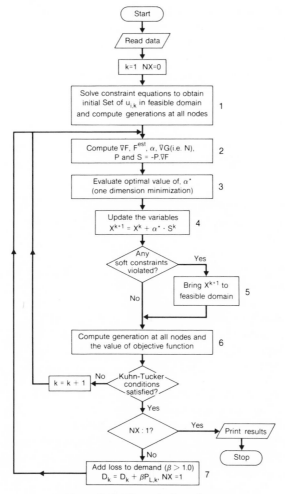

Figure 7.12. The main flow chart.

At stage 5, if the soft constraints are in violation the variables are brought back to their bound (upper or lower) values and a new set of variables is obtained in the feasible domain. At stage 6, the thermal generation can be distributed by equal incremental cost criteria. At stage 7, the losses may be taken into account by a loss coefficient matrix, or if a load flow study is a part of the solution process, they can be directly taken from the load flow study. The former approach is used. The coefficient β is used to augment the losses to correspond to the final values of the system generation. This is necessary because losses are first calculated with the help of a generation schedule obtained without considering the system losses. It is possible to compute the value of β fairly accurately if the loss coefficient matrix is known. In this particular study a value of 1.1 has been used. Since the losses are included at a stage when a suboptimal solution has already been obtained, the convergence of the final stage is extremely fast.

7.3.1.3. Practical Examples

The above algorithm is applied to two practical systems. The data for the first example, which consists of one thermal and one hydroelectric plant, are given by

Maximum water storage	$115 \times 10^4 \, m^3$
Minimum water storage	$0.0 \, m^3$
Basic water head	$20.0 \, m$
Maximum water head	$28.0 \, m$
Maximum water discharge	$130.0 \, m^3/s$
Minimum water discharge	$2.0 \, m^3/s$
Noneffective water discharge	$2.0 \, m^3/s$
Maximum hydroelectric output	$35.0 \, MW$
Maximum thermal output	$50.0 \, MW$
Water inflow in each subinterval	$100 \, m^3/sec$
Cost function $F = 2.5P_S + 0.05P_S^2$	

The above approach is compared with the discrete maximum principle approach of Ref. 7.19. The load duration curve and the results obtained are given in Figure 7.13 for this example. A comparison between the discrete maximum principle and the direct method is shown in the graph.

The initial cost is 174.57 units. The optimal cost by DMP is 168.05 units, whereas by the direct method it is 167.65 units. The number of iterations for convergence were five in the case of DMP and two in the case of the direct method. The computation time by the direct method was about 60% of DMP with insignificant increase in the memory.

In the second example, the system consists of two thermal plants and two hydroelectric plants. The optimization interval was a year. The inflows to the hydroplants are given in Table 7.6, while the incremental fuel costs of thermal plants are given by

$$IC_1 = \frac{\partial F_1(P_1)}{\partial P_1} = 0.80 + 0.04P_1$$

$$IC_2 = \frac{\partial F_2(P_2)}{\partial P_2} = 0.78 + 0.06P_2$$

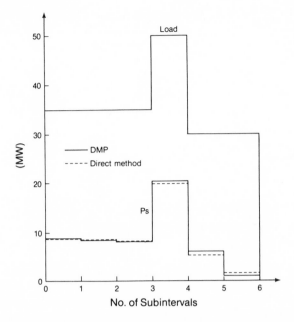

Figure 7.13. The optimal thermal generation. A comparison between discrete maximum principle (DMP) and the direct method.

Table 7.6. The Monthly Inflows to the Hydroplants[a]

Month k	$I_{1,k}$	$I_{2,k}$
1	0.00	0.00
2	0.60	0.00
3	1.20	0.00
4	1.20	1.50
5	1.20	3.00
6	1.80	4.50
7	2.40	4.50
8	1.44	1.50
9	1.20	0.00
10	0.96	0.00
11	0.00	0.00
12	0.00	0.00

[a] Values are per unit

Table 7.7. Hydroelectric Plant Data[a]

	Hydroelectric plant number	
	1	2
h, c, ρ	0.50, 0.05, and 0.2, respectively	0.40, 0.06, and 0.1 respectively
Initial and final storages	30.0	25.0
Minimum and maximum permissible discharges	1.0 and 4.0 respectively	1.0 and 4.0 respectively

[a] Values are per unit.

The load demand for the whole optimization interval is 0.8 p.u. The hydroelectric generation during a period k is given by

$$P_{h_{i,k}} = h_i[1 + 0.5c_i(2x_{i,k-1} - I_{i,k} + u_{i,k})](u_{i,k} - \rho_i)$$

where h_i, c_i, ρ_i for each plant are given in Table 7.7, initial and final storage in per unit, and the maximum and minimum discharges are also given in Table 7.7. The optimal discharges for each plant during each optimization interval are given in Table 7.8 using the direct method and the conjugate gradient method.

Table 7.8. Optimal Discharge for Each Hydroelectric Plant during the Optimization Interval

	Initial outflows		Optimal outflows with conjugate gradient method		Optimal outflows with direct method	
m	$u_{1,k}$	$u_{2,k}$	$u_{2,k}$	$u_{2,k}$	$u_{2,k}$	$u_{2,k}$
1	1.0000	1.2500	0.5000	0.5000	0.6000	0.5000
2	1.0000	1.2500	0.5000	0.5000	0.5000	0.5000
3	1.0000	1.2500	0.5000	0.5000	0.5000	0.5000
4	1.0000	1.2500	0.5000	0.5000	0.5000	0.5000
5	1.0000	1.2500	0.7280	0.5000	0.5000	0.5000
6	1.0000	1.2500	0.8985	0.5000	0.6374	0.6187
7	1.0000	1.2500	0.9880	1.3694	1.0933	2.0087
8	1.0000	1.2500	1.0952	2.0633	1.4903	2.7349
9	1.0000	1.2500	1.2883	2.1534	1.6558	2.5125
10	1.0000	1.2500	1.5202	2.0409	1.7264	1.9743
11	1.0000	1.2500	1.6860	2.0844	1.5095	1.4477
12	1.0000	1.2500	1.7958	2.2805	1.2870	1.2030

Initial cost = 52.185
Optimal cost by conjugate gradient = 48.7512
Optimal cost by direct method = 48.7488

The reduction in cost from the initial set of values is nearly the same, as can be observed from results. The memory required was 10.94 Kbytes and 14.73 Kbytes for the direct method and conjugate gradient method, respectively. The total computation time was 1 min 42 sec and 2 min 3 sec, respectively, on a Riyad 1030 EC computer.

7.3.1.4. Concluding Remarks

From the results of the two examples, the method of feasible direction seems to be better than the alternative method available. The problems that have been studied are simple and small compared to the power system problems that are normally encountered. The main emphasis of the work is, however, on the development of a new technique rather than the improvement or refinement of existing ones. When viewed this way, the proposed method seems to be promising. Within the limited experience that has been gained by solving the two problems mentioned in the paper and a few others, it has been observed that the convergence characteristic of this method is faster than those of the available methods.

The proposed approach needs an initial set of variables in the feasible domain which can be obtained by solving the hard constraints equations. The matrix inversion involved in evaluating P generally requires little extra computation as at any stage only a few constraints are expected to be in violation. When the constraints are in violation the method reduces to a simple steepest descent technique.

Special precautions are necessary to guarantee the convergence, as "zigzagging" or "hemstitching" may occur if the function is ill conditioned. The convergence obtained here was without any difficulties. However, in case of "jamming" a modified version of the method, the "modified feasible direction method," can be used. The commonly used objective function, i.e., cost over a period along with the constraints in a power system, generally do not lead to an ill-conditioned problem and thus the simple version of the method can be used effectively.

The method, though applied in this case to the hydrothermal scheduling problems, is equally suitable for other types of optimization problems in power system studies. Further work is in progress and a future extension will include the analysis of a larger system with more complex optimization objectives and operational constraints.

7.4. Discrete Maximum Principle (Ref. 7.20)

In this section, we discuss the application of the discrete maximum principle to the long-term optimal operation of hydrothermal power systems.

The main features of the discrete maximum principle (DMP) can be summarized as follows.

(1) By introducing adjoint variables, the original problem is converted to a two-point boundary-value problem.

(2) Finding the initial values of adjoint variables is not difficult when using this method and the constraints imposed on state variables add to this difficulty.

(3) Constraints on control variables can easily be handled.

(4) Nonlinearities in system performances and objective function are permissible.

(5) DMP is effective in discrete systems; thus theoretical errors are not introduced in its optimizing processes.

(6) Physical meanings of related variables in the optimizing processes can be obtained very clearly.

All of the above items except (2) are the advantages of the maximum principle. However, although item (2) is the greatest disadvantage, it provides the most important key for the application of this method.

Let us consider the general configuration system given in Figure 7.14. We wish to minimize the total fuel cost of this system, the fuel cost of the thermal plants, over the optimization interval, which is a year. Here we neglect the head variations of the run-of-river plants, as well as the river-flow time lag between hydroplants. In mathematical terms, the problem for the system of Figure 7.14 is to determine the optimal discharge from each hydroplant as well as the thermal generation, which minimizes

$$J = \sum_{k=1}^{K} \left[\sum_{i \in R_s} F_i(P_{s_{i,k}}) \right] \$ \tag{7.37}$$

subject to satisfying the following constraints:

(1) The total active power balance equation should satisfy (i.e., the balance between demand and generation in the k-time interval)

$$\sum_{i \in R_s} P_{s_{i,k}} + \sum_{i \in R_h} P_{h_{i,k}} = D_k \tag{7.38}$$

R_s is the set of thermal plants, while R_h is the set of hydroelectric plants. Here, we neglect the transmission line losses.

(2) The thermal fuel cost function may be approximated by

$$F_i(P_{s_{i,k}}) = \alpha_i + \beta_i P_{s_{i,k}} + \gamma_i (P_{s_{i,k}})^2, \qquad i \in R_s \tag{7.39}$$

(3) The reservoir dynamics may be adequately described by the continuous type difference equation

$$x_{i,k} = x_{i,k-1} + I_{i,k} + \sum_{\sigma \in R_u} (u_{\sigma,k} + s_{\sigma,k}) - u_{i,k} - s_{i,k} \tag{7.40}$$

where R_u is the set of reservoirs immediately upstream from reservoir i.

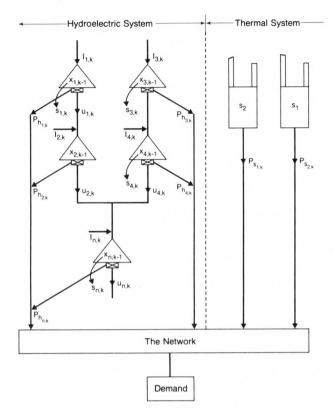

Figure 7.14. Hydrothermal power systems, a general configuration.

(4) Generating output of the ith reservoir type is given by

$$P_{h_{i,k}} = H_{i0}[1 + \tfrac{1}{2}c_i(x_i^{k-1} + x_i^k)](u_{i,k} - u_{i,0}), \qquad i = 1, \ldots, m \qquad (7.41)$$

for run-of-river plants. The generation from these plants is given by

$$P_{h_{j,k}} = H_{j0}(u_{j,k} - u_{j,0}), \qquad j = m + 1, \ldots, n \qquad (7.42)$$

where the following variables are defined: n is the total number of hydro plants; H_{i0} is a base head; c_i is a correction coefficient for head variation; $u_{i,0}$ is a dead discharge; and m is the number of storage reservoirs.

(5) Most of the rivers are multipurpose stream use; to satisfy all the requirements for the multipurpose stream use requirements, the following constraints must be satisfied:

a. Upper and lower bounds on the ith reservoir storage during a period k:

$$x_i^m \leq x_{i,k} \leq x_i^M, \qquad i = 1, \ldots, m \qquad (7.43)$$

b. Upper and lower bounds on the ith reservoir discharge

$$u_{i,k}^m \leq u_{i,k} \leq u_{i,k}^M, \qquad i = 1, \ldots, n \tag{7.44}$$

(6) Upper and lower bounds of thermal plants output

$$P_{s_i}^m \leq P_{s_{i,k}} \leq P_{s_i}^M \tag{7.45}$$

To deal with the inequality constraints of reservoir storages, we introduce the following penalty cost terms:

$$C_{p_{i,k}} = \gamma_i [V_i(x_{i,k})]^2 \tag{7.46}$$

where

$$V_i(x_{i,k}) = [\max(0, x_{i,k} - x_i^M) + \min(0, x_{i,k} - x_i^m)] \tag{7.47}$$

The objective function in equation (7.37) can be written as

$$J = J + \sum_{k=1}^K \sum_{i=1}^m C_{p_{i,k}} \tag{7.48}$$

or, in more detail, equation (7.48) can be written as

$$J = \sum_{k=1}^K \sum_{i \in R_s} [\alpha_i + \beta_i P_{s_{i,k}} + \gamma_i (P_{s_{i,k}})^2] + \sum_{k=1}^K \sum_{i=1}^m \gamma_i [V_i(x_{i,k})]^2 \tag{7.49}$$

Define the Hamiltonian function by

$$H_k = \left\{ \sum_{i \in R_s} [\alpha_i + \beta_i P_{s_{i,k}} + \gamma_i (P_{s_{i,k}})^2] \sum_{i=1}^m \gamma_i' [V_i(x_{i,k})]^2 \right.$$
$$+ \sum_{i=1}^m \lambda_{i,k} [x_{i,k-1} + I_{i,k} + \sum_{\sigma \in R_u} (u_{\sigma,k} + s_{\sigma,k}) - u_{i,k} - s_{i,k}]$$
$$\left. + \mu_{i,k} \left(-\sum_{i \in R_s} P_{s_{i,k}} - \sum_{i \in P_h} P_{h_{i,k}} + D_k \right) \right\} \tag{7.50}$$

In the above problem, we regard the reservoir storage as a state variable, while the thermal plant output and reservoir discharge are control variables.

Applying the discrete maximum principle approach, we obtain the following optimal equations: For thermal power plants we have

$$\beta_i + 2\gamma_i P_{s_{i,k}} - \mu_{i,k} = 0, \qquad i \in R_s \tag{7.51}$$

or

$$P_{s_{i,k}}^* = \frac{\mu_{i,k} - \beta_i}{2\gamma_i}, \qquad i \in R_s \tag{7.52}$$

For the hydro power plants, we have the following costate equation:

$$\lambda_{i,k-1} = \frac{\partial H_k}{\partial x_{i,k-1}}, \qquad i = 1, \ldots, m \tag{7.53}$$

or

$$\lambda_{i,k-1} = \lambda_{i,k} + 2\gamma_i'[V_i(x_{i,k})]$$

$$- \mu_{i,k}H_{i0}c_i(u_{i,k} - u_{i,0}) \qquad i = 1, \ldots, m \qquad (7.54)$$

$$\lambda_{i,K} = 0, \qquad i = 1, \ldots, m \qquad (7.55)$$

Also, we have

$$\frac{\partial H_k}{\partial u_{i,k}} = 0, \ i = 1, \ldots, n \qquad (7.56)$$

which can be written as

$$2\gamma_i'[V_i(x_{i,k})] - \lambda_{i,k} + \sum_{\sigma \in R_d} \lambda_{\sigma,k} - \mu_{i,k}H_{i0}[1 + \tfrac{1}{2}c_i(x_{i,k-1} + x_{i,k})],$$

$$i = 1, \ldots, m \qquad (7.57)$$

and for the run-of-river plant we have

$$\sum_{\sigma \in R_d} \lambda_{\sigma,k} - \lambda_{i,k} - \mu_{i,k}H_{i0} = 0, \qquad i = m+1, \ldots, n \qquad (7.58)$$

where R_d is the set of immediately downstream reservoirs.

Furthermore, we have the following set of equality constraints:

$$x_{i,k} = x_{i,k-1} + I_{i,k} - u_{i,k} - s_{i,k} + \sum_{\sigma \in R_u}(u_{\sigma,k} + s_{\sigma,k}), \qquad i = 1, \ldots, m \quad (7.59)$$

$$0 = I_{i,k} - u_{i,k} - s_{i,k} + \sum_{\sigma \in R_u}(u_{\sigma,k} + s_{\sigma,k}), \qquad i = m+1, \ldots, n \quad (7.60)$$

$$\sum_{i \in R_s} P_{s_{i,k}} + \sum_{i \in R_n} P_{h_{i,k}} = D_k \qquad (7.61)$$

Equations (7.51)-(7.61) give the optimal operation of the system under consideration. The relaxation method is used to solve these equations as is explained in reference 7.20.

The above method is applied to a practical system consisting of five hydroplants ($n = 5$), two of which are reservoir type ($m = 2$) and three run-of-river type ($n - m = 3$) (Figure 7.15). The optimization is done on a monthly time basis for a period of a year. The characteristics of the installation are given in Table 7.9, while Table 7.10 gives the demand powers and natural inflows.

The computational results are shown in Figure 7.16. After three iterations of the relaxation method, almost optimal solutions have been derived.

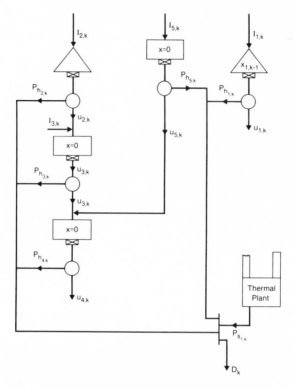

Figure 7.15. Multireservoir power system.

Table 7.9. Constants Characterizing the Model Multireservoir System

		1	2	3	4	5
Maximum storage	x_i^M (m³)	7.2×10^4	125×10^4	65×10^4	90×10^4	26×10^4
Minimum storage	x_i^m (m³)	0	0	0	0	0
Base head	H_{i0} (m)	98	50	115	96	221
Maximum head	h_i^M (m)	142	96.14	117	104.75	225
Maximum discharge	Q_1^M (m³/sec)	73	75.6	64.0	100	24.1
Minimum discharge	Q_i^m (m³/sec)	3.334	4.353	2.151	20	0.588
Dead discharge	Q_{i0} (m³/sec)	3.334	4.353	2.151	0.909	0
Maximum output of hydroelectric plant	$P_{h_i}^M$ (MW)	89.03	57.61	64.72	88.73	42.95
Maximum output of thermal plant	$P_{s_i}^M$ (MW)	240				
Minimum output of thermal plant	$P_{s_i}^m$	0				
Fuel cost function[a]	$F(P_{s_i}) = 720(2{,}500 P_{s_i} - 0.00175 P_{s_i}^2 + 0.00003167 P_{s_i}^3) \times 10^3$ unit					

[a] The fuel cost in this example is assumed to be monotonically increasing.

Table 7.10. Yearly Load and Reservoir Inflow

Time interval	Month	$I_{1,k}$ (m³/sec)	$I_{2,k}$ (m³/sec)	$I_{3,k}$ (m³/sec)	$I_{4,k}$ (m³/sec)	D_k (MW)
1	11	9.8	23.0	3.5	5.6	323
2	12	10.0	17.0	2.6	4.2	273
3	1	11.0	12.0	1.9	3.0	251
4	2	11.9	10.0	1.5	2.5	249
5	3	17.0	20.0	3.1	4.9	295
6	4	18.5	40.0	6.2	9.8	328
7	5	27.6	53.0	8.2	13.0	357
8	6	43.8	63.0	9.7	15.6	365
9	7	56.4	62.0	9.5	15.2	365
10	8	40.3	48.0	7.4	11.8	341
11	9	30.1	55.0	8.5	13.5	359
12	10	16.3	48.0	7.4	11.8	360

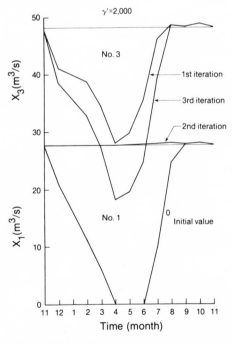

Figure 7.16. Optimal storage for reservoirs 1 and 3.

7.4.1. Concluding Remarks

In this section we discussed the application of the discrete maximum principle to the optimization of hydrothermal power systems for long-term operation. By using this method, the computing time and computing memory can be reduced greatly. The main features of this method are explained earlier in the beginning of this section. However, constraints to state variables can be considered by the penalty cost method. Exact solutions can be obtained by estimating the asymptotes for $\gamma = \infty$.

The maximum principle is used to solve the variational problem after it is converted to a two-point, boundary-value problem. However, in its application it is most difficult to find the initial values of adjoint variables.

7.5. Stochastic Nonlinear Programming (Ref. 7.22)

In the previous sections of this chapter, we discussed the optimal long-term scheduling of hydrothermal power systems and we assumed that the water inflows to the reservoirs and the load demand are known with absolute certainty—that is, we assumed a deterministic problem. In reality, however, this perfect information is not true. In this section, we extend the work to random water inflows and load demand. With these variables being stochastic, the long-term scheduling problem is reduced to the minimization of the expected fuel cost under the constraints of expected water available. The techniques used here are based on first-order gradient methods, where an initial set of control variables is assumed. These variables are adjusted in a direction such that the cost function decreases, thus ensuring the stability of the method.

7.5.1. Optimal Stochastic Control Policy

Given the power system of Figure 7.14, the problem for this system is to schedule the hydro discharge and the thermal generation so that the expected fuel cost of the thermal plants is a minimum. In mathematical terms the stochastic long-term optimal operation can be expressed as follows: Minimize

$$J = E\left[\sum_{k=1}^{K} \sum_{i \in R_s} F_i(P_{s_{i,k}}) \right] \tag{7.62}$$

where $P_{s_{i,k}}$ is the average power output during the optimization interval k subject to satisfying the following constraints.

(1) The thermal fuel-cost equation may be adequately described by a quadratic equation as

$$F(P_{s_{i,k}}) = \alpha_i + \beta_i P_{s_{i,k}} + \gamma_i P_{s_{i,k}}^2 \tag{7.63}$$

(2) The active power balance equation between the demand and the generation is given by

$$\sum_{i \in R_G} P_{i,k} = D_k + P_{L,k} \tag{7.64}$$

In this section, we will discuss how to treat the transmission line losses in the problem formulation. However, with a good approximation, as we said earlier, we can consider these losses as a constant for long-term studies and add them to the load. The transmission-power losses can be expressed through the well-known loss-formula expression as a quadratic function of the power generation:

$$P_{L,k} = \sum_{i \in R_G} \sum_{j \in R_G} P_{i,k} B_{ij} P_{j,k} \tag{7.65}$$

(3) The average hydroelectric generation during the optimization interval depends on the water discharge through the turbine and on the average head, which, in turn, is a function of the storage and thus can be expressed as

$$P_{h_{i,k}} = F_i(x_{i,k}, u_{i,k}), \qquad i \in R_h \tag{7.66}$$

In this study, we will use equation (7.41) for the hydrogeneration; we repeat it here for the reader's convenience:

$$P_{h_{i,k}} = H_{i0}[1 + 0.5 c_i(x_{i,k+1} + x_{i,k})](u_{i,k} - u_{i,0}) \tag{7.67}$$

where H_{i0} is the basic water head and c_i is the water head correction factor to account for variation in head with storage.

(4) The storage at the beginning of period k for the hydro reservoir is given by the continuity type difference equation as

$$x_{i,k+1} = x_{i,k} + I_{ik}, -u_{i,k} + u_{(i-1)k}, \qquad i \in R_h \tag{7.68}$$

The expected storage at the end of the Kth interval, the end of optimization interval, as a functional of the final storage is given by

$$\bar{x}_{i,K+1} = \bar{x}_{i,1} + \sum_{k=1}^{K} \bar{I}_{i,k} - \sum_{k=1}^{K} \bar{u}_{i,k} + \sum_{k=1}^{K} \bar{u}_{(i-1),k} \tag{7.69}$$

where we define the expected value as

$$E[x] = \bar{x} \tag{7.70}$$

(5) There are upper and lower limits, as mentioned in Section 7.3, for the variables:

i. Upper and lower limits on the reservoir storages given by

$$x_i^m \le x_{i,k} \le x_i^M, \qquad i \in R_h \tag{7.71}$$

ii. Upper and lower limits on the discharge through the turbines given by

$$u_{i,k}^m \le u_{i,k} \le u_{i,k}^M, \qquad i \in R_h \tag{7.72}$$

iii. Finally upper and lower limits on the thermal generation

$$P_{s_i}^m < P_{s_{i,k}} \le P_{s_i}^M, \qquad i \in R_s \tag{7.73}$$

The problem now is to minimize the expected value of the fuel cost given by (see Appendix C)

$$J = \sum_{k=1}^K \left\{ \sum_{i \in R_s} [\alpha_i + \beta_i \bar{P}_{s_{i,k}}, \gamma_i (\bar{P}_{s_{i,k}})^2] \right.$$

$$+ \beta_k \sum_{i \in R_h} \sum_{i \in R_h} g_{i,k} g_{j,k} \operatorname{cov}(P_{h_{i,k}}, P_{h_{j,k}})$$

$$\left. + \beta_k \operatorname{var}(D_k)/(e_k)^2 \right\} \tag{7.74}$$

with the initial storage $x_{i,1}$ and the expected final storage $\bar{x}_{i,K+1}$, respectively. Subject to satisfying the equality constraints (7.64), (7.67), (7.68), and (7.69), and the inequality constraints given by equations (7.71)–(7.73).

It has been assumed that the water inflows $I_{i,k}$ into various reservoirs are statistically correlated during the same subinterval, period k, and independent during different subintervals. Their expected values and covariance matrix are assumed to be known from past history. The load demand is assumed to be statistically independent of the water inflows, and its expected value and variance have been preestimated. The probability properties of water storage and hydroelectric generation can be obtained as shown in Appendix C. $x_{i,1}$, $i \in R_h$, can be assumed to be known with probability 1.

Now, we can form an augmented cost functional by augmenting the objective function of equation (7.74) with the equality constraints of equations (7.64), (7.66), (7.68), and (7.69), expressed in terms of expected values, through the dual variables $\lambda_{1,k}$, $\lambda_{2i,k}$, and $\lambda_{3i,k}$, respectively, as follows:

$$\tilde{J} = \sum_{k=1}^K \left\{ \sum_{i \in R_s} (\alpha_i + \beta_i \bar{P}_{s_{i,k}} + \gamma_i \bar{P}_{s_{i,k}}^2) \right.$$

$$+ \beta_k \sum_{i,j \in R_h} g_{i,k} g_{j,k} \operatorname{cov}(P_{h_{i,k}}, P_{h_{j,k}}) + \beta_k \operatorname{var}(D_k)/(e_k)^2$$

$$- \lambda_{1,k} \left(\sum_{i \in R_s} \bar{P}_{s_{i,k}} + \sum_{i \in R_n} \bar{P}_{h_{i,k}} - \bar{P}_{L,k} - \bar{D}_k \right)$$

$$+ \sum_{i \in R_h} \lambda_{2i,k} [P_{h_{i,k}} - F_i(\bar{x}_{i,k}, u_{i,k})]$$

$$\left. + \sum_{i \in R_h} \lambda_{3i,k} (\bar{x}_{i,k+1} - \bar{x}_{i,k} - \bar{I}_{i,k} + u_{i,k} - u_{i-1,k}) \right\}$$

$$+ \sum_{i \in R_n} \tfrac{1}{2} w_i [(\bar{x}_{i,K+1})_{\text{spec}} - \bar{x}_{i,K+1}]^2 \tag{7.75a}$$

Note that the last terms in equation (7.75a) is introduced through a penalty factor w_i, because the amount of water at the last period is specified by equation (7.69). See the example given. During each optimization interval, the control variables are the discharges through the turbines $u_{i,k}$, from which we can calculate the storages and the hydro power generations from equations (7.68) and (7.67), respectively; and therefore these are dependent variables together with the thermal generations. The dual variables $\lambda_{1,k}$, $\lambda_{2i,k}$, and $\lambda_{3i,k}$ are to be determined such that the corresponding equality constraints are satisfied. With $F_i(x_{i,k}, u_{i,k})$ given, applying the principle of differentiation, we obtain the following.

(i) For thermal power plants

$$\left(\frac{\partial \tilde{J}}{\partial P_{s_{i,k}}}\right)_{i \in R_s} = 0 = \beta_i + 2\gamma_i \bar{P}_{s_{i,k}} - \lambda_{1,k}\left(1 - \frac{\partial \bar{P}_{L,k}}{\partial \bar{P}_{s_{i,k}}}\right), \qquad i \in R_s \quad (7.75b)$$

if we define the penalty factor for the thermal power plants as

$$L_{s_{i,k}} = \frac{1}{1 - \partial \bar{P}_{L,k}/\partial \bar{P}_{s_{i,k}}}, \qquad i \in R_s \quad (7.76)$$

then $\lambda_{1,k}$ can be obtained as

$$\lambda_{1,k} = L_{s_{i,k}}(\beta_i + 2\gamma_i \bar{P}_{s_{i,k}}), \qquad i \in R_s \quad (7.77)$$

and the optimal thermal generation is given by

$$\bar{P}_{s_{i,k}} = \frac{\lambda_{1,k} - L_{s_{i,k}}\beta_i}{2\gamma_i L_{s_{i,k}}}, \qquad i \in R_s \quad (7.78)$$

Note that for lossless power systems, we neglect transmission line losses, and $L_{s_{i,k}}$ is equal to one.

(ii) For hydropower plants

$$\frac{\partial \tilde{J}}{(\partial \bar{P}_{h_{i,k}})_{i \in R_h}} = 0 = \lambda_{2i,k} - \lambda_{i,k}\left(1 - \frac{\partial \bar{P}_{L,k}}{\partial \bar{P}_{h_{i,k}}}\right), \qquad i \in R_h \quad (7.79)$$

The penalty factor for the hydro system is defined by

$$L_{h_{i,k}} = \frac{1}{1 - \partial \bar{P}_{L,k}/\partial \bar{P}_{h_{i,k}}}, \qquad i \in R_h \quad (7.80)$$

Thus equation (7.79) can be written as

$$\lambda_{1,k} = \lambda_{2i,k} L_{h_{i,k}} \quad (7.81)$$

Also, we have

$$\left(\frac{\partial \tilde{J}}{\partial \bar{x}_{i,k}}\right)_{i \in R_h} = 0 = \lambda_{3i,k-1} - \lambda_{3i,k} - \lambda_{2i,k} H_{i0} c_i(u_{i,k} - u_{i,0}), \qquad i \in R_h \quad (7.82)$$

or (7.82) can be written as

$$\lambda_{3i,k-1} = \lambda_{3i,k} + \lambda_{2i,k}H_{i0}c_i(u_{i,k} - u_{i,0}), \qquad i \in R_n \qquad (7.83)$$

In order to obtain the above equation, we used the following identity*:

$$\sum_{k=1}^{K} \lambda_{3i,k}\bar{x}_{i,k+1} = (\lambda_{3i,K}x_{i,K+1} - \lambda_{3i,0}x_{i,1}) + \sum_{k=1}^{K} \lambda_{3i,k-1}\bar{x}_{i,k} \qquad (7.84)$$

with the boundary condition given by†

$$\lambda_{3i,K} = w_i[(\bar{x}_{i,K+1})_{\text{spec}} - \bar{x}_{i,K+1}] \qquad (7.85)$$

The gradient vector is given by

$$\left(\frac{\partial \tilde{J}}{\partial u_{i,k}}\right) = 0$$

$$= -\lambda_{3(i+1),k} + \lambda_{3i,k} - \lambda_{2i,k}H_{i0}[1 + 0.5c_i(2\bar{x}_{i,k} + \bar{I}_{i,k} + u_{i-1,k} - u_{i,k})]$$

$$- 0.5H_{i0}c_i(u_{i,k} - u_{i,0})(\lambda_{2(i+1),k} - \lambda_{2i,k})$$

$$+ \{\lambda_{i,k}A_{ii} + (g_{i,k})^2\beta_k\}\frac{d \text{ var}(P_{h_{i,k}})}{du_{i,k}}$$

$$+ \sum_{j \in H} 2\{\lambda_{1,k}A_{ij} + \beta_k g_{i,k}g_{j,k}\}\frac{d \text{ cov}(P_{h_{i,k}}, P_{h_{j,k}})}{du_{i,k}} \qquad (7.86)$$

where $\text{var}(P_{h_{i,k}})$ and $\text{cov}(P_{h_{i,k}}, P_{h_{j,k}})$ are given in Appendix C, and

$$\frac{d \text{ var}(P_{h_{i,k}})}{du_{i,k}} = 2H_{i0}^2 c_i^2(u_{i,k} - u_{i,0}) + [\text{var}(x_{i,k}) + 0.25 \text{ var}(I_{i,k})] \quad (7.87)$$

and

$$\frac{d \text{ cov}(P_{h_{i,k}}, P_{h_{j,k}})}{du_{i,k}} = H_{i0}H_{j0}c_ic_j(u_{i,k} - u_{i,0})[\text{cov}(x_{i,k}, x_{j,k})$$

$$+ 0.25 \text{ cov}(I_{i,k}, I_{j,k})] \qquad (7.88)$$

The upper and lower limits on the control variables can be taken care of by making these variables equal to the respective bound values whenever such limits are violated. For the dependent variables, these limits can be considered by augmenting the cost function through Powell's penalty function as explained in the previous section. Equations (7.75)-(7.88) completely specify the stochastic long-term optimal operation.

Note that in formulating the problem, Ref. 7.22 treated it using a slightly different approach, the author assumed that the reservoirs are hydraulically

---

* The two terms in parentheses are missed in Ref. 7.22, which is the main reference for this approach.

† This equation is not mentioned in Ref. 7.22. It treated the fixed amount of water at the last period studied differently.

independent, and he adjoined equation (7.69) to the cost function through λ_{4i}. Therefore, he obtained the following equations:

$$\left(\frac{\partial \tilde{J}}{\partial P_{s_{i,k}}}\right)_{i \in R_s} = \beta_i + 2\gamma_i \bar{P}_{s_{i,k}} - \lambda_{1,k}\left(1 - \frac{d\bar{P}_{L,k}}{d\bar{P}_{s_{i,k}}}\right) = 0 \qquad (7.89)$$

$$\left(\frac{\partial \tilde{J}}{\partial \bar{P}_{h_{i,k}}}\right)_{i \in R_h} = \lambda_{2i,k} - \lambda_{1,k}\left(1 - \frac{d\bar{P}_{L,k}}{d\bar{P}_{h_{i,k}}}\right) = 0 \qquad (7.90)$$

$$\left(\frac{\partial \tilde{J}}{\partial \bar{x}_{i,k}}\right)_{\substack{i \in R_h \\ k \neq 1}} = \lambda_{3i,k-1} - \lambda_{3i,k} - \lambda_{2i,k}H_{i0}c_i(u_{i,k} - u_{i,0}) = 0 \qquad (7.91)$$

He assumed that $\lambda_{3i,1} = 0$, since the equations associated with the set of dual variables are redundant. The gradient vector is

$$\left(\frac{\partial \tilde{J}}{\partial u_{i,k}}\right)_{i \in R_h} = \lambda_{3i,k} + \lambda_{4i} - \lambda_{2i,k}H_{i0}[1 + 0.5c_i(2\bar{x}_{i,k} - \bar{I}_{i,k} + 2u_{i,k}$$

$$- u_{i,0})] + [\lambda_{1,k}A_{ii} + (g_{i,k})^2\beta_k]\frac{d \text{ var}(P_{h_{i,k}})}{du_{i,k}}$$

$$+ \sum_{j \in R_h} 2(\lambda_{1,k}A_{ij} + \beta_k g_{i,k}g_{j,k})\frac{d \text{ cov}(P_{h_{i,k}}, P_{h_{j,k}})}{du_{i,k}} \qquad (7.92)$$

Reference 7.22 said that the dual variables λ_{4i} are to be adjusted to maximize the Lagrangian function $\tilde{J}$ under the constraints of other optimality conditions. The corresponding gradient vector is

$$\left(\frac{d\tilde{J}}{d\lambda_{4i}}\right)_{i \in R_h} = \bar{x}_{i,K+1} - \bar{x}_{i,1} - \sum_{k=1}^{K} \bar{J}_{i,k} + \sum_{k=1}^{K} u_{i,k} \qquad (7.93)$$

Equations (7.89) to (7.93) with equations (7.87) and (7.88) completely specify the stochastic long-term optimal operation for independent river systems. The next section explains the computer algorithm to solve these equations.

7.5.2. Computer Algorithm

In this section we discuss the computer algorithm used to solve equations (7.89)–(7.93) as stated in Ref. 7.22:

Step 1. Assume a set of λ_{4i} for $i \in R_h$.

Step 2. Set $k = 1$.

Step 3. Assume a set of $u_{i,k}$ for $i \in R_h$.

Step 4. Calculate the expected hydrogeneration from equation (C.18), in Appendix C, and the corresponding covariance matrix from equations (C.19) and (C.20)

Step 5. Assume $\lambda_{1,k}$. Calculate thermal generations from equation (7.89). Calculate matrix A from equation (C.24) and from equation (C.25) after calculating k_i from equations (C.5) and (C.27), e from equation (C.13), and g_i from equation (C.14). Calculate the expected loss from equation (C.26)

Step 6. Check for the power-transfer equation (7.64). If this is not satisfied within a prescribed accuracy, repeat from step 5 for a different $\lambda_{1,k}$. Calculate β from equation (C.29).

Step 7. Calculate $\lambda_{2i,k}$ for $i \in R_h$ from equation (7.90), $\lambda_{3i,k}$ for $i \in R_h$ and $m > 1$ from equation (7.91), and the gradient vector from equation (7.92). If this does not satisfy the optimality conditions with a prescribed accuracy, adjust the water discharges using the conjugate-gradient method and repeat from step 4.

Step 8. Calculate the expected values and covariance matrix of storages from equations (C.15), (C.16), and (C.17).

Step 9. If $k < K$, substitute $k = k + 1$ and repeat from step 3.

Step 10. The solution, although it satisfies the optimality conditions, may not satisfy equation (7.69). Adjust λ_{4i} by the conjugate gradient method with the gradient obtained from equation (7.93) and repeat the procedure from step 2 until this equation is satisfied with a prescribed accuracy.

7.5.3. Practical Example (Ref. 7.22)

The algorithm of the last section is used to solve the long-term planning problem with two thermal generations and two hydroelectric plants. The incremental costs of thermal plants are given by

$$\frac{dF_1(P_{s_1})}{dP_{s_1}} = 0.8 + 0.08 P_{s_1}$$

$$\frac{dF_2(P_{s_2})}{dP_{s_2}} = 0.78 + 0.12 P_{s_2}$$

Data in per unit for the two hydroelectric plants are given in Table 7.11. The optimization is done on a monthly time basis for a year. The water inflows and the standard deviations for the 12 subintervals are given in Table 7.12. The correlation coefficient between the water inputs for the same subinterval is assumed to be 0.5.

Table 7.11. Data for Hydroelectric Plant[a]

	Hydroelectric plant	
	1	2
$H, c, u_{i,0}$	0.50, 0.05, and 0.2, respectively	0.40, 0.06, and 0.1, respectively
Initial and final storages	30.0	25.0
Minimum and maximum permissible discharges	1.0 and 4.0, respectively	1.0 and 4.0, respectively

[a] Values are per unit.

The load demand is 15.0 p.u., and its standard deviation is 2.0. The loss coefficient matrix is

$$B = \begin{bmatrix} 0.002 & 0.001 & 0.001 & 0.0005 \\ 0.001 & 0.003 & 0.0005 & 0.001 \\ 0.001 & 0.0005 & 0.004 & 0.000 \\ 0.0005 & 0.001 & 0.000 & 0.002 \end{bmatrix}$$

In Table 7.13, the available water is divided equally in all the subintervals, and the thermal powers scheduled, for the minimum fuel cost, satisfy equations (7.89) and (7.64). This solution is in the feasible domain, but does not satisfy the optimality conditions. An optimal solution obtained after six iterations for different λ_{4i} is given in Table 7.14. Steps 5 and 6 of the computer algorithm needed 4–5 iterations for different λ_1. The complete

Table 7.12. Water Inflows and Standard Deviations[a]

	k	1	2	3	4	5	6
Hydroelectric plant 1	Expected value	1.00	2.00	2.50	3.00	3.00	4.00
	Standard deviation	0.200	0.400	0.500	0.600	0.600	0.800
Hydroelectric plant 2	Expected value	1.50	1.50	2.00	2.00	2.500	3.00
	Standard deviation	0.300	0.300	0.400	0.400	0.500	0.600
	k	7	8	9	10	11	12
Hydroelectric plant 1	Expected value	3.00	3.00	2.50	2.00	1.00	1.00
	Standard deviation	0.600	0.600	0.500	0.400	0.200	0.200
Hydroelectric plant 2	Expected value	2.50	2.00	2.00	1.50	1.50	1.00
	Standard deviation	0.500	0.400	0.400	0.300	0.300	0.200

[a] Values are per unit.

Table 7.13. Initial Values[a]

Subinterval	Water discharge		Hydroelectric power		Thermal power	
	$u_{1,k}$	$u_{2,k}$	1	2	1	2
1	2.3333	1.9167	2.7022	1.8257	6.3978	4.3876
2	2.3333	1.9167	2.6044	1.8076	6.4689	4.4334
3	2.3333	1.9167	2.5689	1.7821	6.5062	4.4578
4	2.3333	1.9167	2.5689	1.7821	6.5062	4.4578
5	2.3333	1.9167	2.6044	1.7749	6.4889	4.4469
6	2.3333	1.9167	2.6133	1.7894	6.4745	4.4374
7	2.3333	1.9167	2.7289	1.8475	6.3682	4.3682
8	2.3333	1.9167	2.7644	1.8839	6.3242	4.3394
9	2.3333	1.9167	2.8133	1.8875	6.2921	4.3187
10	2.3333	1.9167	2.8356	1.9020	6.2696	4.3041
11	2.3333	1.9167	2.8444	1.8839	6.2753	4.3081
12	2.3333	1.9167	2.7733	1.8766	6.3232	4.3389

[a] Values are per unit. Expected fuel cost = 137.09 per unit

solution needed about 5 min of IBM 1130 digital computer's time, and this is expected to increase proportionally with the number of subintervals.

7.5.4. Concluding Remarks

Because of the tremendous annual fuel cost in thermal plants, a small percentage saving can be considered to be significant, thus justifying the

Table 7.14. Optimal Values[a]

Subinterval	Water discharge		Hydroelectric power		Thermal	
	$u_{1,k}$	$u_{2,k}$	1	2	1	2
1	1.0000	1.0000	1.0000	0.8946	8.0196	5.4423
2	1.0000	1.0000	0.9900	0.9054	8.0192	5.4418
3	1.0000	1.0000	1.0050	0.9108	8.0065	5.4336
4	1.0000	1.0000	1.0300	0.9324	7.9777	5.4147
5	1.0000	1.0000	1.0700	0.9486	7.9429	5.3922
6	1.8295	1.0000	2.2575	0.9756	7.19078	4.9127
7	2.8953	1.4487	3.9499	1.4956	5.8424	4.0411
8	2.6387	2.6754	3.5723	3.0815	5.1032	3.5329
9	3.8455	2.0478	5.4510	2.2844	4.4560	3.1328
10	3.8672	3.1514	5.38389	3.6339	3.6783	2.6064
11	4.0000	3.6535	5.4553	4.1125	3.3470	2.3841
12	4.0000	4.0000	5.1703	4.3514	3.3730	2.2967

[a] Values are per unit. Expected fuel cost = 131.10 per unit

need for more accurate analysis and the consideration of the randomness in the variables and the transmission losses. It is noticed in the numerical example that the optimal power generations of each plant are considerably different in various subintervals, even though the expected load demands are the same in each subinterval.

It can be easily stated that the optimal equations (7.75)–(7.88) can be used for multireservoir power systems which are hydraulically coupled, and have a fixed final storage. These equations can be solved as a TPBVP where the initial states are given and the final costates are given, which was not the case for the problem just described above.

7.6. Aggregation with Stochastic Dynamic Programming Approach (Ref. 7.36)

This section is devoted to the applications of stochastic dynamic programming to the optimal long-term stochastic operation of hydrothermal power systems. The aggregation of hydro plants and reservoir is used to reduce the problem dimensions. The cost of operation for various thermal power plants, the scheduled downtime for various unit maintenance activities, the predicted partial and total forced outages, and the stochastic river inflows to the reservoirs are input requirements to the program which yields the monthly hydro energy scheduling for the most economic loading over a one-year period.

In a time horizon of one year, uncertainties of various types arise in a hydrothermal system. Therefore, in any meaningful representation of a power system the forced outages of the thermal generation, the random inflow to the hydro units, and the stochastic nature of the load must be incorporated in the system modeling.

The stochastic hydrothermal model comprises a load model to compute the expected load energy, a model to represent the controlled hydro generating stations, a thermal model for the thermal units representation and the cost calculation, and a model for the uncontrolled hydro generation.

The system modeling is developed under the following assumptions and considerations:

(i) The stochastic nature of the load is based on a Markov chain model. Line losses are included in the expected load.

(ii) Interchange contracts are considered deterministic in nature. Only long-term capacity interchange is dealt with. This principle is adopted because long-term capacity interchange is dependable and can be regarded as part of the system.

(iii) The power sources available for generation include thermal and nuclear generation plants, and conventional hydro plants.

(iv) Energy is scheduled in monthly blocks over one year. The objective is to allocate monthly blocks of generation from the above sources such that the cost of thermal generation is minimized over one year. The units with smaller average energy costs are loaded first, leaving the high-cost units for the generation of load peaks.

(v) Hydro units are incorporated by using the predicted monthly river flows as a random variable in a stochastic dynamic programming technique to obtain minimum system cost of supplying the required load. The program output is the optimal expected hydro discharge schedule over the desired time frame of one year.

(vi) Nuclear units are assumed to occupy the first portion of the start-up order and to have lower generating costs than either fossil or peaking units.

Any system model should incorporate a consideration of expected load pattern. The load model used here is based on a Markov chain. The loads are represented by daily peak load conditions (Figure 7.17). The sequence of daily peak loads is assumed to be a stationary random process. A reasonable approach to recognize the nonstationary effects of seasonal load changes is to divide the year into intervals of four week duration, or less, so that the load model may be represented in each interval by a stationary stochastic process. Referring to Figure 7.17, the expected value of the load demand is

$$D = \sum_{i=0}^{N} p_i D_i \qquad (7.94)$$

where D_i is the daily peak demand level (MW), $p_i = n_i e / d$ is the availability

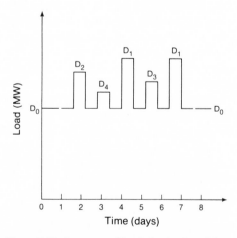

Figure 7.17. Sequence of loads for load model.

of peak load D_i in the interval d, n_i is the expected number of times the demand is at level D_i, e is the expected duration of peak load (fraction of a day), $p_0 = 1 - e$ is the probability that the system is in the state D_0, D_0 is the base load demand level (MW), d is the total interval of time considered (days), and N is the number of daily peak load levels in d.

The N load levels are ordered from highest to lowest. Each peak load lasts e fraction of a day on the average and is always followed by a low load period of mean duration $(1 - e)$ day. The load state transitions occur independently of generation state transitions.

The transition rate downward from D_i is

$$\lambda = \frac{1}{e}$$

and upward from D_0 to D_i is

$$\frac{\beta_i}{1 - e}, \qquad i = 1, \ldots, N$$

where $\beta_i = n_i / d$, $i = 1, 2, \ldots, N$.

It is convenient to distinguish between hydro plants with storage reservoirs, which are called controlled hydro plants, and those without storage reservoirs, which are called uncontrolled hydro plants or run-of-river plants. The hydro system representation comprises a controlled hydro plant model with a procedure to aggregate the reservoirs and hydro generating stations characteristics.

A composite long-term reservoir must exhibit the main characteristics of the multireservoir system that it represents.

To build a composite model, we measure the amount of water at different parts of the system in terms of its equivalent potential energy. The power generated at hydro plant i due to an outflow u can be expressed as

$$P_{h_{i,k}} = a_{1i}q_{i,k}^3 + a_{2i}q_{i,k}h_{i,k} + a_{3i}h_{i,k} + a_{4i} \qquad (7.95)$$

where $q_{i,k}$ is the outflow from reservoir i during a period k in m^3/sec, $h_{i,k}$ is the head at hydro station i during a period k in m, $a_{1i}, \ldots, a_{4i}$ are the known coefficients for station i, and $P_{h_{i,k}}$ is the hydroelectric power generated in MW at station i due to an outflow $u_{i,k}$ (m^3/sec) and head $h_{i,k}(m)$ in period k.

For each hydro plant a conversion factor μ_i is used to measure the number of megawatts generated on plant i by an outflow of $1000\ m^3/sec$. This conversion factor is called the water conversion factor and it is a function of the discharge through the turbines and can be obtained from equation (7.95).

To obtain the equivalent plant of a part of the system, we multiply the volume of water stored in this part by the sum of the conversion factors of the individual plants forming that part. The total energy E_{P_k} stored in the system at the beginning of month k is given by

$$E_{P_k} = \sum_{j=1}^{n} x_{j,k} \mu_j \qquad (7.96)$$

where $x_{j,k}$ is the volume of water stored at the equivalent plant j at the beginning of month k in (m^3), n is the number of equivalent plants in the system, and μ_j is the water conversion factor of the equivalent plant j in $(\text{MW}/\text{m}^3/\text{sec})$. E_{P_k} is measured in MW d (1 MW d is the potential energy accumulated during one day by a continuous inflow of 1 MW).

Similarly, the MW inflow to the system is

$$Q_k = \sum_{j=1}^{n} I_{j,k} \mu_j, \qquad k = 1, \ldots, K \qquad (7.97)$$

where Q_k is the average inflow in the system during month k and $I_{j,k}$ is the average inflow to reservoir i during month k in m^3/sec.

The average inflow $I_{i,k}$ is obtained from the assumption of normally distributed inflows in a given reservoir during a month k.

The MW outflow from the system is

$$u_k = \sum_{j=1}^{n} q_{j,k} \mu_j, \qquad k = 1, \ldots, K \qquad (7.98)$$

where $q_{j,k}$ is the outflow from plant j during month k in m^3/sec and u_k is the outflow from the system during month k.

Equations (7.96)–(7.98) must satisfy the following energy continuity equation:

$$E_{P_{k+1}} = E_{P_k} + (Q_k - u_k) d_k \qquad (7.99)$$

where d_k is the number of days in month k.

7.6.1. Aggregation of Reservoirs and Hydro Storage Characteristics

Long-Term System Description. Only those reservoirs of the long-term hydraulic system that have a storage capacity large enough to allow significant month-to-month storage varation are considered. Such reservoirs are called long-term reservoirs. In a similar fashion, only those hydro plants whose production is influenced by the management policy on midterm reservoirs are considered, and they are called long-term hydro plants.

Some components of the hydro system are omitted from the long-term representation because either their complex contractual constraints are not suitable for mathematical modeling or their electric constraints are not typical. Other components are eliminated if a figure of merit, which is the product of their generating capacity and their storage capacity, is below a specified limit.

Representation of the Long-Term System. The long-term hydroelectric system of a valley can be represented by the following types of elements:

Type I element. Long-term *reservoir*, the storage of which does not influence the downstream plant water head.

Type II element. Long-term *hydro plant*, the water head of which is not influenced by the upstream reservoir storage.

Type III element. Pair of adjacent long-term reservoir and hydro plant, where the plant water head is influenced by the reservoir storage.

Modeling of the Nominal Production Characteristics. Each plant has a certain number of generating units and the flow at the plant can be distributed to the different identical units. The nominal energy production characteristics of a unit (in MWH) as a function of the flow is shown in Figure 7.18.

The operation rule of a plant as its flow increases is to start a generation unit when the production marginal rate of the previously started unit is getting too small, Figure 7.19.

Modeling of the Production Characteristics. For a long-term plant that is adjacent to a long-term reservoir the forebay elevation varies with the reservoir storage, and this has to be taken into account in the production characteristics of the plant. It was found that the latter could be obtained

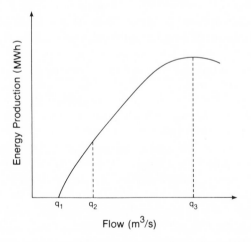

Figure 7.18. Monthly energy production of a generating unit as a function of its monthly average flow.

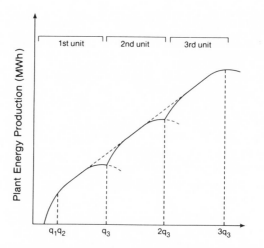

Figure 7.19. Plant operation rule for energy production.

by multiplying the nominal production characteristics by a correction factor α, where

$$\alpha = \frac{A - B}{C - B}$$

Here A is the average forebay elevation, B is the tailwater elevation, and C is the nominal forebay elevation.

Aggregation: Aggregation is desirable because it diminishes problem dimensionality. In the system under consideration, the delay for the water to reach an element from an immediately upstream element in the system ranges from less than an hour to about 72 hr. This is small in comparison with out time unit (month). Therefore, zero delays are assumed in the system.

The previous hypothesis permits us to condense the information on the natural inflows collected along the valley during the month. If u_N is the amount of noncontrollable flow at a plant during the month, the production characteristics in such a plant are shifted to the left by u_N.

The long-term system can be aggregated into a long-term equivalent system by applying the following rules:

Rule 1. Any pair of consecutive or parallel type I element may be aggregated into one type I element.

Rule 2. Any pair of consecutive type II elements may be aggregated into one type II element.

Disaggregation: When disaggregating, the user has the option of splitting the optimal management policy for equivalent reservoirs in the way that suits the system under consideration. For the proposed model, disaggregation of reservoirs is a function of the energy stored in the individual reservoirs at the beginning of each month. Therefore, the total hydro energy expected from an equivalent long-term reservoir will be divided between the individual subreservoirs in proportion to their monthly stored energy.

Thermal System. The thermal system representation comprises a load demand, a generating unit-available capacity model to represent the forced outages of these units, and an equivalent load model to compute the expected value of the thermal energy. Cost calculation is based on the loading of the thermal units.

Load Demand Model. Suppose the load demand curve is given as a continuous curve. Then, the load duration curve will also be continuous (Figure 7.20). Now, let us invert the load duration as shown in Figure 7.20; the inverted load duration curve (ILDC), denoted by (X), is defined as

$$\mathcal{L}(X) = \text{prob}\{D > X\} \qquad (7.100)$$

The cumulative distribution function $F_D(X)$ of the random variable D by definition is the probability that the demand X is greater than or equal to D, i.e.,

$$\mathcal{F}_D(X) \triangleq \text{prob}\{D \le X\} = 1 - \mathcal{L}(X) \qquad (7.101)$$

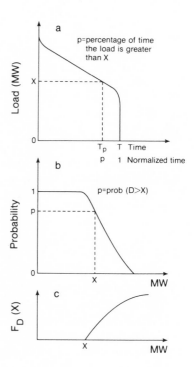

Figure 7.20. Continuous load curves. (a) Typical load duration curve; (b) inverted load duration curve; (c) cumulative distribution of D.

Loading Priority List of Thermal Machines. The production cost of each generating block is approximated by its average proportional costs, neglecting the start-up costs. Then, the blocks are loaded in an ascending order of their average production costs.

Generating-Unit Available-Capacity Model. Let w_i be the rated capacity of generating unit i (MW), c_i be the available capacity (random variable), $\bar{q}_i$ be the forced outage rate, and p_i be the probability that the unit is available. The cumulative distribution function $\mathscr{F}_{c_i}(X)$ is

$$\mathscr{F}_{c_i}(X) = \bar{q}_i u(X) + p_i u(X - w_i) \tag{7.102}$$

and the probability density function f_{c_i} is given by

$$f_{c_i} = \bar{q}_i \delta(X) + p_i \delta(X - w_i) \tag{7.103}$$

where $u(X)$ denotes a unit step and $\delta(X)$ a unit impulse.

Equivalent Load. For computational convenience, the equivalent ILDC, after committing k units, is defined as

$$\mathscr{L}_k(X) = 1 - \mathscr{L}_{T_k}(X), \qquad k = 1, 2, \dots, N \tag{7.104}$$

We thus have the recursive formula:

$$\mathscr{L}_0(X) = \mathscr{L}(X)$$

or

$$\mathscr{L}_k(X) = p_k \mathscr{L}_{k-1}(X - w_k) + \bar{q}_k \mathscr{L}_{k-1}(X) \tag{7.105}$$

Expected Value of Generated Energy. The expected value of P_{TH_k}, i.e., the expected power generation demanded from unit k, $\varepsilon[P_{TH_k}]$, is given by

$$\varepsilon[P_{TH_k}] = \int_0^{w_k} \mathscr{L}_{k-1}(X)\, dX \tag{7.106}$$

The expected power delivered by unit k, obtained by conditioning on the availability of unit k, has the value $P_k \in [P_{TH_k}]$. The expected value of energy delivered by unit, E_{TH_k}, the time period, t, under consideration is given by

$$E_{TH_k} = P_k t \int_0^{w_k} \mathscr{L}_{k-1}(X)\, dX \tag{7.107}$$

Uncontrolled Hydro. Uncontrolled hydro plants utilize the stream flows to generate energy with the amount of energy depending on the flows. The stream flow is considered a random variable having a continuous distribution function as shown in Figure 7.21. These units are base loaded and placed in the very beginning of the loading priority list. Using equation (7.104) we have

$$\mathscr{L}_k(X) = \int_{-\infty}^{\infty} \mathscr{L}_{k-1}(X - z) f_{c_k}(-z) \, dz \tag{7.108}$$

Except for this modification an uncontrolled hydro unit is treated as a thermal unit in all other aspects.

7.6.2. Optimization Problem Formulation

Given the load demand for time period k and the thermal units available for generation, the production cost for this period depends on the amount of energy produced by the controlled-hydro plants. This, in turn, is a function of the variables $q(k)$ or $u(k)$ and $x(k)$, which are the water released from the reservoir and the reservoir content during period k, respectively.

For a given set of values of $q(k)$, $u(k)$, and $x(k)$, the production cost for time period k is evaluated using the thermal probabilistic model. The cost function for this time period is denoted as $F_k[q(k), x(k), I(k)]$, where $I(k)$ is the stochastic flow into the reservoir and is considered as a random

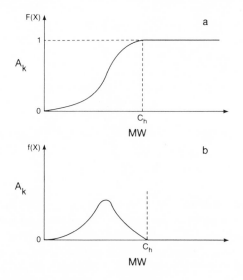

Figure 7.21. Continuous probability. (a) Distribution, and (b) density of available capacity for an uncontrolled hydroelectric unit.

variable. The production cost of the K time period is

$$\sum_{k=1}^{K} F_k[q(k), x(k), I(k)] \tag{7.109}$$

the period k cost, F_k, is computed using the following recursive equations:

$$F_k(x_k) = \min_{x_k} \sum_{I(k)} P(I_k)[f(q(k), x(k), I(k))] \tag{7.110}$$

where

$$f_k(q(k), x(k), I(k)) = c_k(q(k), x(k), I(k)) + F_{k-1}[t(k)(q(k), x(k), I(k))] \tag{7.111}$$

$$F_1(q(1), x(1), I(1)) = C_1(q(1), x(1), I(1)) \tag{7.112}$$

F_k, f_k, and C_k are cost functions for period k and $P(I_k)$ are the probability numbers associated with I_k.

The hydro scheduling problem may now be formulated as a discrete-time optimal control problem as follows:

Find $u(1), \ldots, u(K)$ or $q(1), \ldots, q(K)$ so as to minimize

$$\sum_{k=1}^{K} F_k(u(k), x(k), I(k)) + \psi \tag{7.113}$$

Subject to the constraints

$$x(k-1) = x(k) + I(k) - u(k) \tag{7.114}$$

$$x^m \le x(k) \le x^M \tag{7.115}$$

and

$$u^m(k) \le u(k) \le u^M(k) \tag{7.116}$$

where ψ is the penalty term

$$\psi = \sum_{i=1}^{n} W_i[x_{i,K}^* - x_{i,K}]^2 \tag{7.117}$$

$x_{i,K}^*$ is the desired terminal storage in reservoir i, $x_{i,K}$ is the actual terminal storage in reservoir i, and W_i is the penalty cost associated with equivalent subsystem i.

If, at the end of the cycle (year) x_i^* is not achieved, an extra cost, ψ, is added to the total cost of generation of the system. W_i is the product of the percentage contribution is potential energy of reservoir i during the cycle of one year and the total generation cost during the cycle. The optimization is carried out under the following assumptions and consideration.

(1) In winter, water is drawn from storage to supplement the low river flow. If drawdown is made too soon, the hydrostatic head at the plant drops and the power generated can be severely reduced. At the same time, such drawdown is a consumption of reserves which, if the year turns out dry, can lead to power storage. On the other hand, if the stored water is saved too long, it may never be used, for there is always more water than can be used in spring. Therefore, in the winter months the decisions require a balancing of the benefits of future against immediate water use.

(2) Hydro plants are usually operated near maximum efficiency. Therefore, a logical approach is to optimize in the neighborhood of the outflow at maximum efficiency, which is obtained from the following equation:

$$q_i^M = b_{1i} + b_{2i}h_i + b_{3i}h_i^2 + b_{4i}h_i^3 \tag{7.118}$$

where b_{1i}, b_{2i}, b_{3i}, and b_{4i} are known coefficients for plant i and h_i is the head at plant i.

(3) After a cycle of one year, the reservoirs should return to the same initial volumes. Most possible drawdown occur during peak load months (winter).

7.6.3. Optimization Algorithm

(1) Aggregate the hydro system to a number of equivalent midterm subsystems.

(2) Select a range of outflows for each reservoir in the neighborhood of the outflow at maximum efficiency.

(3) Select an expected set of reservoir volumes for each month to achieve maximum drawdown during winter months and return to the initial level at the end of the annual cycle.

(4) Using equations (7.109)–(7.117) and the information in steps (2) and (3), obtain the minimum cost of supplying the required load with the normally distributed random inflows represented by three probabilities for each month in the explicit dynamic programming procedure.

(5) The generation cost is that of thermal generation and is obtained by loading the thermal units according to a prescribed start-up order after all the hydro units have been loaded. Thermal generation supplies the difference between load energy and hydro energy. A probabilistic model is used to compute the thermal generation costs.

(6) When minimum cost of generation is achieved, the equivalent long-term reservoir outflows, which are the output of the optimization procedure, are disaggregated to yield the individual outflows of the sub-reservoirs.

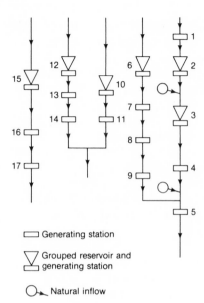

Figure 7.22. Hydroelectric portion of the hydro-thermal system.

7.6.4. Practical Example

The application of the algorithm to a generalized system is presented to illustrate the type of results the proposed technique yields.

Figure 7.22 shows the components of the hydro generating units. In this example, the same generators with the same startup order are used each month during the year.

Table 7.15. Characteristics of the Thermal Generating Units

Unit type	Capacity (MW)	No. of valves	Availability of unit type	Minimum output (MW)
1	495	4	0.8391	0
2	300	4	0.8300	0
3	300	6	0.6810	0
4	300	4	0.7510	0
5	66	1	0.9212	0
6	490	4	0.6040	0
7	490	4	0.7400	0
8	490	1	0.7400	0
9	515	1	0.8000	515
10	208	1	0.8300	208
11	200	4	0.8580	0

Table 7.15 lists the characteristics of the thermal generating units. In this example, the same generators with the same start-up order are used each month during the year.

Table 7.16 gives the start-up order of the thermal generators. Any extra energy needed to supply the load demand is obtained from neighboring utilities.

Table 7.16. The Starting up Order of the Thermal Generators

Generator start-up order	Generator type No.	Valve No.
1	10	1
2	9	1
3	1	1
4	1	2
5	1	3
6	1	4
7	2	1
8	4	1
9	11	1
10	8	1
11	7	1
12	6	1
13	8	2
14	7	2
15	6	2
16	2	2
17	8	3
18	7	3
19	6	3
20	8	4
21	7	4
22	6	4
23	4	2
24	11	2
25	3	1
26	3	2
27	4	3
28	3	3
29	3	4
30	3	2
31	3	5
32	3	6
33	4	4
34	11	3
35	2	4
36	11	4
37	5	1

Table 7.17. Monthly Load Characteristics

Month	Peak load (MW)	Average load (MW)	Energy (MWh)
January	13773	10769	0.8012136×10^7
February	13622	10780	0.7244160×10^7
March	12668	10145	0.7547880×10^7
April	12027	9357	0.6961608×10^7
May	11616	9058	0.6739162×10^7
June	11401	8928	0.6642432×10^7
July	11059	8578	0.6382032×10^7
August	11425	8838	0.6575472×10^7
September	11913	9348	0.6954912×10^7
October	12276	9652	0.7181088×10^7
November	14153	10563	0.7858872×10^7
December	14840	11224	0.8350666×10^7

Table 7.18. Monthly River Inflows and the Associated Probabilistics

Month	Reservoir inflow (m^3/sec)	P	Month	Reservoir inflow (m^3/sec)	P
January	104.50	0.2	July	373.82	0.2
	92.88	0.5		240.72	0.5
	76.46	0.8		154.34	0.8
February	99.12	0.2	August	244.97	0.2
	75.05	0.5		190.59	0.5
	62.30	0.8		124.60	0.8
March	87.79	0.2	September	373.82	0.2
	78.45	0.5		206.17	0.5
	69.38	0.8		73.63	0.8
April	1118.64	0.2	October	543.74	0.2
	580.84	0.5		368.73	0.5
	186.91	0.8		181.25	0.8
May	1670.88	0.2	November	348.90	0.2
	1387.68	0.5		265.64	0.5
	1047.84	0.8		186.91	0.8
June	623.04	0.2	December	181.25	0.2
	514.01	0.5		150.10	0.5
	354.00	0.8		86.88	0.8

Table 7.19. The Characteristics of the Hydroelectric Units

Hydroelectric unit No.	Rating (MW)	Maximum volume (m³)	Minimum volume (m³)
1	136	0.011931×10^6	-0.007564×10^5
2	143	0.021093×10^6	-0.012994×10^6
3	335	0.006374×10^7	0.00
4	110	0.010259×10^7	0.00
5	160	0.004734×10^6	0.016952×10^5
6	173	0.003246×10^7	0.00
7	177	0.008875×10^6	0.00
8	174	0.021484×10^5	0.00
9	40	0.008580×10^5	-0.008498×10^5
10	136	0.006365×10^6	0.00
11	175	0.007792×10^6	0.00
12	121	0.005489×10^7	0.00
13	71	0.017494×10^5	0.00
14	143	0.008829×10^5	0.00
15	132	0.016842×10^5	0.00
16	222	0.024346×10^6	0.00
17	39	0.003475×10^6	0.00

The monthly load characteristics are given in Table 7.17. Standard weekdays, Saturdays, and Sundays are used to represent each month.

Table 7.18 presents the monthly river inflows and the associated probabilistics for one of the reservoirs in the system.

The characteristics of the hydro units are presented in Table 7.19.

The optimal expected hydro discharge from the reservoirs in m³/sec for a period of a year is given in Table 7.20.

7.6.5. Concluding Remarks

A useful mathematical model for optimizing the hydro unit usage over the period of one year is developed. This model yields the expected monthly hydro discharge for a year that minimizes the system production costs.

The proposed model considers the random nature of the river inflows and the partial and total forced outage rates of the thermal generating units.

The application of the model to an existing hydrothermal electrical system has yielded very satisfactory results.

The trade-offs in accuracy in the detailed model are mainly an increase in the computation time and storage.

Table 7.20. Optimal Discharges from the Reservoirs in m³/sec for a Period of a Year

Unit No.

Month	1	2	3	4	5	6	7	8	9	10	11	12	13	14	15	16	17
January	687	7588	8352	9508	11032	1090	1127	1191	1217	1865	1915	1537	1579	1526	694	800	1033
February	740	8233	8831	9633	11259	1102	1116	1186	1880	1880	1953	1451	1458	1422	916	1082	1319
March	620	5951	7109	8154	9461	1001	1011	1062	1059	1616	1650	1761	1785	1714	527	725	892
April	572	4818	6722	10174	14934	1858	2005	2356	2509	1917	2006	1755	1771	1715	125	424	807
May	1909	6694	9080	16091	19755	2201	2232	2422	2487	6890	8941	17887	5591	5592	335	1279	2131
June	845	6388	7482	9276	9726	624	600	633	644	4312	5828	9712	5666	5519	328	717	1013
July	367	4611	5116	6107	6148	259	258	253	268	4185	5147	6657	5062	4859	186	359	546
August	326	3690	4160	4913	4911	189	174	175	190	2165	2427	2169	2271	2126	210	295	546
September	363	3458	3988	4399	4526	280	276	267	287	2235	2729	2615	2773	2621	438	653	993
October	374	4136	5114	6223	6722	457	461	497	527	2081	2402	3077	3152	3051	416	802	1393
November	568	5991	7061	8215	8984	737	751	769	785	2364	2829	4245	4221	4131	604	1343	1943
December	541	7573	8563	10519	12134	1350	1387	1449	1459	1781	1995	2082	2127	2061	492	892	1279

7.7. Aggregation–Decomposition Approach (Ref. 7.53)

The aim of this section is to explain the implementation of the aggregation-decomposition approach for a large hydrothermal power system for long-term optimal operation. The problem (with N reservoirs) is decomposed into S subproblems with two state variables. Each subproblem finds the optimal operating policy for one of the reservoirs as a function of the energy content of that reservoir and the aggregate energy content of the remaining reservoirs. The subproblems are solved by stochastic dynamic programming taking into account the detailed models of the hydro chains as well as the stochasticity and correlation of the hydro inflows.

The reservoir content, inflow, and discharge are modeled in terms of energy in order to allow aggregation of reservoirs. River hydrologies are considered to be the only random variables in the problem. The objective of the long-term operating problem is to find the discharge $u_{i,k}$, $i = 1, .., n$, $k = 1, .., K$ that minimizes the expected thermal fuel cost of supplying the load D_k. In mathematical terms, the problem can be written as

$$\text{Minimize } \sum_{k=1}^{K} E[\text{TC}_k(D_k, u_{1,k}, \ldots, u_{n,k}) + V[x(K)]] \qquad (7.119)$$

where E is the expected value, $\text{TC}_k(\cdot)$ is the thermal cost of supplying the load D_k, and $V[x(K)]$ is the terminal cost as a function of the energy content of the reservoirs, subject to satisfying the following constraints:

$$x(k + 1) = x(k) + I(k) - u(k) - s(k) \qquad (7.120)$$

$$x^m \le x(k) \le x^M \qquad (7.121)$$

$$u^m(k) \le u(k) \le u^M(k) \qquad (7.122)$$

The general stochastic dynamic programming (SDP) solution of this problem assumes that the operating policy is a function of all states: $u_i = u_i(x_1, x_2, \ldots, x_n)$.

The resulting recursive equation thus can be written in the following form:

$$V_k(x_{1,k}, x_{2,k}, \ldots, x_{n,k}) = E_I[\min_{u_1, \ldots, u_n} (T_{c_k}(D_k, u_{1,k}, \ldots, u_{n,k})$$

$$+ V_{k+1}(x_{1,k+1}, \ldots, x_{n,k+1}))] \qquad (7.123)$$

Subject to the constraints (7.120)–(7.122).

7.7.1. Spatial Decomposition

The problem in equation (7.123) is composed of N subproblems, each of two state variables with the following operative policy:

$$u_i = u_i(x_i, x_{ic}) \qquad (7.124)$$

where ic in the above equation denotes the aggregate reservoir associated with i, obtained by aggregation of all reservoirs but i. Its variables are computed as (the same approach was used in Chapter 2)

$$x_{ic}^m = \sum_{j \neq i}^n x_j^m, \qquad x_{ic}^M = \sum_{j \neq i}^n x_j^M, \qquad I_{ic} = \sum_{j \neq i}^n I_j \qquad (7.125)$$

and

$$u_{ic}^m = \sum_{j \neq i}^n u_j^m, \qquad u_{ic}^M = \sum_{j \neq i}^n u_i^M \qquad (7.126)$$

The objective now of each subproblem is to find $u_i(x_i, x_{ic})$ and $u_{ic}(x_i, x_{ic})$ (see Figure 3.2) for each reservoir i during each period k, starting at the end of the planning horizon, so as to minimize the expected cost to go given by

$$V_k(x_{i,k}, x_{ic,k}) = E_I[\min_u (T_{c_k}(D_k, u_i, u_{ic}) + V_{k+1}(x_{i,k+1}, x_{ic,k+1}))] \qquad (7.127)$$

Subject to satisfying the following constraints:

$$x_{i,k+1} = x_{i,k} + I_{i,k} - u_{i,k} - s_{i,k} \qquad (7.128)$$

$$x_{ic,k+1} = x_{ic,k} + I_{ic,k} - u_{ic,k} - s_{ic,k} \qquad (7.129)$$

$$x_i^m \leq x_{i,k} \leq x_i^M \qquad (7.130)$$

$$x_{ic}^m \leq x_{ic,k} \leq x_{ic}^M \qquad (7.131)$$

$$u_{i,k}^m \leq u_{i,k} \leq u_{i,k}^M \qquad (7.132)$$

$$u_{ic,k}^m \leq u_{ic,k} \leq u_{ic,k}^M \qquad (7.133)$$

where the expectation in equation (7.127) is taken over the random variables $I_{i,k}$.

7.7.2. Algorithm of Solution

The following assumptions have been made:

- The load is deterministic with known load duration curve.
- Unit availability is modeled by derating their capacity.
- Stochasticity of hydrologies is treated by considering a large finite sample of hydrological sequences. In each stage, the optimal decision is found for each hydrology and the expected costs are calculated as the average for all samples. Hydrological sequences are precalculated with a synthetic hydrology generator that uses historical series and preserves their serial and spatial correlation. NSQ random sequences are generated.

- Transition cost (thermal and rationing costs) is computed as a function of the total energy release from the reservoirs. A peak shaving algorithm is used, assuming a block dispatch of all hydro plants.

In the design of the algorithm shown in Figure 7.23 special attention was given to its structure in order to reduce the execution time of the program. The main steps and processes are as follows:

Step 1. Preprocessing and initialization.

Step 2. For each period k:
- Load and thermal modeling: build the load duration curve and the thermal system cost curve.
- Modeling of the hydro system: Find the energy model of all the reservoirs and compute the storage limits of the associated reservoirs.

Step 3. For each period and each hydrology:
- Modeling of the hydro system: compute the run of river generation of the chains, the energy inflow to the reservoirs, and the bounds on the reservoir discharges.

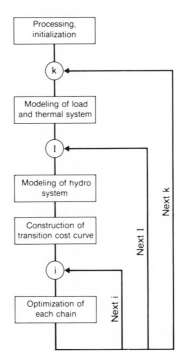

Figure 7.23. Structure of the algorithm.

- Computation of the transition cost: Find the thermal and rationing cost of supplying the load as a function of the total hydro energy generation.
- Optimization of each chain: Find the optimal releases u_i^* and u_{ic}^* as a function of the discrete state of the reservoirs and the corresponding stage cost (S_c):

$$S_{c_k}(x_i, x_{ic}, I_i) = T_{c_k}(D_k, u_i^*, u_i^*) + V_{k+1}(x_i(u_i^*) + x_{ic}(u_{ic}^*))$$

(7.134)

Then, update the tables of the expected cost to go for period k

$$V_k(x_i, x_{ic}) = V_k(x_i, x_{ic}) + Sc_k(x_i, x_{ic}, I_{i,k})p(I_{i,k})$$

(7.135)

where $p(I_{i,k})$ is the probability of occurrence of hydrology I in period k.

Step 4. Stop when steps 2 and 3 are completed for all periods and all hydro sequences.

7.7.3. Stage Optimization, Optimization Procedure

Computation of the optimal release from the reservoirs is the basis of the algorithm since it is repeated for each state (x_i, x_{ic}), each reservoir i, each period k, and each inflow hydrology I_i. The problem is a deterministic nonlinear optimization in the variables u_i, u_{ic} as

$$\min_{u_i, u_{ic}} [T_c(D_k, u_i, u_{ic}) + V_{k+1}(x_{i,k+1}, x_{ic,k+1})]$$

(7.136)

Subject to the following constraints:

$$x_{i,k+1} = x_{i,k} + I_{i,k} - u_{i,k} - s_{i,k}$$

(7.137)

$$x_{ic,k+1} = x_{ic,k} + I_{ic,k} - u_{ic,k} - s_{ic,k}$$

(7.138)

and the following inequality constraints on the variables:

$$x_i^m \le x_{i,k} \le x_i^M$$

(7.139)

$$x_{ic}^m \le x_{ic,k} \le x_{ic}^M$$

(7.140)

$$u_{i,k}^m \le u_{i,k} \le u_{i,k}^M$$

(7.141)

$$u_{ic,k}^m \le u_{ic,k} \le u_{ic,k}^M$$

(7.142)

Reference 7.53 points out three main features characterizing the procedure used in this section, as follows:

First, only controls u_i, u_{ic} that take the reservoirs from a grid point in period k to another grid point in period $k + 1$ are considered. The problem is then to find the optimal state at time $k + 1$ that can be reached from state (x_i, x_{ic}) instead of searching directly over the controls. Consequently, the objective function (7.136) can be written as

$$\min_{x_{i,k+1}, x_{ic,k+1}} [T_c(D_k, u_i(x_{i,k+1}), u_{ic}(x_{ic,k+1})) + V_{k+1}(x_{i,k+1}, x_{ic,k+1})] \qquad (7.143)$$

where u_i and u_{ic} are computed from the dynamic equations (7.137) and (7.138).

Second, the bounds on the controls given by (7.141) and (7.142) are relaxed in such a way that the feasible region is approximated to the nearest grid points. The optimum cost is thus slightly overestimated when the region is reduced and underestimated when the region is enlarged to the nearest grid points. Since the cost is calculated for a large number of hydrologies, on the average the cost will be close to the expected cost.

Third, the optimization algorithm uses a property of the optimal releases in order to reduce the region of search. If the optimal control u^* is known for the state x, then the optimal release $u^* + \Delta u$ associated with the state $x + \Delta x$ satisfies the relation

$$0 \le \left| \sum_{i=1}^{n} \Delta u_i \right| \le \left| \sum_{i=q}^{n} \Delta x_i \right| \qquad (7.144)$$

when the following assumptions apply, as is the case for the stage optimization:

1. T_c is only a function of the component of U:

$$W = \sum_{i=1}^{n} u_i$$

2. $V(x + I - u - s)$ is a continuous convex function of $(x + I - u - s)$.
3. T_c is a nonincreasing continuous convex function of W.

The storage optimization algorithm can be summarized as follows:

Step 1. Find the optimal control for an initial state (x_i, x_{ic}) using an exhaustive search over the feasible states in $k + 1$.

Step 2. The remaining grid points in the row are scanned with x_{ic} fixed. The optimum for each grid point is found efficiently using the band result described in equation (7.144).

Step 3. x_{ic} is incremental Δx_{ic} and the optimum is found within the corresponding band. Steps 2 and 3 are repeated until the whole grid has been scanned.

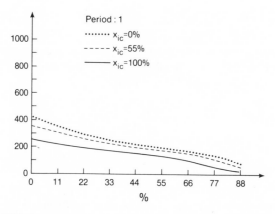

Figure 7.24. Incremental cost curves. Reservoirs with multiannual regulation.

7.7.4. Practical Example

The aggregation–decomposition method was applied to a large system including 10 hydro chains with a wide variety of reservoir sizes. Reservoir regulations were classified as long-term (2–3 years), seasonal (yearly cycle), and short-term reservoirs that can be drawn down in a few weeks. Constraints in the chains include tunnels, aqueduct and irrigation demands, minimum and maximum flows for pollution and flood control. The generating system contained 15 thermal plants and 37 hydro plants.

The results shown correspond to an horizon of 5 years, flat terminal cost, and 100 hydrologies.

Figures 7.24–7.27 show the incremental cost curves of two reservoirs as a function of the energy content of the reservoir for three different levels of its associated reservoir. Period 1 corresponds to the beginning of the dry

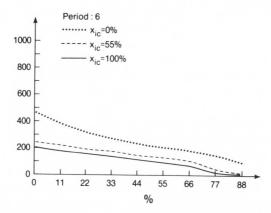

Figure 7.25. Incremental cost curves. Reservoirs with multiannual regulation.

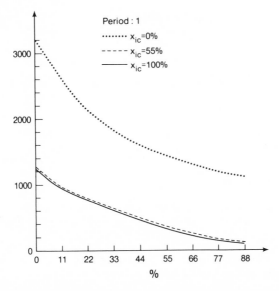

Figure 7.26. Incremental cost curves. Reservoirs with yearly regulation.

season, and the rainy season starts at period 6. One reservoir has a large storage capacity with a multiannual regulation while the other one has a one-year cycle and one third of the storage capacity of the big one.

These incremental cost curves illustrate the following facts:

• The magnitude of the incremental cost depends strongly on the period. As expected, the magnitude of the incremental cost is higher at the beginning of the dry season than at the beginning of the rainy season.

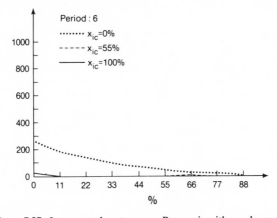

Figure 7.27. Incremental cost curves. Reservoir with yearly regulation.

- The sensitivity of the incremental cost to the content of the associated reservoir depends on the type of reservoir. For example, the multiannual reservoir is the less sensitive. The seasonal reservoir can be sensitive or not depending on the period of the year.
- The results show also that the short-term reservoirs are strongly sensitive to the amount of the energy stored in the rest of the system, in any period of the year.

A five-year case, with an average of 7 hydro chains and a discretization of 10×10 was solved in 73 min on a VAX 11-780.

7.7.5. Concluding Remarks

The aggregation–decomposition approach proved to be an effective tool for solving the long-term generation scheduling in large hydrothermal systems. The results obtained allowed us to draw the following conclusions:

- The value of the water stored in each reservoir is strongly influenced by the total energy stored in the rest of the system.
- The magnitude of this influence varies significantly from one reservoir type to another and from one period of the year to another.

The development of a fast and accurate stage optimization algorithm was a crucial step in achieving reasonable computer times in the application of the AD approach.

7.8. A Minimum Norm Approach, Linear Storage-Elevation Model

This section presents a new approach the authors have used for solving the optimal long-term scheduling problem of a hydrothermal power system for minimum expected fuel cost. The proposed method takes into account the stochasticity of the river inflows and the load demand; we assume that their probability properties are preestimated from past history. The problem is formulated as a minimum norm problem in the framework of functional analytic optimization technique. The cost functional is augmented by using Lagrange and Kuhn–Tucker multipliers to adjoin the equality and inequality constraints. A set of stochastic discrete equations is obtained. The developed approach is efficient in saving the fuel cost and computing time; it can handle large-scale power systems that are hydraulically coupled, which is not the case with most available techniques.

7.8.1. Problem Formulation and the Objective Function

The system under consideration is given in Figure 7.14, where it is composed of many hydroelectric plants in one or more river basins, thermal plants, and interchanges with other systems. The problem is, for this system, to find the active generation of each plant as a function of time over the optimization interval such that the following conditions are satisfied:

(1) The expected total operating costs of the thermal plants over the optimization interval are a minimum.

(2) The operating costs of the ith thermal plant during a month k are approximated by

$$F_i(P_{s_{i,k}}) = \alpha_i + \beta_i P_{s_{i,k}} + \gamma_i(P_{s_{i,k}})^2, \qquad i \in R_s \tag{7.145}$$

(3) The total active generation of the system matches the load plus the transmission line losses during a period k. For long-term study we assume that the transmission losses are constant and added to the demand:

$$\sum_{i \in R_h} P_{h_{i,k}} + \sum_{i \in R_s} P_{s_{i,k}} = D_k, \qquad k = 1, \ldots, K \tag{7.146}$$

(4) The water conversion factor, MW/Mm^3, is a linear function of the storage and is given by

$$WC_{i,k} = (a_i + b_i x_{i,k}) \, MW/Mm^3, \qquad i \in R_h \tag{7.147}$$

In the above equation, we assume that the tail-water elevation is constant independent of the discharge and a_i, b_i are constants; these are obtained by least squares curve fitting to typical plant data available.

(5) The hydroelectric generation for each hydro plant is a function of the head, which in turn is a function of the storage and the discharge. In this part of the section, we will choose $P_{h_{i,k}}$ as

$$P_{h_{i,k}}[a_i u_{i,k} + \tfrac{1}{2} b_i u_{i,k}(x_{i,k} + x_{i,k-1})] \, MW, \qquad i \in R_h \tag{7.148}$$

In the above equation an average of begin-and-end of time storage was used to avoid overestimation for rising water levels, and underestimation for falling water levels.

(6) The reservoirs' dynamics may be adequately described by the following discrete continuity type:

$$x_{i,k} = x_{i,k-1} + I_{i,k} - u_{i,k} - s_{i,k} + \sum_{j \in R_u} (u_{j,k} + s_{j,k}) \tag{7.149}$$

where

$$s_{i,k} = \begin{cases} \left(\left(x_{i,k-1} - x_{i,k} + I_{i,k} + \sum_{j \in R_u} (u_{j,k} + s_{j,k}) \right) \right) - u_{i,k}^M \\ \quad \text{if } \left(x_{i,k-1} - x_{i,k} + I_{i,k} + \sum_{j \in R_u} (u_{j,k} + s_{j,k}) \right) > u_{i,k}^M \\ \quad \text{and} \quad x_{i,k} = x_i^M \\ 0 \qquad \text{otherwise} \end{cases} \tag{7.150}$$

In the above equation $I_{i,k}$ is the inflow to reservoir i during a period k. These are random variables; we assume that they are statistically independent and have a Gaussian distribution, and R_u is the set of reservoirs immediately upstream to the reservoir i.

(7) As we said earlier, most of the rivers are multipurpose streams. To satisfy the multipurpose stream use requirements, there are upper and lower bounds on the storage and the discharge. These are given by

$$x_i^m \leq x_{i,k} \leq x_i^M, \qquad i \in R_h \tag{7.151}$$

$$u_{i,k}^m \leq u_{i,k} \leq u_{i,k}^M, \qquad i \in R_h \tag{7.152}$$

These inequality constraints simply state that the reservoir storage may not exceed a maximum level, nor be lower than a minimum level. For the former, this is determined by the elevation of the spillway crest of the top of the spillway gates. The minimum level may be fixed by the elevation of the lowest outlet in the dam or by conditions of operating efficiency for the turbines.

(8) The water left in storage at the end of optimization interval is given by

$$x_{i,K} = x_{i,0} + \sum_{k=1}^{K} I_{i,k} - \sum_{k=1}^{K} (u_{i,k} + s_{i,k})$$
$$+ \sum_{k=1}^{K} \left[\sum_{j \in R_u} (u_{j,k} + s_{j,k}) \right], \qquad i \in R_h \tag{7.153}$$

and this is a specified amount of water (specified energy) at the last period studied.

If we substitute from equation (7.149) into equation (7.148) for $x_{i,k}$, we obtain the following hydro-power generation

$$P_{h_{i,k}} = \left[b_{i,k} u_{i,k} + b_i u_{i,k} x_{i,k-1} + \tfrac{1}{2} b_i u_{i,k} \left(\sum_{j \in R_h} u_{j,k} - u_{i,k} \right) \right], \qquad i \in R_h \tag{7.154}$$

where

$$q_{i,k} = I_{i,k} - s_{i,k} + \sum_{j \in R_h} s_{j,k}, \qquad i \in R_h \tag{7.155a}$$

$$b_{i,k} = a_i + \tfrac{1}{2} b_i q_{i,k} \tag{7.155b}$$

In mathematical terms, the object of the optimizing computation is to find $P_{h_{i,k}}$ and $P_{s_{i,k}}$ that minimize

$$J = E \left[\sum_{k=1}^{K} \sum_{i \in R_s} (\alpha_i + \beta_i P_{s_{i,k}} + \gamma_i (P_{s_{i,k}})^2) \right] \tag{7.156}$$

Subject to satisfying the equality constraints given by equations (7.146), (7.149), and (7.153) and the inequality constraints given by equations (7.151) and (7.152).

To take care of the storage at the last period studied, in equation (7.153), the cost function of equation (7.156) can be modified to

$$J = E = \left[\sum_{i \in R_h} \tfrac{1}{2} w_i (x_{i,K}^* - x_{i,K})^2 + \sum_{k=1}^{K} \sum_{i \in R_s} (\alpha_i + \beta_i P_{s_{i,k}} + \gamma_i (P_{s_{i,k}})^2) \right] \quad (7.157)$$

where w_i is a weighting factor, $x_{i,k}^*$ is the specified amount of water given by equation (7.153) and $x_{i,K}$ is the calculated storage at the last period during the optimization procedures.

7.8.2. A Minimum Norm Formulation

The augmented cost function can be obtained by adjoining the equality constraint (7.146) using the unknown dual variable θ_k, the equality constraints (7.149) via Lagrange multiplier $\lambda_{i,k}$, and the inequality constraints (7.151) and (7.152) via Kuhn-Tucker multipliers, so that a modified cost function is obtained:

$$\tilde{J} = E\left[\sum_{i \in R_h} \tfrac{1}{2} w_i (x_{i,K}^* - x_{i,K})^2 + \sum_{k=1}^{K} \left[\sum_{i \in R_s} \{ (\alpha_i + (\beta_i - \theta_k) P_{s_{i,k}} + \gamma_i (P_{s_{i,k}})^2) \} \right. \right.$$

$$+ \sum_{i \in R_h} \left\{ -\theta_k (b_{i,k} u_{i,k} + b_i u_{i,k} x_{i,k-1} + \tfrac{1}{2} b_i u_{i,k} \left(\sum_{i \in R_h} u_{j,k} - u_{i,k} \right) \right.$$

$$+ \lambda_{i,k} \left(-x_{i,k} + x_{i,k-1} + q_{i,k} - u_{i,k} + \sum_{i \in R_h} u_{j,k} \right) + e_{i,k}^m (x_i^m - x_{i,k})$$

$$\left. \left. \left. + e_{i,k}^M (x_{i,k} - x_i^M) + g_{i,k}^m (u_{i,k}^m - u_{i,k}) + g_{i,k}^M (u_{i,k} - u_{i,k}^M) \right\} \right] \right] \quad (7.158)$$

Employing the discrete version of integration by parts and dropping constant terms, one obtains

$$\tilde{J} = E\left[\sum_{i \in R_h} \{ \tfrac{1}{2} w_i (x_{i,K}^* - x_{i,K})^2 + \lambda_{i,K} x_{i,K} - \lambda_{i,0} x_{i,0} \} \right.$$

$$+ \sum_{k=1}^{K} \left[\sum_{i \in R_s} \{ \delta_{i,k} P_{s_{i,k}} + \gamma_i (P_{s_{i,k}})^2 \} \right.$$

$$+ \sum_{i \in R_h} \left\{ B_{i,k} u_{i,k} + \beta_{i,k} u_{i,k} x_{i,k-1} + \tfrac{1}{2} \beta_{i,k} u_{i,k} \left(\sum_{j \in R_h} u_{j,k} - u_{i,k} \right) \right.$$

$$+ (\lambda_{i,k} - \lambda_{i,k-1} + \mu_{i,k}) x_{i,k-1} + \lambda_{i,k} \left(\sum_{j \in R_h} u_{j,k} - u_{i,k} \right)$$

$$\left. \left. \left. + \psi_{i,k} u_{i,k} + \mu_{i,k} \left(\sum_{j \in R_h} u_{j,k} - u_{j,k} \right) \right\} \right] \right] \quad (7.159)$$

where

$$\delta_{i,k} = \beta_i - \theta_k, \ i \in R_s \tag{7.160}$$

$$B_{i,k} = -\theta_k b_{i,k}, \ i \in R_h \tag{7.161}$$

$$\beta_{i,k} = -\theta_k b_i, \ i \in R_h \tag{7.162}$$

$$\mu_{i,k} = e_{i,k}^M - e_{i,k}^m, \ i \in R_h \tag{7.163}$$

$$\psi_{i,k} = g_{i,k}^M - g_{i,k}^m, \ i \in R_h \tag{7.164}$$

If one defines the following vectors for thermal plants

$$\delta(k) = \text{col}(\delta_{i,k}; i \in R_s) \tag{7.165}$$

$$\gamma = \text{diag}(\gamma_i; i \in R_s) \tag{7.166}$$

and if one also defines the following vectors for hydro plants

$$B(k) = \text{col}(B_{i,k}, i \in R_n) \tag{7.167}$$

$$\lambda(k) = \text{col}(\lambda_{i,k}, i \in R_h) \tag{7.168}$$

$$\mu(k) = \text{col}(\lambda_{i,k}, i \in R_h) \tag{7.169}$$

$$\psi(k) = \text{col}(\psi_{i,k}, i \in R_h) \tag{7.170}$$

$$x(k) = \text{col}(x_{i,k}, i \in R_h) \tag{7.171}$$

$$u(k) = \text{col}(u_{i,k}, i \in R_h) \tag{7.172}$$

and if one furthermore defines the following vectors

$$w = \text{diag}(w_i, i \in R_h) \tag{7.173}$$

$$\beta(k) = \text{diag}(\beta_{i,k}, i \in R_h) \tag{7.174}$$

Then the augmented cost functional in equation (7.159) can be written in the following vector form:

$$\tilde{J} = E\left[\tfrac{1}{2}(x^*(K) - x(K))^T w(x^*(K) - x(K)) + \lambda^T(K)x(K) - \lambda^T(0)x(0) \right.$$

$$+ \sum_{k=1}^{K} \{\delta^T(k)P_s(k) + P_s^T(k)\gamma P_s(k) + B^T(k)u(k)$$

$$+ \tfrac{1}{2}x^T(k-1)B(k)u(k) + \tfrac{1}{2}u^T(k)B(k)x(k-1)$$

$$+ \tfrac{1}{2}u^T(k)\beta(k)Mu(k) + (\lambda^T(k) - \lambda^T(k-1) + \mu^T(k))$$

$$\left. \times x(k-1) + (\lambda^T(k)M + \mu^T(k)M + \psi^T(k))u(k)\} \right] \tag{7.175}$$

In the above equation M is a lower triangular matrix that contains the topological arrangement of the upstream reservoirs.

If we define the augmented vectors as

$$X^T(k) = [P_s^T(k), x^T(k-1), u^T(k)] \tag{7.176}$$

$$R^T(k) = [(\delta(k))^T, (\lambda(k) - \lambda(k-1) + \mu(k))^T, (B(k) + M^T\lambda(k)$$
$$+ M^T\mu(k) + \psi(k))^T] \tag{7.177}$$

and

$$L(k) = \begin{bmatrix} \gamma & 0 & 0 \\ 0 & 0 & \frac{1}{2}\beta(k) \\ 0 & \frac{1}{2}\beta(k) & \frac{1}{2}\beta(k)M \end{bmatrix} \tag{7.178}$$

then equations (7.175) can be written as

$$\tilde{J} = E[\tfrac{1}{2}(x^*(K) - X(K))^T w(x^*(K) - x(K)) + \lambda^T(K)x(K) - \lambda^T(0)x(0)]$$
$$+ e\left[\sum_{k=1}^{K} \{X^T(k)L(k)X(k) + R^T(k)X(k)\} \right] \tag{7.179}$$

Define the vector $V(k)$ as

$$V(k) = L^{-1}(k)R(k) \tag{7.180}$$

then, the second part of equation (7.179) can be written in the following form similar to completing the squares:

$$\tilde{J} = E[\tfrac{1}{2}(x^*(K) - x(K))^T w(x^*(K) - X(K)) + \lambda^T(K)x(K) - \lambda^T(0)x(0)]$$
$$+ E\left[\sum_{k=1}^{K} (X(k) + \tfrac{1}{2}V(k))^T L(k)(X(k) + \tfrac{1}{2}V(k)) - \tfrac{1}{4}V^T(k)L(k)V(k) \right] \tag{7.181}$$

It can be noticed that $\tilde{J}$ in equation (7.181) is composed of two parts, the boundary part and the discrete integral part, which are independent of each other. The discrete integral part defines a norm in the Hilbert space; then equation (7.181) can be written as

$$\min_{[x(K), X(k)]} \tilde{J} = \min_{x(K)} E[\tfrac{1}{2}(x^*(K) - X(K))^T w(x^*(k) - x(K))$$
$$+ \lambda^T(K)x(K) - \lambda^T(0)x(0)] + \min_{X(k)} E[\|X(k) + \tfrac{1}{2}V(k)\|]_{L(k)} \tag{7.182}$$

because $V(k)$ is constant independent of $X(k)$.

7.8.3. Optimal Solution

To minimize $\tilde{J}$ in equation (7.182), one minimizes each term separately. The minimization of the boundary term is clearly achieved when

$$E[w(x(K) - x^*(K)) - \lambda(K)] = 0 \qquad (7.183)$$

because $x(0)$ is constant and $\delta x(K)$ is free, or in a component form

$$E[\lambda_{i,K}] = E[w_i(x_{i,K} - x_{i,K}^*)] \qquad (7.184)$$

The above equation gives the values of Lagrange multiplier at the last period studied.

The discrete integral part of equation (7.182) is a minimum when the norm of this part is equal to zero:

$$E[X(k) + \tfrac{1}{2}V(k)] = 0 \qquad (7.185)$$

Substituting for $V(k)$ from equation (7.180) into equation (7.185), one obtains

$$E[R(k) + 2L(k)X(k)] = 0 \qquad (7.186)$$

Also, if one substitutes from equations (7.176)–(7.178) into equation (7.186), one obtains

$$E[\delta(k) + 2\gamma P_s(k)] = 0 \qquad (7.187)$$

$$E[\lambda(k) - \lambda(k-1) + \mu(k) + \beta(k)u(k)] = 0 \qquad (7.188)$$

$$E[B(k) + M^T\lambda(k) + M^T\mu(k) + \psi(k) + \beta(k)x(k-1) + \beta(k)Mu(k)] = 0$$

$$(7.189)$$

Writing the above equations in component form, and adding the equality constraints given by equations (7.146), (7.149), (7.154), and (7.155), one obtains the following set of optimal equations:

$$E[P_{s_{i,k}}] = E\left[\frac{\theta_k - \beta_i}{2\gamma_i}\right], \qquad i \in R_s \qquad (7.190)$$

$$E[x_{i,k}] = E\left[x_{i,k-1} + I_{i,k} + q_{i,k} - u_{i,k} + \sum_{j \in R_u} u_{j,k}\right], \qquad i \in R_h \quad (7.191)$$

$$E[P_{h_{i,k}}] = E\left[b_{i,k}u_{i,k} + b_iu_{i,k}x_{i,k-1} + \tfrac{1}{2}b_iu_{i,k}\left(\sum_{j \in R_h} u_{j,k} - u_{i,k}\right)\right], \qquad i \in R_h$$

$$(7.192)$$

$$E\left[\sum_{i \in R_h} P_{h_{i,k}} + \sum_{i \in R_s} P_{s_{i,k}}\right] = E[D_k] \qquad (7.193)$$

$$E[\lambda_{i,k-1}] = E[\lambda_{i,k} + \mu_{i,k} + \beta_{i,k}u_{i,k}], \qquad i \in R_h \qquad (7.194)$$

$$E\left[B_{i,k} + \sum_{\sigma \in R_d} \lambda_{\sigma,k} - \lambda_{i,k} + \sum_{\sigma \in R_d} \mu_{\sigma,k} - \mu_{i,k} + \beta_{i,k} x_{i,k-1} + \psi_{i,k} \right.$$

$$\left. + \beta_{i,k}\left(\sum_{j \in R_u} u_{j,k} - u_{i,k} \right) \right] = 0, \qquad i \in R_h \qquad (7.195)$$

where

$$q_{i,k} = I_{i,k} + \sum_{j \in R_u} s_{j,k} - s_{i,k}, \qquad i \in R_h \qquad (7.196)$$

$$b_{i,k} = a_i + \tfrac{1}{2} b_i q_{i,k}, \qquad i \in R_h \qquad (7.197)$$

and R_d is the set of the immediately downstream reservoirs to reservoir i.

Besides the above equations, one has the following exclusion Kuhn–Tucker equations:

$$e_{i,k}^m (x_i^m - x_{i,k}) = 0, \qquad i \in R_h \qquad (7.198)$$

$$e_{i,k}^M (x_{i,k} - s_i^M) = 0, \qquad i \in R_h \qquad (7.199)$$

$$g_{i,k}^m (u_{i,k}^m - u_{i,k}) = 0, \qquad i \in R_h \qquad (7.200)$$

$$g_{i,k}^M (u_{i,k} - u_{i,k}^M) = 0, \qquad i \in R_h \qquad (7.201)$$

Also, we have the following limits on the variables, $i \in R_h$:

$$\text{if } x_{i,k} > x_i^M, \qquad \text{then we put } x_{i,k} = x_i^M \qquad (7.202)$$

$$\text{if } x_{i,k} < x_i^m, \qquad \text{then we put } x_{i,k} = x_i^m \qquad (7.203)$$

$$\text{if } u_{i,k} > u_{i,k}^M, \qquad \text{then we put } u_{i,k} = u_{i,k}^M \qquad (7.204)$$

$$\text{if } u_{i,k} < u_{i,k}^m, \qquad \text{then we put } u_{i,k} = u_{i,k}^m \qquad (7.205)$$

Equations (7.190)–(7.205) completely specify the optimal long-term operation of hydro thermal power systems. During each optimization interval, the control variables are the water discharges through the turbines—$u_{i,k}$, $i \in R_h$ of the hydroelectric plants. The storage at the end of period k and the hydroelectric power generations can be obtained from equations (7.191) and (7.192), respectively, and therefore these are dependent variables. Irrespective of the actual hydroelectric generations, the thermal generations satisfy equation (7.190) and the power transfer equation (7.193). These are also taken as dependent variables. The next section explains the algorithm of solution of these optimal equations.

7.8.4. Algorithm of Solution

Assume we are given the natural inflows $I_{i,k}$, the demand D_k with their probability properties during the optimization interval, the initial storages $x_{i,0}$, and the constants for thermal and hydro electric power plants.

Step 1. Assume initial guess for $u_{i,k}$ such that

$$u_{i,k}^m \leq u_{i,k}^0 \leq u_{i,k}^M, \qquad i \in R_h$$

Step 2. Calculate $x_{i,k}$ by solving equation (7.191) forward in stages with $x_{i,0}$, $i \in R_h$, given

Step 3. Calculate the hydroelectric generation from equation (7.192).

Step 4. Assume θ_k such that

$$\theta_k \geq \beta_i^M \qquad i \in R_s$$

where β_i^M is the maximum values for the β's thermal fuel cost coefficient.

Step 5. Calculate the expected value of the thermal generations from equation (7.190).

Step 6. Check the active power balance equation, equation (7.193). If this is not satisfied with a prescribed accuracy, repeat from step 4, for different values of θ_k using the conjugate gradient method as

$$E(\theta_k)_{\text{new}} = E[(\theta_k)_{\text{old}} + \alpha_k \Delta \theta_k]$$

where

$$E[\Delta \theta_k] = E\left[\sum_{i \in R_h} P_{h_{i,k}} + \sum_{i \in P_s} P_{s_{i,k}} - D_k \right]$$

and α_k is a positive scalar, which is chosen with consideration given to such factors as convergence. Otherwise go to step 7.

Step 7. Solve equation (7.194) backward in stages with the boundary condition given by equation (7.184).

Step 8. Calculate the gradient $\Delta u_{i,k}$ given by equation (7.195). If this does not satisfy the optimality condition, $\Delta u_{i,k} = 0$, with a prescribed accuracy, adjust the water discharge $u_{i,k}$, using the conjugate gradient method as

$$E[u_{i,k}]_{\text{new}} = E[[u_{i,k}]_{\text{old}} + \alpha_k \Delta u_{i,k}]$$

and go to step 9.

Step 9. Repeat the calculation starting from step 2. Continue until the states $P_{s_{i,k}}$, $x_{i,k}$ and the control $u_{i,k}$ do not change significantly from iteration to iteration and the expected total fuel cost during the optimization period is a minimum.

The above steps are described briefly in the flow chart given in Figure 7.28.

7.8.5. Concluding Remarks

The method developed in this section has the ability to deal with large-scale coupled power systems. Most of the current approaches either use composite reservoirs or do not consider the coupling of reservoirs. The proposed method takes into account the stochasticity of the water inflows and the load demand; we assume that their probability properties are preestimated from past history.

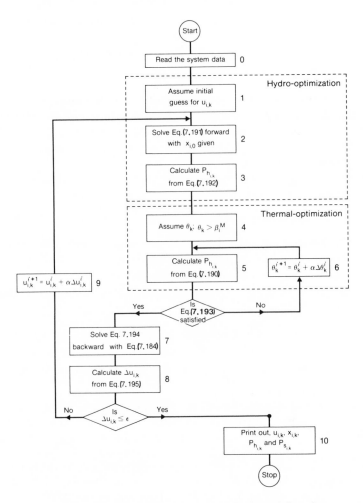

Figure 7.28. The optimal long-term hydrothermal flow chart.

7.9. A Minimum Norm Approach, Nonlinear Storage-Elevation Curve

In Section 7.8, the model for the storage-elevation curve was a linear model; for the reservoirs with a small storage this model may be adequate, but for the reservoirs with a large storage, we need a more accurate model. In this section the model used for the storage-elevation curve is a quadratic model. The hydroelectric generating function is a highly nonlinear function. To cast this function into a quadratic one, a set of pseudo-state-variables are defined. The optimization is done on a monthly time basis for a period of a year so that the total thermal fuel cost is a minimum. We repeat here, for the reader's convenience, the problem formulation and the objective function.

7.9.1. Problem Formulation and Objective Functional

The problem of the power system of Figure 7.14 is to find the thermal power generation as well as the discharge from each reservoir so that the following requirements are satisfied:

(1) The expected total fuel cost for thermal power plants over the optimization interval is a minimum.

(2) The fuel cost for a thermal power plant, input–output characteristics, may be adequately described by a quadratic model as

$$F_i(P_{s_{i,k}}) = \alpha_i + \beta_i(P_{s_{i,k}}) + \gamma_i(P_{s_{i,k}})^2, \qquad i \in R_s \qquad (7.206)$$

(3) The balance between the active power generated and the load demand should be satisfied. In other words, the total power generated should match the load and the losses on the system:

$$\sum_{i \in R_h} P_{h_{i,k}} + \sum_{i \in R_s} P_{s_{i,k}} = D_k + P_{L,k} \qquad (7.207)$$

(4) The transmission line losses may be adequately described by the general loss formula given by

$$P_{L,k} = K_{L_0} + \sum_{i \in R_G} B_{i0}P_{i,k} + \sum_{i,j \in R_G}\sum P_{i,k}B_{ij}P_{j,k} \qquad (7.208)$$

In the above equations, R_G is the set of the generation nodes. In this section we will discuss the effect of the transmission losses on the problem formulation, which was not the case in Section 7.8. K_{L_0}, B_{i0}, and B_{ij} are the loss coefficients.

(5) The storage-elevation curve for the reservoirs may be described by a quadratic model as

$$h_{i,k} = [a_i' + b_i'x_{i,k} + c_i'(x_{i,k})^2], \qquad i \in R_h \qquad (7.209)$$

In this equation, we assume that the tailwater elevation is constant independent of the discharge. Note that the MW/mm^3, which is the water conversion factor, is proportional to the head, i.e., equation (7.209) scales the water conversion factor as

$$WC_{i,k} = [a_i + b_i x_{i,k} + c_i(x_{i,k})^2], \qquad i \in R_h \ MW/Mm^3 \qquad (7.210)$$

where a_i, b_i, and c_i are constants. These are obtained by least-squares curve fitting to typical plant data available.

(6) The reservoir dynamics may be adequately described by the following discrete difference continuity type equation: as

$$x_{i,k} = x_{i,k-1} + I_{i,k} - u_{i,k} + \sum_{j \in R_u} (u_{j,k} + s_{j,k}) - s_{i,k} \qquad (7.211)$$

where $s_{i,k}$ is given by

$$s_{i,k} = \begin{cases} \left(x_{i,k-1} - x_{i,k} + I_{i,k} + \sum_{j \in R_u} (u_{j,k} + s_{j,k}) - u_{i,k} - s_{i,k} \right) - u_{i,k}^M. \\ \quad \text{if } x_{i,k} = x_i^M \quad \text{and} \\ \left(x_{i,k-1} - x_{i,k} + I_{i,k} + \sum_{j \in R_u} (u_{j,k} + s_{j,k}) - u_{i,k} - s_{i,k} \right) > u_{i,k}^M \\ 0 \quad \text{otherwise} \end{cases} \qquad (7.212)$$

(7) The hydroelectric generation of a hydropower plant is a function of the discharge and the head, which itself is a function of the storage:

$$P_{h_{i,k}} = [a_i + \tfrac{1}{2}b_i(x_{i,x} + x_{i,k-1}) + \tfrac{1}{2}c_i(x_{i,k} + x_{i,k-1})^2]u_{i,k}, \qquad i \in R_h \quad (7.213)$$

In the above equation, an average of begin and end of time storage is used to avoid overestimation in the production for rising water levels and underestimation for falling water levels. Equation (7.213) is obtained by multiplying equation (7.210) by the discharge $u_{i,k}$ through the turbine.

If one substitutes for $x_{i,k}$ from equation (7.211) into equation (7.213), we obtain

$$P_{h_{i,k}} = \left[b_{i,k}u_{i,k} + d_{i,k}u_{i,k}x_{i,k-1} + f_{i,k}u_{i,k}\left(\sum_{j \in R_u} u_{j,k} - u_{i,k} \right) + c_i u_{i,k} y_{i,k} \right.$$
$$\left. + \tfrac{1}{4}c_i u_{i,k} z_{i,k} + c_i r_{i,k}\left(\sum_{j \in R_u} u_{j,k} u_{i,k} \right) \right], \qquad i \in R_h \qquad (7.214)$$

where

$$q_{i,k} = I_{i,k} - s_{i,k} + \sum_{j \in R_u} s_{j,k}, \qquad i \in R_h \qquad (7.215a)$$

$$b_{i,k} = a_i + \tfrac{1}{2}b_i q_{i,k} + \tfrac{1}{4}c_i(q_{i,k})^2, \qquad i \in R_h \tag{7.215b}$$

$$d_{i,k} = b_i + c_i q_{i,k}, \qquad i \in R_h \tag{7.215c}$$

$$f_{i,k} = \tfrac{1}{2}b_i + \tfrac{1}{2}c_i q_{i,k}, \qquad i \in R_h \tag{7.215d}$$

and the following are pseudo-state-variables.

$$y_{i,k} = (x_{i,k-1})^2, \qquad i \in R_h \tag{7.216}$$

$$z_{i,k} = \left(\sum_{j \in R_u} u_{j,k} - u_{i,k} \right)^2, \qquad i \in R_h \tag{7.217}$$

$$r_{i,k} = u_{i,k} x_{i,k-1}, \qquad i \in R_h \tag{7.218}$$

(8) The expected storage at the end of the optimization period can be obtained from the initial storage as follows:

$$x_{i,K} = x_{i,0} - \sum_{k=1}^{K} (u_{i,k} + s_{i,k}) + \sum_{k=1}^{K} I_{i,k} + \sum_{k=1}^{K} \sum_{j \in R_u} (u_{j,k} + s_{j,k}) \tag{7.219}$$

(9) The following inequality constraints must be satisfied at the optimum to satisfy multipurpose stream use requirements.

   (a) Upper and lower limits on the storage

$$x_i^m \le x_{i,k} \le x_i^M, \qquad i \in R_h \tag{7.220}$$

   (b) Upper and lower limits on the discharge

$$u_{i,k}^m \le u_{i,k} \le u_{i,k}^M, \qquad i \in R_h \tag{7.221}$$

In mathematical terms, the objective of the computational optimization is to find the discharges $u_{i,k}$, $i \in R_h$ and $P_{s_{i,k}}$, $i \in R_s$ that minimize the total expected fuel cost, over the optimization period, as given by

$$J = E\left[\sum_{k=1}^{K} \sum_{i \in R_s} (\alpha_i + \beta_i P_{s_{i,k}} + \gamma_i P_{s_{i,k}}^2) \right], \qquad i \in R_s \tag{7.222}$$

subject to satisfying the equality constraints (7.206), (7.207), (7.211), (7.214), (7.216), (7.217), and (7.218) and the inequality constraints (7.220) and (7.222). In equation (7.222), E stands for the expected; the expectation is taken over the random variables $I_{i,k}$ and the load demand on the system D_k.

To take equation (7.219) into account, the cost function in equation (7.222) is modified to

$$J = E\left[\sum_{i \in R_h} \tfrac{1}{2} w_i (x_{i,K}^* - x_{i,K})^2 + \sum_{k=1}^{K} \sum_{i \in R_s} (\alpha_i + \beta_i P_{s_{i,k}} + \gamma_i P_{s_{i,k}})^2 \right] \tag{7.223}$$

where w_i is a weighting factor, $x_{i,K}^*$ is the specified amount of water at the last period studied given by equation (7.219), and $x_{i,K}$ is the calculated storage at the last period during the optimization procedures.

7.9.2. A Minimum Norm Formulation

We can now form an augmented cost function by adjoining equation (7.207) to the cost functional using the unknown dual variable θ_k, equation (7.211) using the Lagrange multiplier $\lambda_{i,k}$, $i \in R_h$, equation (7.214) using Lagrange multiplier $\nu_{i,k}$, $i \in R_h$, and equations (7.216)–(7.218) using Lagrange multipliers $\mu_{i,k}$, $\phi_{i,k}$, $\delta_{i,k}$ $i \in R_h$, respectively. Also, by adjoining the inequality constraints (7.220) and (7.221) via Kuhn–Tucker multipliers, we obtain

$$
\begin{aligned}
\tilde{J} = E\Bigg[&\sum_{i \in R_h} \tfrac{1}{2} w_i (x_{i,K}^* - x_{i,K})^2 \\
&+ \sum_{k=1}^{K} \Bigg[\sum_{i \in R_s} \{\alpha_i + (\beta_i + \theta_k c_i) P_{s_{i,k}} + \gamma_i P_{s_{i,k}}^2\} \\
&- \sum_{i,j \in R_G} P_{i,k} \theta_k B_{ij} P_{j,k} \\
&+ \sum_{i \in R_h} \Bigg\{ \theta_k c_i P_{h_{i,k}} + \nu_{i,k}\bigg(-P_{h_{i,k}} + b_{ik} u_{i,k} + d_{i,k} u_{i,k} x_{i,k-1} \\
&+ f_{i,k} u_{i,k}\bigg(\sum_{j \in R_u} u_{j,k} - u_{i,k} \bigg) + c_i u_{i,k} y_{i,k} \\
&+ \tfrac{1}{4} c_i u_{i,k} z_{i,k} + c_i r_{i,k}\bigg(\sum_{j \in R_u} u_{j,k} - u_{i,k} \bigg)\bigg) \\
&+ \lambda_{i,k}\bigg(-x_{i,k} + x_{i,k-1} + q_{i,k} - u_{i,k} + \sum_{j \in R_u} u_{j,k} \bigg) \\
&+ \mu_{i,k}(-y_{i,k} + (x_{i,k-1})^2) + \phi_{i,k}\bigg(-z_{i,k} + \bigg(\sum_{j \in R_u} u_{j,k} - u_{i,k} \bigg)^2 \bigg) \\
&+ \delta_{i,k}(-r_{i,k} + u_{i,k} x_{i,k-1}) + e_{i,k}^m(x_i^m - x_i^k) + e_{i,k}^M(x_i^k - x_i^M) \\
&+ g_{i,k}^m(u_{i,k}^m - u_{i,k}) + g_{i,k}^M(u_{i,k} - u_{i,k}^M) \Bigg\} \Bigg] \Bigg]
\end{aligned}
\tag{7.224}
$$

where

$$
C_i = 1 - B_{i0}, \qquad i \in R_G
\tag{7.225a}
$$

Define the following variables:

$$\beta_{i,k} = \beta_i + C_i\theta_k, \qquad i \in R_s \tag{7.225b}$$

$$B_{ij,k} = B_{ij}\theta_k, \qquad i \in R_G \tag{7.225c}$$

$$C_{i,k} = \theta_k C_i - \nu_{i,k}, \qquad i \in R_h \tag{7.225d}$$

$$B_{i,k} = b_{i,k}\nu_{i,k}, \qquad i \in R_h \tag{7.225e}$$

$$D_{i,k} = \nu_{i,k}d_{i,k}, \qquad i \in R_h \tag{7.225f}$$

$$F_{i,k} = \nu_{i,k}f_{i,k}, \qquad i \in R_h \tag{7.225g}$$

$$E_{i,k} = \nu_{i,k}c_i, \qquad i \in R_h \tag{7.225h}$$

$$h_{i,k} = e_{i,k}^M - e_{i,k}^m, \qquad i \in R_h \tag{7.225i}$$

$$l_{i,k} = g_{i,k}^M - g_{i,k}^m, \qquad i \in R_h \tag{7.225j}$$

Then, the cost functional in equation (7.224) can be written as

$$
\begin{aligned}
\tilde{J} = E\Bigg[&\sum_{i \in R_h} \tfrac{1}{2}w_i(x_{i,K}^* - x_{i,K})^2 + \sum_{k=1}^{K}\Bigg[\sum_{i \in R_s}(\beta_{i,k}P_{s_{i,k}} + \gamma_i P_{s_{i,k}}^2) \\
&+ \sum_{i,j \in R_G} P_{i,k}B_{ij,k}P_{j,k} + \sum_{i \in R_h}\Bigg\{ C_{i,k}P_{h_{i,k}} + B_{i,k}u_{i,k} + D_{i,k}u_{i,k}x_{i,k-1} \\
&+ F_{i,k}u_{i,k}\Bigg(\sum_{j \in R_u} u_{j,k} - u_{i,k}\Bigg) + E_{i,k}u_{i,k}y_{i,k} \\
&+ \tfrac{1}{4}E_{i,k}u_{i,k}z_{i,k} + E_{i,k}r_{i,k}\Bigg(\sum_{j \in R_u} u_{j,k} - u_{i,k}\Bigg) \\
&+ \lambda_{i,k}\Bigg(-x_{i,k} + x_{i,k-1} - u_{i,k} + \sum_{j \in R_u} u_{j,k}\Bigg) + \mu_{i,k}(-y_{i,k} + (x_{i,k-1})^2) \\
&+ \phi_{i,k}\Bigg(-z_{i,k} + \Bigg(\sum_{j \in R_u} u_{j,k} - u_{i,k}\Bigg)^2\Bigg) \\
&+ \delta_{i,k}(-r_{i,k} + u_{i,k} + u_{i,k}x_{i,k-1}) \\
&+ h_{i,k}\Bigg(x_{i,k-1} + \sum_{j \in R_u} u_{j,k} - u_{i,k}\Bigg) + l_{i,k}u_{i,k}\Bigg\}\Bigg]\Bigg]
\end{aligned}
\tag{7.226}
$$

We will use the following identity:

$$\sum_{k=1}^{K} \lambda_{i,k}x_{i,k} = \lambda_{i,K}x_{i,K} - \lambda_{i,0}x_{i,0} + \sum_{k=1}^{K}\lambda_{i,k-1}x_{i,k-1}, \qquad i \in R_h \tag{7.227}$$

The above equation can be written in the following vector form, assuming for simplicity that the loss coefficient matrix is a diagnal matrix, and dropping constant terms:

$$\tilde{J} = E[\{(x^*(K) - x(K))^T W(x^*(K) - x(K)) - \lambda^T(K)x(K) + \lambda^T(0)x(0)]$$

$$+ \sum_{k=1}^{K} [\beta^T(k)P_s(k) + P_s^T(k)\Gamma(k)P_s(k) + P_h^T(k)B(k)P_h(k)$$

$$+ C^T(k)P_h(k) + B_1^T(k)u(k) + u^T(k)D(k)x(k-1)$$

$$+ u^T(k)F(k)Mu(k) + u^T(k)E(k)y(k) + \tfrac{1}{4}u^T(k)E(k)z(k)$$

$$+ r^T(k)E(k)Mu(k) + (\lambda^T(k) - \lambda^T(k-1) + h^T(k))x(k-1)$$

$$+ \mu^T(k)(-y(k) + x^T(k-1)Hx(k-1)) + \phi^T(k)$$

$$\times (-z(k) + u^T(k)M^THMu(k))$$

$$+ \delta^T(k)(-r(k) + u^T(k)Hx(k-1))$$

$$+ (\lambda^T(k)M + l^T(k) + h^T(k)M)u(k)]] \tag{7.228}$$

where

$$\Gamma_i = \gamma_i + B_{ij,k}, \qquad i, j \in R_s \tag{7.229}$$

In the above equation we defined the following variables:

$$P_s(k) = \text{col}(P_{s_{i,k}}, i \in R_s) \tag{7.230}$$

$$P_h(k) = \text{col}(P_{h_{i,k}}, i \in R_h) \tag{7.231}$$

$$x(k) = \text{col}(x_{i,k}, i \in R_h) \tag{7.232}$$

$$u(k) = \text{col}(u_{i,k}, i \in R_h) \tag{7.233}$$

$$y(k) = \text{col}(y_{i,k}, i \in R_h) \tag{7.234}$$

$$z(k) = \text{col}(z_{i,k}, i \in R_h) \tag{7.235}$$

$$r(k) = \text{col}(r_{i,k}, i \in R_h) \tag{7.236}$$

$$\lambda(k) = \text{col}(\lambda_{i,k}, i \in R_h) \tag{7.237}$$

$$\mu(k) = \text{col}(\mu_{i,k}, i \in R_h) \tag{7.238}$$

$$\phi(k) = \text{col}(\phi_{i,k}, i \in R_h) \tag{7.239}$$

$$\delta(k) = \text{col}(\delta_{i,k}, i \in R_h) \tag{7.240}$$

$$\beta(k) = \text{col}(\beta_{i,k}, i \in R_s) \tag{7.241}$$

$$C(k) = \text{col}(C_{i,k}, i \in R_h) \tag{7.242}$$

$$B_1(k) = \text{col}(B_{i,k}, i \in R_h) \tag{7.243}$$

$$h(k) = \text{col}(h_{i,k}, i \in R_h) \tag{7.244}$$

$$l(k) = \text{col}(l_{i,k}, i \in R_h) \tag{7.245}$$

Furthermore, the following diagonal matrices are defined:

$$\Gamma(k) = \text{diag}(\Gamma_i, i \in R_s) \tag{7.246}$$

$$B(k) = \text{diag}(B_{ij,k}, i, j \in R_h) \tag{7.247}$$

$$D(k) = \text{diag}(D_{i,k}, i \in R_h) \tag{7.248}$$

$$F(k) = \text{diag}(F_{i,k}, i \in R_h) \tag{7.249}$$

$$E(k) = \text{diag}(E_{i,k}, i \in R_h) \tag{7.250}$$

$$W = \text{diag}(w_i, i \in R_h) \tag{7.251}$$

In equation (7.228) M is a lower triangular matrix that contains the topological management of the upstream reservoirs, and $\mathbf{H}$ is $n \times n$ vector matrix whose index varies from 1 to n while the matrix dimension of $\mathbf{H}$ is $n \times n$.

Define the following component vectors:

$$X^T(k) = [P_s^T(k), P_h^T(k), x^T(k-1), u^T(k), y^T(k), z^T(k), r^T(k)] \tag{7.252}$$

$$R^T(k) = [\beta^T(k), C^T(k), (\lambda(k) - \lambda(k-1) + h(k))^T, (B_1(k) + l(k)$$
$$+ M^T h(k) + M^T \lambda(k))^T - \mu^T(k), -\phi^T(k), -\delta^T(k)] \tag{7.253}$$

and

$$L(k) = \begin{bmatrix} \Gamma(k) & 0 & 0 & 0 & 0 & 0 & 0 \\ 0 & B(k) & 0 & 0 & 0 & 0 & 0 \\ 0 & 0 & \mu^T(k)\mathbf{H} & (\frac{1}{2}\delta^T(k)\mathbf{H} + \frac{1}{2}D(k) & 0 & 0 & 0 \\ 0 & 0 & (\frac{1}{2}\delta^T(k)\mathbf{H} + \frac{1}{2}D(k) & F(k)M + \phi^T(k)M^T\mathbf{H}M & \frac{1}{2}E(k) & \frac{1}{8}E(k) & \frac{1}{2}M^TE(k) \\ 0 & 0 & 0 & \frac{1}{2}E(k) & 0 & 0 & 0 \\ 0 & 0 & 0 & \frac{1}{8}E(k) & 0 & 0 & 0 \\ 0 & 0 & 0 & \frac{1}{2}E(k)M & 0 & 0 & 0 \end{bmatrix}$$

$$\tag{7.254}$$

Then, the cost function in equation (7.228) can be written as

$$\tilde{J} = E\bigg[\{(x^*(K) - x(K))^T W(x^*(K) - x(K)) - \lambda^T(K)x(K) + \lambda^T(0)x(0)\}$$

$$+ \sum_{k=1}^{K} \{X^T(k)L(k)X(k) + R^T(k)X(k)\}\bigg] \tag{7.255}$$

Define the vector $V(k)$ as

$$V(k) = L^{-1}(k)R(k) \tag{7.256}$$

then, the discrete integral part of equation (7.255) can be written in the following form by a process similar to completing the squares:

$$\tilde{J} = E[(x^*(K) - x(K))^T W(x^*(K) - x(K)) - \lambda^T(K)x(K) + \lambda^T(0)x(0)]$$

$$+ E\left[\sum_{k=1}^{K} \{(X(k) + \tfrac{1}{2}V(k))^T L(k)(X(k) + \tfrac{1}{2}V(k))\right.$$

$$\left. - \tfrac{1}{4}V^T(k)L(k)V(k)\}\right] \qquad (7.257)$$

since $x(0)$ is always given at the beginning of the optimization interval and $V(k)$ is constant independent of $X(k)$. Then, equation (7.257) can be written as

$$\tilde{J} = E[(x^*(K) - x(K))^T W(x^*(K) - x(K)) - \lambda^T(K)x(K)]$$

$$+ E\left[\sum_{k=1}^{K} \{(X(k) + \tfrac{1}{2}V(k))^T L(k)(X(k) + \tfrac{1}{2}V(k))\}\right] \qquad (7.258)$$

7.9.3. Optimal Solution

It can be noticed from equation (7.258) that $\tilde{J}$ is composed of two terms that are independent of each other. To minimize $\tilde{J}$, one minimizes each term separately:

$$\min_{[x(K),X(k)]} \tilde{J} = \min_{x(K)} E[(x^*(K) - x(K))^T W(x^*(K) - x(K)) - \lambda^T(K)x(K)]$$

$$+ \min_{X(k)} E\left[\sum_{k=1}^{K} \{(X(k) + \tfrac{1}{2}V(k))^T L(k)(X(k) + \tfrac{1}{2}V(k))\}\right]$$

$$(7.259)$$

The minimization of the boundary term is clearly achieved when

$$E[\lambda(K)] = E[W(x(K) - x^*(K))] \qquad (7.260)$$

the above equation can be written in the component form as

$$E[\lambda_{i,K}] = E[w_i(x_{i,K} - x^*_{i,K})] \qquad (7.261)$$

Equation (7.261) gives the values of Lagrange multipliers at the last period studied.

The discrete integral part in equation (7.259) defines a norm in Hilbert space. This part can be written as

$$\min_{X(k)} J_2 = \min_{X(k)} E\|X(k) + \tfrac{1}{2}V(k)\|_{L(k)} \qquad (7.262)$$

The minimization of equation (7.262) is clearly achieved when the norm of this equation is zero:

$$E[X(k) + \tfrac{1}{2}V(k)] = 0 \tag{7.263}$$

Substituting for $V(k)$ from equation (7.256) into equation (7.263), one obtains

$$E[R(k) + 2L(k)X(k)] = 0 \tag{7.264}$$

Writing equation (7.264) explicitly, by substituting from equations (7.252)-(7.254) into (7.264) one obtains the following optimal equations:

$$E[\beta(k) + 2\Gamma(k)P_s(k)] = 0 \tag{7.265}$$

$$E[C(k) + 2B(k)P_h(k)] = 0 \tag{7.266}$$

$$E[\lambda(k) - \lambda(k-1) + h(k) + 2\mu^T(k)\mathbf{H}x(k-1) + \delta^T(k)\mathbf{H}u(k)$$
$$+ D(k)u(k)] = 0 \tag{7.267}$$

$$E[M^T\lambda(k) + B_1(k) + l(k) + M^Th(k) + \delta^T(k)\mathbf{H}x(k-1)$$
$$+ D(k)x(k-1) + 2F(k)Mu(k) + 2\phi^T(k)M^T\mathbf{H}\,Mu(k)$$
$$+ E(k)y(k) + \tfrac{1}{4}E(k)z(k) + M^TE(k)r(k)] = 0 \tag{7.268}$$

$$E[-\mu(k) + E(k)u(k)] = 0 \tag{7.269}$$

$$E[-\phi(k) + \tfrac{1}{4}E(k)u(k)] = 0 \tag{7.270}$$

$$E[-\delta(k) + E(k)Mu(k)] = 0 \tag{7.271}$$

Writing the above equations in component form, and adding the equality constraints (7.207), (7.208), (7.211), (7.214), and (7.216)-(7.218), one obtains

(1) For thermal power plants

$$E[P_{s_{i,k}}] = E\left[\frac{\theta_k(B_{i0} - 1) - \beta_i}{2(\gamma_i - \theta_k B_{ii})}\right], \qquad i \in R_s \tag{7.272}$$

(2) For hydro power plants

$$E[x_{i,k}] = \left[x_{i,k-1} + I_{i,k} + q_{i,k} - u_{i,k} + \sum_{j \in R_u} u_{j,k}\right] \qquad i \in R_h \tag{7.273}$$

$$E[P_{h_{i,k}}] = E\left[b_{i,k}u_{i,k} + d_{i,k}u_{i,k}x_{i,k-1}\right.$$
$$+ f_{i,k}u_{i,k}\left(\sum_{j \in R_u} u_{j,k} - u_{i,k}\right) + c_i u_{i,k}y_{i,k} + \tfrac{1}{4}c_i u_{i,k}z_{i,k}$$
$$\left. + c_i r_{i,k}\left(\sum_{j \in R_u} u_{j,k} - u_{i,k}\right)\right], \qquad i \in R_h \tag{7.274}$$

$$E[\nu_{i,k}] = E[\theta_k(1 - B_{i0}) - 2\theta_k B_{ii}P_{h_{i,k}}], \qquad i \in R_h \qquad (7.275)$$

$$E\left[\sum_{i \in R_h} P_{h_{i,k}} + \sum_{i \in R_s} P_{s_{i,k}}\right] = E[D_k + P_{L,k}] \qquad (7.276)$$

$$P_{L,k} = K_{L_0} + \sum_{i \in R_G} B_{i0}P_{i,k} + \sum\sum_{i,j \in R_G} P_{i,k}B_{ij}P_{j,k} \qquad (7.277)$$

$$E[\lambda_{i,k} - \lambda_{i,k-1} + h_{i,k} + \nu_{i,k}d_{i,k}u_{i,k} + 2\mu_{i,k}x_{i,k-1} + \delta_{i,k}u_{i,k}] = 0, \qquad i \in R_h \qquad (7.278)$$

$$E\left[\nu_{i,k}b_{i,k} + \sum_{\sigma \in R_d} \lambda_{\sigma,k} - \lambda_{i,k} + \sum_{\sigma \in R_d} h_{\sigma,k} - h_{i,k} + l_{i,k} + \nu_{i,k}d_{i,k}x_{i,k-1}\right.$$

$$+ 2\nu_{i,k}f_{i,k}\left(\sum_{j \in R_u} u_{j,k} - u_{i,k}\right) + c_i\nu_{i,k}y_{i,k} + \tfrac{1}{4}c_i\nu_{i,k}z_{ik}$$

$$\left. + c_i\nu_{i,k}\left(\sum_{\sigma \in R_d} r_{\sigma,k} - r_{i,k}\right)\right] = 0, \qquad i \in R_h \qquad (7.279)$$

$$E[-\mu_{i,k} + c_i\nu_{i,k}u_{i,k}] = 0, \qquad i \in R_h \qquad (7.280)$$

$$E[-\phi_{i,k} + \tfrac{1}{4}c_i\nu_{i,k}u_{i,k}] = 0, \qquad i \in R_h \qquad (7.281)$$

$$E\left[-\delta_{i,k} + c_i\nu_{i,k}\left(\sum_{j \in R_u} u_{j,k} - u_{i,k}\right)\right] = 0, \qquad i \in R_h \qquad (7.282)$$

In the above equations R_d denotes the set of immediately downstream reservoirs from reservoir i.

Besides the above equations, we have the following Kuhn-Tucker exclusion equations:

$$e_{i,k}^m(x_i^m - x_{i,k}) = 0, \qquad i \in R_h \qquad (7.283)$$

$$e_{i,k}^M(x_{i,k} - x_i^M) = 0, \qquad i \in R_h \qquad (7.284)$$

$$g_{i,k}^m(u_{i,k}^m - u_{i,k}) = 0, \qquad i \in R_h \qquad (7.285)$$

$$g_{i,k}^M(u_{i,k} - u_{i,k}^M) = 0, \qquad i \in R_h \qquad (7.286)$$

One also has the following limits on the variables:

$$\begin{aligned}
&\text{if } x_{i,k} < x_i^m, &&\text{then we put } x_{i,k} = x_i^m \\
&\text{if } x_{i,k} > x_i^M, &&\text{then we put } x_{i,k} = x_i^M \\
&\text{if } u_{i,k} < u_{i,k}^m, &&\text{then we put } u_{i,k} = u_{i,k}^m \\
&\text{if } u_{i,k} > u_{i,k}^M, &&\text{then we put } u_{i,k} = u_{i,k}^M
\end{aligned} \qquad (7.287)$$

Equations (7.272)–(7.287) with equation (7.261) completely specify the optimal long-term operation of hydrothermal power systems. In the next section we discuss the algorithm of solution of these equations.

7.9.4. Algorithm of Solution

Assume we are given the natural inflows $I_{i,k}$, the load demand D_k with their probability properties during the optimization interval, the initial storages $x_{i,0}$, $i \in R_h$ and the constants for thermal and hydroelectric power plants.

Step 1. Assume initial guess for $u_{i,k}$ such that

$$u_{i,k}^m \leq u_{i,k} \leq u_{i,k}^M. \qquad i \in R_h$$

Step 2. Solve equations (7.273), (7.280), (7.281), and (7.282) forward in stages with $x_{i,0}$ given.

Step 3. Calculate the hydroelectric generations, $P_{h_{i,k}}$, $i \in R_h$, from equation (7.274).

Step 4. Assume θ_k such that the thermal generation is always positive.

Step 5. Calculate the thermal generations from equation (7.272).

Step 6. Calculate the transmission losses from equation (7.278).

Step 7. Check the active power balance equation, equation (7.276). If this is not satisfied with a prescribed accuracy, repeat from step 4 for different values of θ_k using the conjugate gradient method:

$$E[\theta_k]_{\text{new}} = E[(\theta_k)_{\text{old}} + \alpha_k \Delta \theta_k]$$

where

$$E[\Delta \theta_k] = E\left[\sum_{i \in R_h} P_{h_{i,k}} + \sum_{i \in R_s} P_{s_{i,k}} - D_k - P_{L,k}\right]$$

and α_k is a positive scalar, which is chosen with consideration given to such factors as convergence. Otherwise, go to step 8.

Step 8. Calculate $\nu_{i,k}$ from solving equation (7.275) forward in stages.

Step 9. Solve equation (7.278) backward in stages with equation (7.261) as a boundary condition.

Step 10. Calculate the gradient $\Delta u_{i,k}$ given by equation (7.279). If this does not satisfy the optimality condition, $\Delta u_{i,k} \cong 0$, with a prescribed accuracy, adjust the water discharges $u_{i,k}$ using the conjugate gradient:

$$E[u_{i,k}]_{new} = E[(u_{i,k})_{old} + \alpha_k u_{i,k}]$$

and go to step 11.

Step 11. Repeat the calculation starting from step 2. Continue until the states $P_{s_{i,k}}$, $P_{h_{i,k}}$ and $x_{i,k}$ and the control $u_{i,k}$ do not change significantly from iteration to iteration and the expected fuel cost during the optimization period is a minimum.

7.9.5. Concluding Remarks

In the previous section we have discussed the application of the minimum norm problem to the optimization of the hydrothermal power system for long-term operations, where the objective is to minimize the expected total fuel cost of thermal plants. The model used in that section is more accurate than that obtained in Section 7.8 since the water conversion factor is a quadratic function of the storage. The transmission line losses are included through use of the general loss formula; we assumed that the loss coefficient matrix is a diagonal for the sake of simplicity. At the end of the section we discussed the algorithm for solution.

7.10 Nuclear, Hydrothermal Power Systems (Refs. 7.63–7.81)

At present, owing to economic considerations, most of the nuclear generating stations throughout the world are operated at base load, and the length of the reactor refueling cycles for most reactors is of the order of several years. Thus it is recognized that the load scheduling of mixed power systems with nuclear plants is essentially an optimization problem of a long-term nature.

With nuclear plants becoming an increasing part of power-generating systems, it is expected that in the near future the operating role of the nuclear plants will change. It is likely that in the absence of suitable energy storage methods the nuclear plants will be required to operate in a load-following mode in order to achieve the flexibility needed to match the power demand.

The multiyear length of the optimization interval makes it virtually impossible to consider the hourly or even the daily load variations of the power demand.

The purposes of this section are (1) to formulate the problem of optimal long-term operation of hydrothermal-nuclear power systems, (2) to obtain the solution by using a functional analytic optimization technique, and (3) to propose an algorithm suitable for implementing the optimal solution.

7.10.1. Problem Formulation

The power system under consideration consists of N_n nuclear plants; we shall denote this set by R_n, N_s fossil fueled plants denoted by R_s and N_h multichain hydrogenerating plants located on different river valleys. The problem is to determine the optimal long-term schedule such that the total operating cost of the system during the optimization interval is a minimum, under the following conditions:

(1) The fossil fuel cost of the thermal plants is assumed to be of the form

$$F_{i,k}(P_{s_{i,k}}) = \alpha_{s_i} + \beta_{s_i} P_{s_{i,k}} + \gamma_{s_i} P^2_{s_{i,k}}, \qquad i \in R_s \qquad (7.288)$$

(2) The nuclear fuel cost as a function of the generated power during a period k is given by

$$F_{i,k}(P_{n_{i,k}}) = \alpha_{n_i} + \beta_{n_i} P_{n_{i,k}} + \gamma_{n_i} P^2_{n_{i,k}}, \qquad i \in R_n \qquad (7.289)$$

where $P_{s_{i,k}}$ and $P_{n_{i,k}}$ are the electric power generations at the ith thermal and nuclear plants, respectively. The coefficients α_i, β_i, and γ_i are known constants. These are obtained by least-squares curve fitting to typical plant data available.

(3) The total active power generation in the system matches the predicted power demand $P_{d,k}$:

$$\sum_{i \in R_s} P_{s_{i,k}} + \sum_{i \in R_n} P_{n_{i,k}} + \sum_{i \in R_h} P_{h_{i,k}} = P_{d,k} \qquad (7.290)$$

where $P_{h_{i,k}}$ denotes the hydropower generation of reservoir i during a period k and R_h is the set of hydroreservoirs. In equation (7.290), the transmission line losses are considered as a part of the demand on the system.

(4) The hydropower generation of reservoir i during a period k is given by

$$P_{h_{i,k}} = a_i u_{i,k} + \tfrac{1}{2} b_i u_{i,k}(x_{i,k} + x_{i,k+1}), \qquad i \in R_h \qquad (7.291)$$

In the above equation an average of begin and -end -of time storage is used to avoid overestimation in the production for rising water levels and underestimation for falling water levels; the tail water elevation is assumed constant, independent of the discharge. Furthermore, we assume the storage-elevation curve is linear and the water turbines have constant efficiency. The coefficients a_i and b_i are obtained by fitting equation (7.291) to the plant capability curves.

(5) The reservoirs dynamics can be adequately described by the continuity type difference equation as

$$x_{i,k+1} = x_{i,k} + y_{i,k} - u_{i,k} - s_{i,k} + \sum_{v \in R_u} (u_{v,k} + s_{v,k}), \qquad i \in R_h$$

$$(7.292)$$

where $x_{i,k}$ is the storage of reservoir i during the period k in m^3; $y_{i,k}$ is the natural inflows to reservoir i during period k in m^3; (these are statistically independent random variables; we assume that their probability properties are preestimated from the past history); $u_{i,k}$ is the discharge from reservoir i during the period k in m^3; and $s_{i,k}$ is the spill from reservoir i during the period k in m^3; water is spilt from reservoir i when the reservoir i is filled to capacity and the inflow to the reservoir exceeds $u_{i,k}^M$. It is given by

$$
s_{i,k} = \begin{cases} \left(x_{i,k} + y_{i,k} - x_{i,k+1} + \sum_{v \in R_u} (u_{v,k} + s_{v,k}) \right) - u_{i,k}^M, \\ \quad \text{if} \left(x_{i,k} + y_{i,k} - x_{i,k+1} + \sum_{v \in R_u} (u_{v,k} + s_{v,k}) \right) > u_{i,k}^M \\ \quad \text{and } x_{i,k} = x_i^M \\ 0 \quad \text{otherwise} \end{cases} \tag{7.293}
$$

R_u is the set of reservoirs immediately upstream from reservoir i.

If one substitutes from equation (7.292) into equation (7.291) for $x_{i,k}$, then one obtains

$$
P_{h_{i,k}} = a_{i,k} u_{i,k} + b_i u_{i,k} x_{i,k} + \tfrac{1}{2} b_i u_{i,k} \left(\sum_{v \in R_u} u_{v,k} - u_{i,k} \right), \qquad i \in R_h \tag{7.294}
$$

where we define

$$
q_{i,k} = y_{i,k} - s_{i,k} + \sum_{v \in R_u} s_{v,k}, \qquad i \in R_h \tag{7.295}
$$

and

$$
a_{i,k} = a_i + \tfrac{1}{2} b_i q_{i,k}, \qquad i \in R_h \tag{7.296}
$$

(6) Most hydro rivers are multipurpose streams used for such things as flood control, fishing, navigation, irrigation, and other purposes if any. To satisfy multipurpose stream use requirements the following limits are imposed on the variables magnitude:

a. Upper and lower limits on the storage given by

$$
x_i^m \leq x_{i,k} \leq x_i^M, \qquad i \in R_h \tag{7.297}
$$

b. Upper and lower limits on the water discharge

$$
u_{i,k}^m \leq u_{i,k} \leq u_{i,k}^M \tag{7.298}
$$

The first set of the inequality constraints simply states that the reservoir storage (or elevation) may not exceed a maximum level, nor be lower than a minimum level. For maximum level this is determined by the elevation of the spillway crest or the top of the spillway gates. The minimum level may be fixed by the elevation of the lowest outlet in the dam or by conditions of operating efficiency for the turbines. The second set is determined by the discharge capacity of the power plant as well as its efficiency.

(7) The discrete changes of the average xenon and iodine concentrations in the ith core, which are induced by power level changes, are accounted for by the following difference equations:

$$x_{i,k+1} = \lambda'_X X_{i,k} + \Lambda_I I_{i,k} + \gamma'_X P_{n_{i,k}} - \Gamma'_X X_{i,k} P_{n_{i,k}}, \qquad i \in R_n \qquad (7.299)$$

$$I_{i,k+1} = \lambda'_I I_{i,k} + \gamma'_I P_{n_{i,k}}, \qquad i \in R_n \qquad (7.300)$$

where

$$\lambda'_I = 1 - \Delta t \lambda_I \qquad (7.301a)$$

$$\gamma'_I = \Delta t \frac{\gamma_I}{e_i G} \qquad (7.301b)$$

$$\lambda'_X = 1 - \Delta t \lambda_X \qquad (7.301c)$$

$$\Lambda_I = \Delta t \lambda_I \qquad \left.\right\} i \in R_n \qquad (7.301d)$$

$$\gamma'_X = \Delta t \frac{\gamma_X}{e_i G} \qquad (7.301e)$$

$$\Gamma'_X = \frac{\Delta t \Gamma_X}{e_i \sum_{f_i} G} \qquad (7.301f)$$

and $I_{i,k}$, $X_{i,k}$ denote the iodine and xenon average concentrations during period k, respectively. λ_I and λ_X represent the iodine and xenon radioactive decay constants, and $\sum_{f_i}$ is the effective fission macroscopic cross section of the ith core. Γ_X denotes the xenon's thermal neutron microscopic absorption cross section. G is the energy released per fission. e_i is the plant's overall efficiency. γ_I and γ_X denote the iodine and xenon fission yield, and Δt is the incremental time used for discretization.

(8) The external reactivity ρ_{c_i}, provided by the control devices whose compensating action against variations of power level and xenon concentration maintains the reactor critical, is given by

$$\rho_{c_{i,k}} = \rho_{o_i} - \alpha_{d_i} P_{n_{i,k}} - \alpha_{X_i} X_{i,k}, \qquad i \in R_n \qquad (7.302)$$

(9) Only limited xenon override capability is available in most reactors. This is accounted for by upper and lower magnitude constraints on the external control reactivity ρ_{c_i}; we assume that

$$\rho_{c_i}^m \leq \rho_{c_{i,k}} \leq \rho_{c_i}^M, \qquad i \in R_n \qquad (7.303)$$

(10) Physical limitations impose the magnitude constraints on the generation given by

$$P_{n_i}^m \leq P_{n_{i,k}} \leq P_{n_i}^M, \qquad i \in R_n \qquad (7.304)$$

$$P_{s_i}^m \leq P_{s_{i,k}} \leq P_{s_i}^M, \qquad i \in R_s \qquad (7.305)$$

In mathematical terms, the optimal long-term operating problem aims to find the generation of the system which minimizes

$$J = E\left[\sum_{k=0}^{K-1} \left\{ \sum_{i \in R_s} (\alpha_{s_i} + \beta_{s_i} P_{s_{i,k}} + \gamma_{s_i} P^2_{s_{i,k}}) + \sum_{i \in R_n} (\alpha_{n_i} + \beta_{n_i} P_{n_{i,k}} + \gamma_{n_i} P^2_{n_{i,k}}) \right\} \right]$$

(7.306)

Subject to satisfying the equality constraints given by equations (7.290), (7.292), (7.294), (7.299), (7.300), (7.302) and the inequality constraints given by equations (7.297), (7.298), (7.303), (7.304), and (7.305). The symbol E stands for the expected value; the expectation is taken over the random variables $y_{i,k}$.

7.10.2. A Minimum Norm Formulation

We can form an augmented cost functional by adjoining the equality constraints of equation (7.290) to the cost functional of equation (3.306) via the unknown variable θ_k, the equality constraints (7.292), (7.299), (7.300), and (7.302) via Langrange multipliers $\lambda_{i,k}$, $\phi_{i,k}$, $\psi_{i,k}$, and the inequality constraints via Kuhn–Tucker multipliers:

$$
\begin{aligned}
\tilde{J} = E \sum_{k=0}^{K-1} \Bigg[& \sum_{i \in R} (\alpha_{s_i} + \beta_{s_i} P_{s_{i,k}} \gamma_{s_i} P^2_{s_{i,k}}) + \sum_{i \in R_n} (\alpha_{n_i} + \beta_{n_i} P_{n_{i,k}} + \gamma_{n_i} P^2_{n_{i,k}}) \\
& - \theta_k \Bigg[-P_{d,k} + \sum_{i \in R_s} P_{s_{i,k}} + \sum_{i \in R_n} \Bigg(a_{i,k} u_{i,k} + b_i u_{i,k} x_{i,k} \\
& + b_i u_{i,k} \Bigg(\sum_{v \in R_u} u_{v,k} - u_{i,k} \Bigg) \Bigg) + \sum_{i \in R_n} P_{n_{i,k}} \Bigg] \\
& + \sum_{i \in R_h} \lambda_{i,k+1} \Bigg(-x_{i,k+1} + x_{i,k} + q_{i,k} + \sum_{v \in R_u} u_{v,k} - u_{i,k} \Bigg) \\
& + \sum_{i \in R_n} \{ \phi_{i,k+1}(-I_{i,k+1} + \gamma'_I P_{n_{i,k}} + \lambda'_I I_{i,k}) \\
& + \psi_{i,k+1}(-X_{i,k+1} + \lambda'_X X_{i,k} + \Lambda_I I_{i,k} + \gamma'_X P_{n_{i,k}} - \Gamma'_X X_{i,k} P_{n_{i,k}}) \\
& + \mu_{i,k}(-\rho_{c_{i,k}} + \rho_{o_i} - \alpha_{d_i} P_{n_{i,k}} - \alpha_{X_i} X_{i,k}) \} \\
& + \sum_{i \in R_h} \{ e^m_{i,k}(x^m_i - x_{i,k}) + e^M_{i,k}(x_{i,k} - x^M_i) \\
& + g^m_{i,k}(u^m_{i,k} - u_{i,k}) + g^M_{i,k}(u_{i,k} - u^M_{i,k}) \} \\
& + \sum_{i \in R_n} \{ f^m_{i,k}(\rho^m_{c_i} - \rho_{c_{i,k}}) + f^M_{i,k}(\rho_{c_{i,k}} - \rho^M_{c_i}) \\
& + l^m_{i,k}(P^m_{n_i} - P_{n_{i,k}}) + l^M_{i,k}(P_{n_{i,k}} - P^M_{n_i}) \} \\
& + \sum_{i \in R_s} \left\{ h^m_{i,k}(P^m_{s_i} - P_{s_{i,k}}) + h^M_{i,k}(P_{s_{i,k}} - P^M_{s_i}) \right\} \Bigg]
\end{aligned}
$$

(7.307)

If we use the identity

$$\sum_{k=0}^{K-1} \lambda_{i,k+1} x_{i,k+1} = \lambda_{i,K} x_{i,K} - \lambda_{i,0} x_{i,0} + \sum_{k=0}^{K-1} \lambda_{i,k} x_{i,k} \tag{7.308}$$

and drop constant terms, then the augmented cost function of equation (7.307) can be written as

$$
\begin{aligned}
\tilde{J} = E\Bigg[&\sum_{i \in R_h} (\lambda_{i,0} x_{i,0} - \lambda_{i,K} x_{i,K}) \\
&+ \sum_{i \in R_n} (\phi_{i,0} I_{i,0} - \phi_{i,K} I_{i,K} - \psi_{i,K} x_{i,k} + \psi_{i,0} X_{i,0}) \Bigg] \\
&+ E \sum_{k=0}^{K-1} \Bigg[\sum_{i \in R_s} \{(C_{s_{i,k}} + h_{i,k}) P_{s_{i,k}} + \gamma_{s_i} P_{s_{i,k}}^2\} \\
&+ \sum_{i \in R_h} \Bigg\{ (g_{i,k} - \theta_k a_{i,k}) u_{i,k} - b_i \theta_k u_{i,k} x_{i,k} \\
&- \tfrac{1}{2} b_i \theta_k u_{i,k} \Bigg(\sum_{v \in R_u} u_{v,k} - u_{i,k} \Bigg) + (\lambda_{i,k+1} - \lambda_{i,k} + e_{i,k}) x_{i,k} \\
&+ \lambda_{ik+1} \Bigg(\sum_{v \in R_u} u_{v,k} - u_{i,k} \Bigg) \Bigg\} \\
&+ \sum_{i \in R_n} \{(C_{n_{i,k}} + \gamma_I' \phi_{i,k+1} + \gamma_x' \psi_{i,k+1} + l_{i,k} - \alpha_{d_i} \mu_{i,k}) P_{n_{i,k}} \\
&+ \gamma_{n_i} P_{n_{i,k}}^2 + (\lambda_I' \phi_{i,k+1} - \phi_{i,k} + \Lambda_I \psi_{i,k+1}) I_{i,k} \\
&+ (\lambda_x' \psi_{i,k+1} - \psi_{i,k} - \mu_{i,k} \alpha_{X_i}) X_{i,k} - \Gamma_x' \psi_{i,k+1} X_{i,k} P_{n_{i,k}} \\
&+ (f_{i,k} - \mu_{i,k}) \rho_{c_{i,k}} + a_{c_i} I_{i,k}^2 + b_{c_i} \rho_{c_{i,k}}^2 \} \Bigg]
\end{aligned}
\tag{7.309}
$$

where

$$C_{s_{i,k}} = \beta_{s_i} - \theta_k, \qquad i \in R_s \tag{7.310a}$$

$$C_{n_{i,k}} = \beta_{n_i} - \theta_k, \qquad i \in R_n \tag{7.310b}$$

and a_{c_i} and b_{c_i} are small numbers to satisfy Hilbert space requirements.

Let us define the following vectors:

(i) For thermal power plants

$$P_s(k) = \text{col}(P_{s_{i,k}}, i \in R_s) \tag{7.311a}$$

$$C_s(k) = \text{col}(C_{s_{i,k}}, i \in R_s) \tag{7.311b}$$

$$h(k) = \text{col}(h_{i,k}, i \in R_s) \tag{7.311c}$$

$$\gamma_s = \text{diag}(\gamma_{s_i}, i \in R_s) \tag{7.311d}$$

(ii) For hydro power plants

$$x(k) = \text{col}(x_{i,k}, i \in R_h) \qquad (7.312\text{a})$$

$$u(k) = \text{col}(u_{i,k}, i \in R_h) \qquad (7.312\text{b})$$

$$\lambda(k) = \text{col}(\lambda_{i,k}, i \in R_h) \qquad (7.312\text{c})$$

$$g(k) = \text{col}(g_{i,k}, i \in R_h) \qquad (7.312\text{d})$$

$$e(k) = \text{col}(e_{i,k}, I \in R_h) \qquad (7.312\text{e})$$

$$A_{i,k} = -\theta_k a_{i,k}, \; i \in R_h \qquad (7.312\text{f})$$

$$A(k) = \text{col}(A_{i,k}, i \in R_h) \qquad (7.312\text{g})$$

$$B_{i,k} = -\theta_k b_i, \; i \in R_h \qquad (7.312\text{h})$$

$$B(k) = \text{diag}(B_{i,k}, i \in R_h) \qquad (7.312\text{i})$$

$$b = \text{diag}(b_i, i \in R_h) \qquad (7.312\text{j})$$

(iii) For nuclear power plants

$$P_n(k) = \text{col}(P_{n_{i,k}}, i \in R_n) \qquad (7.313\text{a})$$

$$I(k) = \text{col}(I_{i,k}, i \in R_n) \qquad (7.313\text{b})$$

$$X(k) = \text{col}(X_{i,k}, i \in R_n) \qquad (7.313\text{c})$$

$$\phi(k) = \text{col}(\phi_{i,k}, i \in R_n) \qquad (7.313\text{d})$$

$$\psi(k) = \text{col}(\psi_{i,k}, i \in R_n) \qquad (7.313\text{e})$$

$$l(k) = \text{col}(l_{i,k}, i \in R_n) \qquad (7.313\text{f})$$

$$\mu(k) = \text{col}(\mu_{i,k}, i \in R_n) \qquad (7.313\text{g})$$

$$\rho_c(k) = \text{col}(\rho_{c_{i,k}}, i \in R_n) \qquad (7.313\text{h})$$

$$f(k) = \text{col}(f_{i,k}, i \in R_n) \qquad (7.313\text{i})$$

$$\alpha_{i,k} = \mu_{i,k} \alpha_{d_i}, \qquad i \in R_n \qquad (7.313\text{j})$$

$$\alpha(k) = \text{col}(\alpha_{i,k}, i \in R_n) \qquad (7.313\text{k})$$

$$C_n(k) = \text{col}(C_{n_{i,k}}, i \in R_n) \qquad (7.313\text{l})$$

$$\beta_{i,k} = \alpha_{x_i} \mu_{i,k}, \qquad i \in R_n \qquad (7.313\text{m})$$

$$B(k) = \text{col}(\beta_{i,k}, i \in R_n) \qquad (7.313\text{n})$$

Then, the cost function in equation (22) can be written as

$$\tilde{J} = E[\lambda^T(0)x(0) - \lambda^T(K)x(K) + \phi^T(0)I(0) - \phi^T(K)I(K)$$

$$+ \psi^T(0)x(0) - \psi^T(K)X(K)] + E \sum_{k=0}^{K-1} [(C_s(k) + h(k))^T P_s(k)$$

$$+ P_s^T(k)\gamma_s P_s(k) + (A(k) + g(k))^T u(k)$$

$$+ u^T(k)B(k)x(k) + \tfrac{1}{2}u^T(k)bMu(k) + (\lambda(k+1) - \lambda(k)$$

$$+ e(k))^T x(k) + (M^T\lambda(k+1))^T u(k)$$

$$+ (C_n(k) + \gamma_I'\phi(k+1) + \gamma_X'\psi(k+1) + l(k) - \alpha(k))^T P_n(k)$$

$$+ P_n^T(k)\gamma_n P_n(k) + (\lambda_I'\phi(k+1) - \phi(k) + \Lambda_I\psi(k))^T I(k)$$

$$+ I^T(k)a_c I(k) + (\lambda_X\psi(k+1) - \psi(k) - \beta(k))^T X(k)$$

$$- \Gamma_X' X^T(k)\psi^T(k+1)\mathbf{H}P_n(k) + (f(k) - \mu(k))^T \rho_c(k)$$

$$+ \rho_c^T(k)b_c\rho_c(k)] \tag{7.314}$$

In the above equation M is an $N_h \times N_h$ matrix which contains the topological arrangement for the system and $\mathbf{H}$ is a vector matrix whose vector index varies from 1 to N_n, while the matrix dimension of $\mathbf{H}$ is $N_n \times N_n$.

If one defines the following vectors

$$U(k) = \text{col}(P_s(k), x(k), u(k), P_n(k), X(k), I(k), \rho_c(k)) \tag{7.315}$$

$$R^T(k) = [(C_s(k) + l(k))^T, (\lambda(k+1) - \lambda(k) + e(k))^T, (A(k) + M^T\lambda(k+1)$$

$$+ g(k))^T, (C_n(k) + \gamma_I'\phi(k+1) + \gamma_X'\psi(k+1) + h(k) - \alpha(k))^T,$$

$$\times (\lambda_X'\psi(k+1) - \psi(k) - \beta(k))^T, (\lambda_I'\phi(k+1) - \phi(k)$$

$$+ \Lambda_I\psi(k))^T, (f(k) - \mu(k))^T] \tag{7.316}$$

and

$$L(k) = \begin{bmatrix} \gamma_s & 0 & 0 & 0 & 0 & 0 & 0 \\ 0 & 0 & \tfrac{1}{2}B(k) & 0 & 0 & 0 & 0 \\ 0 & \tfrac{1}{2}B(k) & \tfrac{1}{2}bM & 0 & 0 & 0 & 0 \\ 0 & 0 & 0 & \gamma_n & \tfrac{1}{2}[\Gamma_X'\psi^T(k+1)\mathbf{H}] & 0 & 0 \\ 0 & 0 & 0 & \tfrac{1}{2}[\Gamma_X'\psi^T(k+1)\mathbf{H}] & 0 & 0 & 0 \\ 0 & 0 & 0 & 0 & 0 & a_c & 0 \\ 0 & 0 & 0 & 0 & 0 & 0 & b_c \end{bmatrix} \tag{7.317}$$

then the cost functional in equation (7.314) can be written as

$$\tilde{J} = E[\lambda^T(0)x(0) - \lambda^T(K)x(K) + \phi^T(0)I(0)$$

$$- \phi^T(K)I(K) + \psi^T(0)X(0) - \psi^T(K)X(K)]$$

$$+ E\left[\sum_{k=0}^{K-1} \{U^T(k)L(k)U(k) + R^T(k)U(k)\}\right] \tag{7.318}$$

It can be noticed that the cost functional in equation (31) is composed of two parts, the boundary part and the discrete integral part, which are independent of each other. $\tilde{J}$ in (7.318) can be written as

$$\tilde{J} = J_1(K) + J_2(k) \tag{7.319}$$

where

$$J_1(K) = E[\lambda^T(0)x(0) - \lambda^T(K)x(K) + \phi^T(0)I(0)$$
$$- \phi^T(K)I(K) + \psi^T(0)X(0) - \psi^T(K)X(K)] \tag{7.320}$$

and

$$J_2(k) = E\left[\sum_{k=0}^{K-1} \{U^T(k)L(k)U(k) + R^T(k)U(k)\}\right] \tag{7.321}$$

If one defines $V(k)$ as

$$V(k) = L^{-1}(k)R(k) \tag{7.322}$$

then the discrete integral part of equation (7.321) can be written as

$$J_2(k) = E\left[\sum_{k=0}^{K-1} \{(U(k) + \tfrac{1}{2}V(k))^T L(k)(U(k) + \tfrac{1}{2}V(k))\right.$$
$$\left. - \frac{-1}{4} V^T(k)L(k)V(k)\}\right] \tag{7.323}$$

The last term in equation (7.323) is a constant independent of $U(k)$, equation (7.323) can be written as

$$J_2(k) = E\left[\sum_{k=0}^{K-1} \{(U(k) + \tfrac{1}{2}V(k))^T L(k)(U(k) + \tfrac{1}{2}V(k))\}\right] \tag{7.324}$$

Equation (7.324) defines a norm in Hilbert space; it can be written as

$$J_2(k) = \| U(k) + \tfrac{1}{2}V(k)\|_{L(k)} \tag{7.325}$$

To minimize $\tilde{J}$ in equation (7.319) one minimizes $J_1(K)$ and $J_2(k)$ separately.

7.10.3. The Optimal Solution

The boundary part in equation (7.320) is minimized when

$$E[\lambda(K)] = 0 \tag{7.326}$$

$$E[\phi(K)] = 0 \tag{7.327}$$

and

$$E[\psi(K)] = 0 \tag{7.328}$$

because $x(0)$, $I(0)$, and $X(0)$ are constants and $\delta x(K)$, $\delta I(K)$, and $\delta X(K) \neq 0$.

The above three equations can be written in components form as

$$E[\lambda_{i,k}] = 0, \qquad i \in R_h \tag{7.329}$$

$$E[\phi_{i,k}] = 0, \qquad i \in R_n \tag{7.330}$$

$$E[\psi_{i,k}] = 0, \qquad i \in R_n \tag{7.331}$$

The minimum of the discrete integral part is clearly achieved when the norm of equation (7.325) is equal to zero:

$$E[U(k) + \tfrac{1}{2}V(k)]_{L(k)} = 0 \tag{7.332}$$

Substituting from equation (7.322) into equation (7.332), one obtains

$$E[R(k) + 2L(k)U(k)] = 0 \tag{7.333}$$

Writing equation (7.333) explicitly, one obtains
(1) For thermal power plants

$$E[P_s(k)] = \frac{-1}{2\gamma_s} E[C_s(k) + h(k)] \tag{7.334}$$

(2) For hydropower plants

$$E[\lambda(k)] = E[\lambda(k+1) + B(k)u(k) + e(k)] \tag{7.335}$$

$$E[A(k) + M^T(k) + g(k) + B(k)x(k) + bMu(k)] = 0 \tag{7.336}$$

(3) For nuclear power plants

$$E[C_n(k) + \gamma'_I\phi(k+1) + \gamma'_X\psi(k+1) + l(k) - \alpha(k) + 2\gamma_n P_n(k)$$
$$+ \Gamma'_X\psi^T(k+1)\mathbf{H}X(k)] = 0 \tag{7.337}$$

$$E[\psi(k)] = E[\lambda'_X\psi(k+1) - \beta(k) + \Gamma'_X\psi^T(k+1)\mathbf{H}P_n(k)] \tag{7.338}$$

$$E[\phi(k)] = E[\lambda'_I\phi(k+1) + \Lambda_I\psi(k) + 2a_c I(k)] \tag{7.339}$$

$$E[\mu(k)] = E[f(k) + 2b_c \rho_c(k)] \tag{7.340}$$

If we write the above equation in components form, and add the system equality constraints, we obtain the following set of optimal equations:
(1) For thermal power plants, $I \in R_s$

$$E[P_{s_{i,k}}] = \frac{1}{2\gamma_s} E[\theta_k - \beta_{s_i} - (h_{i,k}^M - h_{i,k}^m)], \qquad i \in R_s \tag{7.341}$$

(2) For hydro power plants, $I \in R_h$

$$E[x_{i,k+1}] = E\left[x_{i,k} + q_{i,k} - u_{i,k} + \sum_{v \in R_u} u_{v,k} \right] \tag{7.342}$$

$$E[P_{h_{i,k}}] = E[a_i u_{i,k} + \tfrac{1}{2} b_i u_{i,k}(x_{i,k} + x_{i,k+1})] \tag{7.343}$$

$$E[\lambda_{i,k}] = E[\lambda_{i,k+1} - b_i \theta_k u_{i,k} + (e_{i,k}^M - e_{i,k}^m)] \tag{7.344}$$

$$E\left[-\theta_k a_{i,k} + (g_{i,k}^M - g_{i,k}^m) + \sum_{v \in R_d} \lambda_{v,k} - \lambda_{i,k} \right.$$
$$\left. - b_i \theta_k x_{i,k} + b_i \sum_{j \in R} u_{j,k} - b_i u_{i,k} \right] = 0 \tag{7.345}$$

(3) For nuclear power plants, $i \in R_n$

$$E[I_{i,k+1}] = E[\lambda_I I_{i,k} + \gamma_I P_{n_{i,k}}] \tag{7.346}$$

$$E[X_{i,k+1}] = E[\lambda_X' X_{i,k} + \Lambda_I I_{i,k} + \gamma_X' P_{n_{i,k}} - \Gamma_X' X_{i,k} P_{n_{i,k}}] \tag{7.347}$$

$$E[\psi_{i,k}] = E[\lambda_X' \psi_{i,k+1} - \mu_{i,k} \alpha_{X_i} + \Gamma_X' \psi_{i,k+1} P_{n_{i,k}}] \tag{7.348}$$

$$E[\phi_{i,k}] = E[\lambda_I' \phi_{i,k+1} + \Lambda_I \psi_{i,k} + 2 a_{c_i} I_{i,k}] \tag{7.349}$$

$$E[\rho_{c_{i,k}}] = E[\rho_{o_i} - \alpha_{d_i} P_{n_{i,k}} - \alpha_{X_i} X_{i,k}] \tag{7.350}$$

$$E[\mu_{i,k}] = E[(f_{i,k}^M - f_{i,k}^m) + 2 b_{c_i} \rho_{c_{i,k}}] \tag{7.351}$$

$$\sum_{i \in R_s} P_{s_{i,k}} + \sum_{i \in R_h} P_{h_{i,k}} + \sum_{i \in R_n} P_{n_{i,k}} = P_{d,k} \tag{7.352}$$

$$E[\beta_{n_i} - \theta_k + \gamma_I' \phi_{i,k+1} + \gamma_X' \psi_{i,k+1} + (l_{i,k}^M - l_{i,k}^m)$$
$$- \mu_{i,k} \alpha_{d_i} + 2 \gamma_{n_i} P_{n_{i,k}} + \Gamma_X' \psi_{i,k+1} X_{i,k}] = 0 \tag{7.353}$$

Besides the above equations, one has the following limits on the variables:

$$\left. \begin{array}{ll} \text{if } P_{s_{i,k}} < P_{s_i}^m, & \text{then we put } P_{s_{i,k}} = P_{s_i}^m \\ \text{if } P_{s_{i,k}} > P_{s_i}^M, & \text{then we put } P_{s_{i,k}} = P_i^{Ms} \end{array} \right\}, \quad i \in R_s \tag{7.354}$$

$$\left. \begin{array}{ll} \text{if } x_{i,k} < x_i^m, & \text{then we put } x_{i,k} = x_i^m \\ \text{if } x_{i,k} > x_i^M, & \text{then we put } x_{i,k} = x_i^M \\ \text{if } u_{i,k} < u_{i,k}^m, & \text{then we put } u_{i,k} = u_{i,k}^m \\ \text{if } u_{i,k} > u_{i,k}^M, & \text{then we put } u_{i,k} = u_{i,k}^M \end{array} \right\}, \quad i \in R_h \tag{7.355}$$

$$\left. \begin{array}{ll} \text{if } \rho_{c_{i,k}} < \rho_{c_i}^n, & \text{then we put } \rho_{c_{i,k}} = \rho_{c_i}^m \\ \text{if } \rho_{c_{i,k}} > \rho_{c_i}^M, & \text{then we put } \rho_{c_{i,k}} = \rho_{c_i}^M \\ \text{if } p_{n_{i,k}} < P_{n_i}^m, & \text{then we put } P_{n_{i,k}} = P_{n_i}^m \\ \text{if } P_{n_{i,k}} > P_{n_i}^M, & \text{then we put } P_{n_{i,k}} = P_{n_i}^M \end{array} \right\}, \quad i \in R_n \tag{7.356}$$

One also has the following exclusion equations, Kuhn–Tucker exclusion equations, which must be satisfied at the optimum:

$$\left.\begin{aligned} h_{i,k}^m(P_{s_i}^m - P_{s_{i,k}}) &= 0\\ h_{i,k}^M(P_{s_{i,k}} - P_{s_i}^M) &= 0 \end{aligned}\right\}, \qquad i \in R_s \qquad (7.357)$$

$$\left.\begin{aligned} e_{i,k}^m(x_i^m - x_{i,k}) &= 0\\ e_{i,k}^M(x_{i,k} - x_i^M) &= 0\\ g_{i,k}^m(u_{i,k}^m - u_{i,k}) &= 0\\ g_{i,k}^M(u_{i,k}^M - u_{i,k}) &= 0 \end{aligned}\right\}, \qquad i \in R_h \qquad (7.358)$$

$$\left.\begin{aligned} l_{i,k}^m(P_{n_i}^m - P_{n_{i,k}}) &= 0\\ l_{i,k}^M(P_{n_{i,k}} - P_{n_i}^M) &= 0\\ f_{i,k}^m(\rho_{c_i}^m - \rho_{c_{i,k}}) &= 0\\ f_{i,k}^M(\rho_{c_{i,k}} - \rho_{c_i}^M) &= 0 \end{aligned}\right\}, \qquad i \in R_n \qquad (7.359)$$

In the above equations $h_{i,k}^m$, $h_{i,k}^M$, $e_{i,k}^m$, $e_{i,k}^M$, $g_{i,k}^m$, $g_{i,k}^M$, $l_{i,k}^m$, $l_{i,k}^M$, $f_{i,k}^m$, and $f_{i,k}^M$ are Kuhn–Tucker multipliers. These are equal to zero if the constraints do not violate their limits and greater than zero if the variables violate their limits. Equations (7.341)–(7.359) with equations (7.329)–(7.331) completely specify the optimal long-term operation of a nuclear-hydrothermal power system. In the next section we discuss an algorithm proposed to solve these optimal equations.

7.10.4. Algorithm of Solution, Computer Logic

In this section, we discuss the algorithm of solution for the optimal equations obtained in the previous section. We assume that the initial storage $x(0)$ is given, the expected natural inflows for the reservoirs are given, and the operating data for the system are also given.

Step 1. Assume an initial guess for θ_k such that $\theta_k > \beta_{s_i}^M$, $\beta_{s_i}^M$ is the largest β coefficient for the thermal plants and assume that $P_{s_{i,k}}$ is within the limits.

Step 2. Find the optimal thermal generation from

$$P_{s_{i,k}} = \frac{\theta_k - \beta_{s_i}}{2\gamma_{s_i}}, \qquad i \in R_s$$

If one of these variables violates the limits, put it to its limit and decrease the load $P_{d,k}$ by this amount.

Step 3. Assume an initial guess for $u(k)$ such that

$$u^m(k) \leq (k) \leq u^M(k)$$

Step 4. Solve equation (7.342) forward in stages with $x(0)$ given, and calculate the spill if any.

Step 5. Calculate the hydropower generations from equation (7.343).

Step 6. Solve equation (7.344) backward in stages with equation (7.329) as a boundary condition.

Step 7. Check the gradient given by (7.345). If it is satisfied within a prescribed limit, go to step 8; otherwise update $u(k)$, using the conjugate gradient method and go to step 4, where

$$E[u^{\text{new}}(k)] = E[u^{\text{old}}(k) + \alpha_k \delta u(k)]$$

where $\delta u(k)$ is given by equation (7.345) and α_k is a positive scalar with consideration given to such factors as convergence.

Step 8. Assume an initial guess for $P_{n_{i,k}}$. This can be obtained from steps 2 and 5 and the demand on the system $P_{d,k}$.

Step 9. Solve equations (7.346), (7.347), (7.350), and (7.351) forward in stages and equations (7.348) and (7.349) backward in stages with equations (7.330) and (7.331) as boundary conditions.

Step 10. Check equation (7.353); if it is satisfied within a prescribed limit go to step 11; otherwise update $P_{n_{i,k}}$ as

$$E[P_{n_{i,k}}^{\text{new}}] = E[P_{n_{i,k}}^{\text{old}} + \alpha_k' \delta P_{n_{i,k}}]$$

where $\delta P_{n_{i,k}}$ is given by equation (7.353), using the conjugate gradient and go to step 8.

Step 11. Check the active power balance equation given by (7.352). If it is satisfied within a prescribed limit, terminate the iteration; otherwise update θ_k using the conjugate gradient as

$$\theta_k^{\text{new}} = \theta_k^{\text{old}} + \beta_k \delta \theta_k$$

where $\delta\theta$ is given by equation (7.352), and repeat the calculation starting from step 2. Continue until the variations in the variable are small from iteration to iteration and J in equation (7.306) is a minimum.

7.10.5. Conclusion

In this section we discuss the problem of the optimal discrete long-term operation of nuclear-hydrothermal power systems and present an algorithm for solving the problem on a computer. We have discretized the problem, since this, in our experience, leads to an efficient computer algorithm for the long-term problem.

We have formulated the problem in a general manner in that an arbitrary number of each kind of plant has been included; hence the algorithm presented here can easily deal with large power systems. This is also true when equality and/or inequality constraints are present for the various plants. It is well known that the xenon and iodine concentrations in a nuclear reactor can play a major role in the dynamic operation of the reactor; therefore the differential equations describing these concentrations have also been discretized and are included in the optimization problem.

7.11. General Comments

Long-term scheduling is an important as well as a difficult task for hydrothermal power systems with significant percentages of hydro generation. Substantial reductions in operating cost and in the risk of energy curtailments can be achieved by appropriate management of the energy stored in the various reservoirs.

Long-term multireservoir management is a complex problem for the following reasons:

- It is dynamic: present decisions (reservoir releases) for one reservoir have an impact on future decisions for all reservoirs.
- The optimal operating strategy for one reservoir depends not only on its own energy content but also on the corresponding content of each one of the remaining reservoirs.
- It is a highly stochastic problem in which the major uncertainties are associated with the reservoir inflows, the load, and the unit availability. Spatial and time correlation among hydro inflows is often high and must be modeled.
- It is a nonlinear problem: thermal operating costs, rationing costs, the head effects, etc.

Stochastic dynamic programming is, in principle, the optimization technique most suited to solve this kind of problem in the sense that it can, theoretically at least, handle all these characteristics. Computing time and storage requirement, however, make it impractical for systems with more than three reservoirs since they grow exponentially with the number of state variables. Various types of approximations have been proposed in the past in order to cope with the dimensionality problem:

- *The deterministic approach* ignores the stochastic nature of the hydro inflows, load, etc. This approximation is considered in general to be too unrealistic, particularly for predominantly hydro power systems.
- *The aggregation approach* can be applied to systems with reservoirs that have similar characteristics. Otherwise, local reservoir constraints cannot be guaranteed to be met and consequently hydro generation may be overestimated.
- *Statistic dynamic programming with successive approximations* (DPSA) uses a local feedback policy in which each reservoir is optimized independently assuming an expected operation (state or release) of the rest of the reservoirs. The procedure iterates until convergence is found. Detailed representation of each hydro chain can be used (random inflows, serial correlation, local constraints, etc). The major drawback of this approach is that it ignores the dependence of the operating policy of one reservoir on the actual energy content of other reservoirs. The method may give good results if the actual operation of the rest of the reservoirs is close to expected value (state or release).
- *In the aggregation–decomposition* (AD) approach the optimization of a system of N reservoirs is broken into N subproblems in which one reservoir is optimized knowing the total energy content of the rest of the reservoirs. The global feedback characteristic of the problem is thus retained and the technique can potentially handle all the uncertainties as well as the local constraints in each hydro chain. Furthermore, the computational requirements in this method grow linearly with the number of reservoirs.

A comparison of the last two approaches, DPSA and AD, has been made on a simulation basis for a six-reservoir system. The results indicate that the AD method may give small thermal operating costs and spillage with a computer burden similar for both methods.

The application of the minimum norm theorem to the optimization of the long-term operation of hydrothermal power systems is discussed in the last two sections of this chapter. In Section 7.8, the model used for the water conversion factor was a linear model. This model is adequate for small storage reservoirs, when the variation of the head with the storage is small. A set of optimal equations is obtained; these equations constitute a two-point boundary-value problem, TPBVP; we solve these equations forward and backward in time.

For the power systems in which the water heads vary by a considerable amount, the linear model used in Section 7.8 for the water conversion factor is not adequate. In Section 7.9, we used a quadratic model for the water conversion factor as a function of storage. The cost functional obtained in this section is a highly nonlinear function; we defined a set of pseudo-state-variables to cast the problem into a quadratic model. The optimal equations obtained are solved forward and backward in time.

In Section 7.10, we discuss the problem of the optimal discrete long-term operation of nuclear-hydrothermal power systems and present an algorithm for solving the problem on a computer. We have discretized the problem, since this, in our experience, leads to an efficient computer algorithm for the long-term problem.

We have formulated the problem in a general manner in that an arbitrary number of each kind of plant has been included, hence the algorithm presented here can easily deal with large power systems. This is also true when either equality and/or inequality constraints are present for the various plants. It is well known that the xenon and iodine concentrations in a nuclear reactor can play a major role in the dynamic operation of the reactor; therefore the differential equations describing these concentrations have also been discretized and are included in the optimization problem.

Appendix A: One Dimension Minimization (Ref. 7.17)

To derive the optimum value of α, α^*, so that the function is minimized, cubic interpolation technique has been used.

For the function

$$Z(\alpha) = f(x + \alpha S) \tag{A.1}$$

we have

$$Z'(\alpha) = S^T \cdot \nabla f \tag{A.2}$$

$Z'(\alpha)$ is examined for $\alpha = 0, 2\alpha, 4\alpha, \ldots, \alpha a, \alpha b$ until there is a change in the sign of $Z'(\alpha)$. Then α^* lies in the range αa and αb, call them α_1 and α_2, the exact value of α^* is given by

$$\alpha^* = \alpha_2 - \frac{(\alpha_2 - \alpha_1)[Z'(b) + u - \sigma]}{Z'(b) - Z'(a) + 2u} \tag{A.3}$$

where

$$\sigma = 3\left[\frac{f(b) - f(a)}{\alpha_2 - \alpha_1}\right] + Z'(a) + Z'(b) \tag{A.4}$$

and

$$u = [\sigma^2 - Z'(b) \cdot Z'(a)]^{1/2} \tag{A.5}$$

Appendix B: Projection Matrix P

The feasible direction is given as

$$S = -\nabla f - \sum_{j \in J} l_j \cdot \nabla g_j \tag{B.1}$$

where g_j belongs to a set of active constraints only. In matrix form

$$S = B - NL \tag{B.2}$$

where

$$B = -\nabla F$$

$$\nabla g_j = A_j = (r_{1j}, \ldots, n_{nj})$$

N is defined as $n \times r$ matrix:

$$N = (n_{ij}) = (A_1, A_2, \ldots, A_r) \tag{B.3}$$

This means r constraints are active, where n is the number of variables.
Equation (7.33) can be written as

$$S^T \nabla g_j = 0 \tag{B.4}$$

which can be written as

$$A_i^T \cdot S = 0, \qquad i = 1, 2, \ldots, r \tag{B.5}$$

Substituting for S from equation (7.33), we obtain

$$A_i^T B - \sum_{j=1}^{r} l_j A_i^T A_j = 0 \tag{B.6}$$

from which we obtain

$$N^T NL = N^T B$$

Thus

$$L = (N^T N)^{-1} N^T B \tag{B.7}$$

by substituting equation (B.7) back into equation (B.2), we obtain

$$S = B - N(N^TN)^{-1}N^TB$$

$$S = [I - N(N^TN)^{-1}N^T]B \qquad \text{(B.8)}$$

or

$$S = -P\nabla F$$

where

$$P = I - N(N^TN)^{-1}N^T \qquad \text{(B.9)}$$

P is called the projection matrix.

Appendix C: Some Probability Characteristics of Electric Power Systems

C.1. Approximate Evaluation of Expected Value and Variance of $g(x)$ (Ref. 7.21)

If x has an expected value of $\bar{x}$ and variance σ^2, then, on substituting $x = \bar{x} + \Delta x$, with $\overline{\Delta x} = 0$, and $\overline{\nabla x^2} = 2$, we obtain

$$g(x) = g(\bar{x}) + g'(\bar{x})x + g''(\bar{x})\frac{\Delta x^2}{2} + \cdots \qquad \text{(C.1)}$$

or

$$\overline{g(x)} = g(x) + g''(\bar{x})\frac{\sigma^2}{2} \qquad \text{(C.2)}$$

$$\text{Var}[g(x)] \cong [g'(x)]^2\sigma^2 \qquad \text{(C.3)}$$

C.2. Variances of Thermal Power Generation

For optimal operation at any instant, the thermal powers satisfy the following relation:

$$\lambda = \left(\frac{\beta_i + 2\gamma_i^P s_{i,k}}{1 - dP_{L,k}/dP_{s_{i,k}}}\right), \quad i \in R_s \qquad \text{(C.4)}$$

a perturbation of equation (C.4) gives

$$2\left(1 - \frac{dP_{L,k}}{dP_{s_{i,k}}}\right)\gamma_i\Delta P_{s_{i,k}} + 2(\beta_i + 2\gamma_i P_{s_{i,k}}) \sum_{j \in R_G} B_{ij}P_j = \left(1 - \frac{dP_{L,k}}{dP_{s_{i,k}}}\right)^2 \Delta\lambda \qquad \text{(C.5)}$$

ΔP_s can thus be obtained from the following matrix equation:

$$D\nabla P_s = C \tag{C.6}$$

where

$$D_{ij} = 2(\beta_i + 2\gamma_i P_{s_{i,k}})B_{ij} i \neq j$$

$$D_{ii} = 2(\beta_i + 2\gamma_i P_{s_{i,k}})B_{ii} + 2\left(1 - \frac{dP_{L,k}}{dP_{s_{i,k}}}\right)\gamma_i$$

$$C_i = \left(1 - \frac{dP_{L,k}}{dP_{s_{i,k}}}\right)^2 \nabla\lambda$$

The solution of equation (C.6) gives

$$\nabla P_{s_{i,k}} = k_i \Delta P_s \qquad \text{with } k_s = 1 \tag{C.7}$$

$$\{\text{Var}(P_{s_{i,k}})\}_{i \in R_s} = k_i^2 \text{Var}(P_s) \tag{C.8}$$

$$\{\text{Cov}(P_i, P_j)\}_{i,k \in R_s} = k_i k_j \text{Var}(P_s) \tag{C.9}$$

From the perturbation of equation (7.64), we have

$$\sum_{i \in R_G}\left(1 - \frac{dP_{L,k}}{dP_i}\right)\Delta P_i = \Delta D_k \tag{C.10}$$

$$\Delta P_s = -\sum_{i \in R_h} g_i \Delta P_i + \frac{\Delta D}{e} \tag{C.11}$$

$$\text{Var}(P_s) = \sum_{i \in R_h}\sum_{j \in R_h} g_i g_j \text{Cov}(P_i, P_j) + \frac{\text{Var}(D)}{e^2} \tag{C.12}$$

where

$$e = \sum_{j \in R_s} k_j\left(1 - \frac{dP_L}{dP_j}\right) \tag{C.13}$$

and

$$g_i = \frac{(1 - dP_L/dP_i)}{e} \tag{C.14}$$

C.3. Probability Properties of Hydroelectric Generation

Assuming the water inflows into reservoirs of hydroelectronics plants to be statistically correlated at the same optimization interval but independent at different subintervals, from equation (7.68) we have

$$x_{i,k+1} = x_{i,k} + I_{i,k} - u_{i,k} + u_{(i-1),k} \tag{C.15}$$

$$\text{Var}(x_{i,k+1}) = \text{Var}(x_{i,k}) + \text{Var}(I_{i,k}) \tag{C.16}$$

$$\text{Cov}(x_{i,k+1}, x_{j,k+1}) = \text{Cov}(x_{i,k}, x_{j,k}) + \text{Cov}(I_{i,k}, I_{j,k}) \tag{C.17}$$

From equation (7.67) we have

$$(P_{h_{i,k}})_{i \in R_h} = H_{i0}[1 + 0.5c_i(2\bar{x}_{i,k} + \bar{I}_{i,k} - u_{i,k} + u_{i-1,k})](u_{i,k} - u_{i,0}) \quad \text{(C.18)}$$

$$[\text{Var}(P_{h_{i,k}})]_{i \in R_h} = H_{i0}^2(u_{i,k} - u_{i,0})^2 c_i^2 \cdot [\text{Var}(x_{i,k}) + 0.25\,\text{Var}(I_{i,k})] \quad \text{(C.19)}$$

$$[\text{Cov}(P_{h_{i,k}}, P_{h_{j,k}})]_{i,j \in R_h} = H_{i0}H_{j0}(u_{i,k} - u_{i,0})(u_{j,k} - u_{j,0})$$
$$\times\, c_i c_j[\text{Cov}(x_{i,k}, x_{j,k}) + 0.25\,\text{Cov}(I_{i,k} + I_{j,k})]$$

$$\text{(C.20)}$$

C.4. Expected Transmission Losses

Substituting $P_i = \bar{P}_i + \Delta P_i$, with $\overline{\Delta P}_i = 0$, into equation (7.65), we obtain

$$\bar{P}_L = \sum\sum_{i,j \in R_G} \bar{P}_i B_{ij}\bar{P}_j + \overline{\Delta P}_L$$

where

$$\overline{\Delta P}_L = E[\nabla P^T B \nabla P]$$

$$= E\begin{bmatrix} \Delta P_s \\ \Delta P_h \end{bmatrix}^T \begin{bmatrix} B_{ss} & B_{sh} \\ B_{hs} & B_{hh} \end{bmatrix} \begin{bmatrix} \Delta P_s \\ \Delta P_h \end{bmatrix} \quad \text{(C.21)}$$

and subscripts s and h refer to thermal and hydroelectric plants, respectively.

From equations (C.7) and (C.11)

$$\Delta P_s = -kg^T \Delta P_h + k\frac{\Delta D}{e} \quad \text{(C.22)}$$

where k and g are column vectors with their elements defined by equations (C.7) and (C.14)

$$\overline{\Delta P}_L = E(\Delta P_h^T A \Delta P_h) + F\,\text{Var}(D) \quad \text{(C.23)}$$

where the matrix

$$A = B_{hh} + gk^T B_{ss}kg^T - B_{hs}kg^T - gk^T B_{sh} \quad \text{(C.24)}$$

and

$$F = \frac{1}{e^2}k^T B_{ss}k \quad \text{(C.25)}$$

$$E(P_L) = \sum\sum_{i,j \in R_G} \bar{P}_i B_{ij}\bar{P}_j + \sum_{i \in R_h}\sum_{j \in R_h} A_{ij}\,\text{Cov}(P_i, P_j) + F\,\text{Var}(D) \quad \text{(C.26)}$$

C.5. Expected Cost of Thermal Generation

$$F_i(P_{s_i}) = \beta_i P_{s_i} + \gamma_i P_{s_i}^2 \tag{C.27}$$

$$\overline{F_i(P_{s_i})} = \beta_i \overline{P_{s_i}} + \gamma_i (\overline{P_{s_i}})^2 + \gamma_i \, \mathrm{Var}(P_{s_i}) \tag{C.28}$$

$$\overline{\sum_{i \in R_s} F_i(P_{s_i})} = \sum_\varepsilon [\beta_i \overline{P_{s_i}} + \gamma_i (\overline{P_{s_i}})^2]$$

$$+ \beta \sum_{i,j \in R_h} \sum g_i g_j \, \mathrm{Cov}(P_i, P_j) + \beta \, \mathrm{Var}(D)/e^2 \tag{C.29}$$

where

$$\beta = \sum_{i \in R_s} \beta_i k_i^2 \tag{C.30}$$

References

7.1. BOSCH, P. P. J., "Optimal Static Dispatch with Linear, Quadratic and Nonlinear Functions of the Fuel costs," *IEEE Transactions on Power Apparatus and Systems* **PAS-104**(12), 3402–3408 (1985).

7.2. SHOULTS, R. R., VENKTATESH, S. V., HELMICK, S. D., WARD, G. L., and LOLLAR, M. J., "A Dynamic Programming Based Method for Developing Dispatch Curves When Incremental Heat Rate Curves are Non-Monotinically Increasing," *IEEE Transactions on Power Systems* **PWRS-1**(1), 10–16 (1986).

7.3. KUSIC, G. L., *Computer-Aided Power Systems Analysis*, Prentice-Hall, Englewood Cliffs, New Jersey, 1986.

7.4. HAPP, H. H., "Optimal Power Dispatch—A Comprehensive Survey," *IEEE Transactions on Power Apparatus and Systems* **PAS-96**(3), 841–854 (1977).

7.5. VAN DEN BOSCH, P. P. J., and HONDERD, G., "A Solution of the Unit Commitment Problem via Decomposition and Dynamic Programming," Preprints IEEE PES Summer Conference, paper 84 SM 609-4, Seattle, July 1984.

7.6. VAN DEN BOSCH, P. P. J., "Optimal Dynamic Dispatch Owing to Spinning-Reserve and Power-Rate Limits," Preprints IEEE PES Winter Conference, New York, January 1985.

7.7. OTTENHOF, F. A. "Economische Optimalisatie van een Hierarchisch Electriciteits-Productie Systeem" M.Sc. thesis, Delft University of Technology, Laboratory for Control Engineering, 1978.

7.8. MURTY, K. G., *Linear and Combinatorial Programming*, John Wiley and Sons, New York, 1976.

7.9. VAN DEN BOSCH, P. P. J., "Short-Term Optimization of Thermal Power Systems," Ph.D. thesis, Delft University of Technology, 1983.

7.10. VAN DEN BOSCH, P. P. J., and LOOTMA, F. A., "Large-Scale Electricity-Production Scheduling via Nonlinear Optimization," Report 84-07 of the Department of Mathematics and Information, Delft University of Technology 1984, Submitted for possible publication in *Mathematical Programming*.

7.11. VIVIANNI, G. L., "Practical Optimization," IEEE PES Summer Meeting, Paper 84 SM 613-6, Seattle, 1984.

7.12. VENKATESH, S. V., "A New Approach for Performing the Economic Dispatch Calculation Based upon the Principle of Dynamic Programming," Master's Thesis, The University of Texas at Arlington, August 1984.

7.13. WOOD. A. J., and WOLLENBERG, B. F., *Power Generation, Operation, and Control*, Wiley, New York, 1984.

7.14. HILLER, F. S., and LIEBERMAN, F. J., *Introduction to Operations Research*, Holden-Day Inc., San Francisco, 1967.

7.15. DRAPER, N. R., and SMITH, H., *Applied Regression Analysis*, Wiley, New York, 1981, pp. 85–89, 122.

7.16. BELLMAN, R., and ROTH, "Curve Fitting by Segmented Straight Lines," *American Statistical Association Journal* **64**(327), 1079–1084 (1967).

7.17. SAHA, T. N., and KHAPARDE, S. A., "An Application of a Direct Method to the Optimal Scheduling of Hydrothermal Systems," *IEEE Transactions on Power Apparatus and Systems* **PAS-97**(3), 977–983 (1978).

7.18. SOARES, S., LYRA, C., and TAVARES, H., "Optimal Generation Scheduling of Hydrothermal Power Systems," *IEEE Transactions on Power Apparatus and Systems* **PAS-99**(3), 1107–1115 (1980).

7.19. NARITA, S., OH, Y., HANO, I., and TAMURA, Y., "Optimum System Operation by Discrete Maximum Principle," Proceedings of the PICA Conference, 1967, pp. 189–207.

7.20. BERNHOLTZ, B., and GRAHAM, L. J., "Hydrothermal Economic Scheduling," *AIEE, Transactions Pt. III Power Apparatus and Systems* **79**, 921–932 (1960).

7.21. OH, Y. N., "An Application of the Discrete Maximum Principle to the Most Economical Power-System Operation," *Electrical Engineering Japan* **4**, 17–28 (1967).

7.22. AGARWAL, S. K., "Optimal Stochastic Scheduling of Hydrothermal Systems," *Proceedings of the IEEE*, **120**(6), 674–678 (1973).

7.23. AGARWAL, S. K., and NAGRATH, I. J., "Optimal Scheduling of Hydothermal Systems," *Proceedings of the IEEE* **119**(2), 169–173 (1971).

7.24. GUPTA, P. C., "Statistical and Stochastic Techniques for Peak Power Demand Forecasting," Ph.D. thesis, Purdue University, Lafayette, Indiana, 1970.

7.25. BENJAMIN, J. R., and CORNELL, C. A., *Probability Statistics and Decision for Civil Engineers*, McGraw-Hill, New York, 1970.

7.26. POWELL, M. J. D., "A Method for Nonlinear Constraints in Minimization Problem," presented at the conference on optimization, Keele University, Keele, England, 1968.

7.27. KUSHNER, H. J., *Stochastic Stability and Control*, Academic Press, New York, 1967.

7.28. AOKI, M., *Optimization of Stochastic Systems*, Academic Press, New York, 1967.

7.29. NARITA, S., OH, Y., HANO, L., and TAMURA, Y., "Optimum System Operation by Discrete Maximum Principle," Proceedings of PICA conference, Pittsburg, Pennsylvania, 1967, pp. 189–207.

7.30. PAPOULIS, A., *Probability, Random Variables, and Stochastic Process*, McGraw-Hill, New York, 1965.

7.31. FELDMAN, A. A., *Optimum Control System*, Academic Press, New York, 1965.

7.32. SKORKHOD, A. V., *Studies in the Theory of Random Processes*, Addison-Wesley, Reading, Massachusetts, 1965.

7.33. FLETCHER, R., and REEVES, C. M., "Function Minimization by Conjugate Gradients," *Computer Journal* **7**, 149–154 (1964).

7.34. HADLEY, G., *Nonlinear and Dynamic Programming*, Addison-Wesley, Reading, Massachusetts, 1964.

7.35. KIRCHMAYER, L. K., *Economic Operation of Power Systems*, Wiley, New York, 1958.

7.36. QUINTANA, V. H., and CHIKHANI, A. Y., "A Stochastic Model for Mid-Term Operation Planning of Hydro-Thermal Systems with Random Reservoir Inflows," *IEEE Transactions on Power Apparatus and Systems* **PAS-100**(3), 1119–1127 (1981).

7.37. CROLEY II, T. E., "Sequential Stochastic Optimization for Reservoir System," *Journal of the Hydraulic Division, HYL*, **10263**, 201–219 (1974).

7.38. BALERIAUX, J., JAMOULE, E., and DEGUERTECHIN, R. L., *Simulation de l'Exploitation d'un Parc de Machines Thermiques de Production d'Electricite Couple a des Stations de Pompage*, Revue T., Edition SBRE, Vol. V, No. 7, 1967.

7.39. BOOTH, R. R., "Power System Simulation Model Based on Probability Analysis," *IEEE Transactions on Power Apparatus and Systems* **PAS-91**, 62–69 (1972).

7.40 BOOTH, R. R., "Optimal Generation Planning Considering Uncertainty," *IEEE Transactions on Power Apparatus and Systems* **PAS-91**, 70–71 (1972).

7.41. JOY, D. S., and JENKINS, R. T., "A Probabilistic Model for Estimating the Operating Cost of an Electrical Power Generating System," Oak Ridge National Laboratory, Report No. ORNL-TM-3549, Oak Ridge, Tennessee, 1971.

7.42. WU, F., and GROSS, C., "Probabilistic Simulation of Power System Operation for Production Cost and Reliability Evalutation," Special Session on Power Systems, IEEE International Symposium on Circuit and Systems, Phoenix, Arizona, 1977.

7.43. VIRAMONTES, F. A., and HAMILTON, H. B., "Optimal Long Range Hydro Scheduling in the Integrated Power System," IEEE PES Winter Meeting, New York, Paper No. F-77-112-6, 1977.

7.44. RINGLEE, R. J., and WOOD, A. J., "Frequency and Duration Methods for Power Reliability Calculations: II—Demand Model for Capacity Reserve Model," *IEEE Transactions on Power Apparatus and Systems* **PAS-84**, 61–78 (1965).

7.45. LUENBERGER, D. G., *Introduction to Linear and Nonlinear Programming*, Addison-Wesley, Reading, Massachusetts, 1973.

7.46. GAGNON, C. R., HICKS, R. H., JACOBY, S. L. S., and KOWALIK, J. S., "A Nonlinear Programming Approach to a Very Large Hydroelectric System Optimization," *Mathematical Programming* **6**, 28–41 (1974).

7.47. ARVANTIDIS, N. V., and ROSING, J., "Composite Representation of a Multireservoir Hydroelectric Power System," *IEEE Transactions on Power Apparatus and Systems* **PAS-89**(2), 319–326 (1970).

7.48. HANSCOM, M., and LAFOND, L., "Modeling and Resolution of the Deterministic Midterm Energy Production Problem for the Hydro-Quebec System," IREQ Report No. 1453, Project 01570-57351-503, Varennes, P.Q., Canada, 1976.

7.49. NEMHAUSER, G. L., *Introduction to Dynamic Programming*, John Wiley, New York, pp. 149–179, 1966.

7.50. LITTLE, J. D., "The Use of Storage Water in a Hydroelectric System," *Journal of the Operations Research Society of America* **III**, 187–197 (1955).

7.51. SULLIVAN, R. L., *Power System Planning*, McGraw-Hill, New York, 1976.

7.52. HILSON, D. W., SULLIVAN, R. L., and WILSON, J. A., "Theory and Application of the Power System Probabilistics Simulation Method," IEEE Summer Power Meeting, Paper No. A78 530-8, Los Angeles, July 1978.

7.53. DURAN, H., *et al.*, "Optimal of Multireservoir Systems Using an Aggregation–Decomposition Approach," *IEEE Transaction on Power Apparatus and Systems* **PAS-104**(8), 2086–2092 (1985).

7.54. REES, F. J., and LARSON, R. E., "Computer-Aided Dispatching and Operations Planning for an Electric Utility with Multiple Types of Generation," *IEEE Transactions on Power Apparatus and Systems* **PAS-10**(2), (1971).

7.55. ARVANTIDIS, N., and ROSING, J., "Optimal Operation of Multireservoir Systems using a Composite Representation," *IEEE Transactions on Power Apparatus and Systems* **PAS-89**, (1970).

7.56. PRONOVOST, R., and BOULVA, J., "Long-Range Operation Planning of a Hydro-Thermal System, Modeling and Optimization," Canadian Electrical Association, Toronto, Ontario, March, 1978.

7.57. TURGEON, A., "Optimal Operation of Multireservoir Power Systems with Stochastic Inflows," *Water Resources Research* **16**(2), 275–283 (1980).

7.58. LEDERER, P., TORRION, PH., and BOUTTES, J. P., "Overall Control of an Electricity Supply and Demand System: A Global Feedback for the French System," 11th IFIP Conference on System Modeling and Optimization, Copenhagen, July 1983.

7.59. DAVIS, R., and PRONOVOST, R., "Two Stochastic Dyanmic Programming Procedures for Long-Term Reservoir Management," IEEE Summer Power Meeting, San Francisco, July 1972.

7.69. DURAN, H., QUERUBIN, R., CUERVO, G., and RENGIFO, A., "A Model for Planning Hydrothermal Power Systems,": 9th Power Industry Computer Applications Conference, June 1975.

7.61. DELEBECQUE, F., and QUADRAT, J. P., "Contribution of Stochastic Control Singular Perturbation Averaging and Team Theories to an Example of Large-Scale System Management of Hydropower Production," *IEEE Transaction on Automatic Control* **AC-23**(2), (1978).

7.62. SOLIMAN, S. A., and CHRISTENSEN, G. S., "Discrete Stochastic Optimal Long-Term Scheduling of Hydro-Thermal Power Systems," Applied Simulation and Modeling. IASTED Proceedings Conference, ASM'86, Vancouver, B.C., Canada. June 4–6, 1986, pp. 103–106.

7.63. KIEFER, W. M., and KONCEL, E. F., "Scheduling Generations on Systems with Fossil and Nuclear Units," *Transactions of the American Nuclear Society* **13**, 768 (1970).

7.64. HOSKINS, R. E., and REES, F. J., "Power Systems Optimization Approach to Nuclear Fuel Management," *Transactions of the American Nuclear Society* **13**, 768 (1970).

7.65. GROSSMAN, L. M., and REINKING, A. G., "Fuel Management and Load Optimization of Nuclear Units in Electric Systems," *Transactions of the American Nuclear Society* **20**, 391 (1975).

7.66. CHOU, W. B., "Characteristics and Maneuverability of Candu Nuclear Power Stations Operated for Base-Load and Load Following Generation," *IEEE Transactions on Power Apparatus and Systems* **PAS-94**(3), 792–801 (1975).

7.67. EL-WAKIL, M. M., *Nuclear Power Engineering*, McGraw-Hill, New York, 1962.

7.68. YASUKAWA, S., "An Analysis of Continuous Rector Refueling," *Nuclear Science and Engineering* **24**, 253–260 (1966).

7.69. MILLAR, C. H., "Fuel Management in Candu Reactors," *Transactions of the American Nuclear Society* **20**, 350 (1975).

7.70. EL-HAWARY, M. E., and CHRISTENSEN, G. S., *Optimal Economic Operation of Electric Power Systems*, Academic, New York, 1979.

7.71. PORTER, W. A., *Modern Foundations of Systems Engineering*, Macmillan, New York, 1966.

7.72. HAMILTON, E. P., and LAMONT, I. W., "An Improved Short Term Hydrothermal Coordination Model," Paper No. A77 518-4, Institute of Electrical and Electronics Engineers Summer Power Meeting, Mexico City, 1977.

7.73. ISBIN, H. S., *Introductory Nuclear Reactor Theory*, Reinhold, New York, 1963.

7.74. SHAMALY, A., *et al.*, "A Transformation for Necessary Optimality Conditions for Systems with Polynomial Nonlinearities," *IEEE Transactions on Automatic Control* **AC-24**, 983–985 (1979).

7.75. MAHMOUD, M. S., "Multilevel Systems Control and Applications: A Survey," *IEEE Transactions on Systems, Man and Cybernetics* **7**(3), 125–143 (1977).

7.76. NIEVA, R., CHRISTENSEN, G. S., and EL-HAWARY, M. E., "Functional Optimization of Nuclear-Hydro-Thermal Systems," Proceedings, CEC, Toronto, 1978.

7.77. NIEVA, R., CHRISTENSEN, G. S., and EL-HAWARY, M. E., "Optimum Load Scheduling of Nuclear-Hydro-Thermal Power Systems," *Optimization Theory and Applications* **35**(2), 261–275 (1981).

7.78. TSOURI, N., and ROOTENBERG, J., "Optimal Control of A Large core Reactor in Presence of Xenon," *IEEE Transactions on Nuclear Science* NS-22, 702–710 (1975).

7.79. LIN, C., and GROSSMAN, L. M., "Optimal Control of a Boiling Water Reactor in Load-Following Via Multilevel Methods," *Nuclear Science and Engineering* 92, 531–544 (1986).

7.80. CHRISTENSEN, G. S., EL-HAWARY, M. E., and SOLIMAN, S. A., *Optimal Control Applications in Electric Power Systems*, Plenum Press, New York, 1987.

7.81. CHAUDHURI, S. P., "Distributed Optimal Control in a Nuclear Reactor," *International Journal of Control* 16(5), 927–937 (1972).

8

Conclusion

8.1. Summary

The aims of this book are to discuss the applications of optimal control to the long-term operation of electric power systems. It is clear that such problems can only now be solved numerically because of the advent of large-scale digital computers. Chapter 1 reviews the historical developments in the field and discusses briefly all techniques used to solve the problem. At the end of this chapter we offer in detail the modeling of hydroplants for long-term studies, where all the system equations become discrete equations; by using this model the problem may be simplified, in contrast to using the continuous differential equations.

Some mathematical background is offered in the beginning of Chapter 2, which is helpful for the reader, while the rest of the chapter discusses the optimization techniques used throughout the book, including calculus of variations, dynamic programming, and the discrete maximum principle. We review in this chapter the minimum norm application to the optimal long-term problem. A number of solved examples are included to illustrate these techniques.

In Chapter 3 we discuss the optimal long-term operation of multireservoir power systems connected in cascade on a river (series). In this chapter, we compare different techniques used to solve the problem. In the beginning of the chapter we discuss the decomposition approach with dynamic programming used to solve the problem. This approach is justified only for similar reservoirs; otherwise, the solution obtained by this method is only a suboptimal operating policy. In the middle of the chapter, we discuss the minimum norm approach, developed by the authors, to solve the problem. We use, in the beginning of this part, a linear model for the water conversion factor; at the same time a constant water conversion factor is used to model the amount of water at the end of the year. It was found that this algorithm can deal with a large-scale power system with stochastic inflows. We compare the minimum norm approach with the dynamic programming and decomposition approach; we obtain increased benefits, using a smaller computing time, for the same system.

For power systems in which the water heads vary by a small amount, the assumption of constant water conversion factor is adequate, but for power systems in which this variation is large using this assumption is not adequate. On the other hand, the linear storage-elevation curve used in modeling the reservoirs is adequate only for small-capacity reservoirs of rectangular cross section area, which is generally not the case in practice in that most reservoirs are nearly trapezoidal in cross section. The last section of the chapter discusses the long-term optimal operating problem of power systems having a variable water conversion factor and a nonlinear storage elevation curve. The optimal solution is obtained using the minimum norm formulation in the framework of functional analysis. A practical example is presented at the end of the section.

The intent of Chapter 4 is to solve the long-term optimal operating problem of a multichain power system. The rivers in this chapter may or may not be independent of each other. In this chapter we discuss different approaches used to solve the problem, including the stochastic dynamic programming, aggregation–decomposition approach; we compare this approach with a well-known approach called "The one–at-a-time method," the discrete maximum principle, and finally we discuss the application of the minimum norm formulation, developed by the authors, to solve the problem. Each technique is applied to a practical example to demonstrate the main features of this technique. Different models are used for each approach.

There is a period in the water management during which the inflows to the reservoirs are the lowest on record, and the reservoirs should be drawn down from full to empty at the end of the period. This period is referred to as the "critical period" and the stream flows that occur during the critical period are called the critical stream flows. We discuss in Chapter 5 the optimal operation during this period. The first section of this chapter concerns maximizing the total benefits from the system during this period, but the load shape on the system is not take into account in this section. The second section of the chapter discusses the maximization of the total generation from the system during this period, but here the load shape on the system is taken into account. The nature of this load is that it depends on the total generation at the end of each year of the critical period, i.e., the monthly load is equal to a certain percentage of the total generation at the end of each year. The minimum norm formulation technique is used in this chapter. A practical example for a real system in operation is provided in each section.

Chapter 6 discusses the operation of hydroelectric power systems for a maximum hydro energy capability during the critical period, where the firm energy is defined as the difference between the total generation during a certain period and the load on the system during that period; the minimum norm formulation of functional analysis is used to solve the problem. In the first section of this chapter a linear storage-elevation curve is used, while in the second section a nonlinear storage-elevation curve is used. The resulting cost function is a highly nonlinear function. A set of pseudostate variables are defined to cast the problem into a quadratic problem.

Most of the utility electric power systems contain a combination of hydro and thermal power plants to supply the required load on the system. The aim of Chapter 7 is to study the optimal long-term operation of this combination, where the objective function, is to minimize the thermal fuel cost, and at the same time satisfying the hydro constraints on the system. In this chapter we discuss different techniques used to solve the problem, including the stochastic dynamic programming, aggregation–decomposition approach, nonlinear programming, discrete maximum principle, and finally the minimum norm approach developed by the authors. Practical examples are provided for each technique to demonstrate the main features of the technique used.

8.2. Future Work

In looking toward future research needs, it is clear that an example is needed for a combination of hydrothermal and nuclear power system for optimal long-term operation where the model used is a discrete model. By using this model the computing time and computer storage requirements are reduced greatly.

The problem discussed in Chapter 6 needs a practical example (which is the current research of the authors), to indicate how powerful the minimum norm formulation is when dealing with a large-scale power system. Also, the minimum norm application to the optimization of the hydrothermal power system for long-term operation, which we discuss at the end of Chapter 7 for both the linear and the quadratic hydro model, should be applied to a practical system in operation. This is also the current research of the authors.

Index